I0072240

Molecular Genetics

Molecular Genetics

Edited by **Erik Pierre**

SYRAWOOD
PUBLISHING HOUSE

New York

Published by Syrawood Publishing House,
750 Third Avenue, 9th Floor,
New York, NY 10017, USA
www.syrawoodpublishinghouse.com

Molecular Genetics
Edited by Erik Pierre

© 2016 Syrawood Publishing House

International Standard Book Number: 978-1-68286-099-1 (Hardback)

This book contains information obtained from authentic and highly regarded sources. Copyright for all individual chapters remain with the respective authors as indicated. All chapters are published with permission under the Creative Commons Attribution License or equivalent. A wide variety of references are listed. Permission and sources are indicated; for detailed attributions, please refer to the permissions page and list of contributors. Reasonable efforts have been made to publish reliable data and information, but the authors, editors and publisher cannot assume any responsibility for the validity of all materials or the consequences of their use.

The publisher's policy is to use permanent paper from mills that operate a sustainable forestry policy. Furthermore, the publisher ensures that the text paper and cover boards used have met acceptable environmental accreditation standards.

Trademark Notice: Registered trademark of products or corporate names are used only for explanation and identification without intent to infringe.

Printed in the United States of America.

Contents

Preface

This book has been an outcome of determined endeavour from a group of educationists in the field. The primary objective was to involve a broad spectrum of professionals from diverse cultural background involved in the field for developing new researches. The book not only targets students but also scholars pursuing higher research for further enhancement of the theoretical and practical applications of the subject.

Molecular genetics deals with the study of structure of genes on a molecular level. This book provides comprehensive insights into interaction and biosynthesis of proteins, structure of RNA and DNA, polymerase chain reaction, cloning, proteinase activities, and risk factors of hyper-dyslipidemia, etc. The topics included in this book provide a significant overview of the various genetic structures and gene transfer mechanisms. Coherent flow of topics, student-friendly language and extensive use of examples make this book an invaluable source of knowledge for students and academicians alike.

It was an honour to edit such a profound book and also a challenging task to compile and examine all the relevant data for accuracy and originality. I wish to acknowledge the efforts of the contributors for submitting such brilliant and diverse chapters in the field and for endlessly working for the completion of the book. Last, but not the least; I thank my family for being a constant source of support in all my research endeavours.

Editor

Evaluation of genetic heterogeneity in glutamate carboxypeptidase II (H475Y) and reduced folate carrier (SLC19A1) gene variants increased risk factor for the development of neural tube defects in eastern region of India

Ajit Kumar Saxena[1]*, S. Pandey[1] and L. K. Pandey[2]

[1]Human Molecular Cytogenetic Laboratory, Centre of Experimental Medicine and Surgery, Institute of Medical Sciences, Banaras Hindu University, Varanasi-221005, India.
[2]Department of Obstetrics and Gynecology, Institute of Medical Sciences, Banaras Hindu University, Varanasi-221005, India.

In humans, neural tube failure to close during the 4th week of gestation leads to the development of severe congenital malformations of the central nervous system because of an error in maternal folate metabolism associated gene variants. The frequency of genotypic variants of GCP II (H475Y) and folate carrier RFCl (SLC19A1) gene polymorphism (80 G →A) were evaluated as potential candidate gene(s) and also assess their clinical association to increase "risk" in neural tube defects (NTDs). In the present study, blood samples (0.5 ml) were collected from NTD cases, mother and their respective controls and genomic DNA was isolated to evaluate the impact of GCP II and *RFCl* genotypic variants as risk factor using polymerase chain reaction (PCR) based restriction fragment length polymorphism (RFLP) analysis. Significant differences ($p<0.05$) were observed between case mothers and control for GCP II genotype using Fischer's exact two tailed probability test. The odd ratio was calculated to determine the risk factors at 95% C.I. (1.56-87.60), which seems to be very high, suggesting significant involvement of GCP II gene in the development of NTDs. The significant ($p = 0.03$) risk factor was also calculated (OR=4.85: 95%, C.I. 1.33-17.36) for *RFCl* gene between heterozygote (GA) and homozygote (AA) mothers having NTDs child. The present finding strongly suggests that genotype variants of GCPII and RFCl gene, in heterozygous condition, are responsible for increasing as independent risk factor for the development of NTDs like meningomyelocele (MMC) susceptibility in this region.

Key words: Glutamate carboxy peptidase, reduced folate carrier, gene polymorphism, neural tube defects, meningomyelocele.

INTRODUCTION

Neural tube defects (NTDs) are severe form of congenital malformation of central nervous system including brain and spinal cord where neural tube fails to close in early embryonic development. Such "birth defects" are multifactorial in origin including genetics and environmental factors (Finnell et al., 2000; Cabrera et al., 2004; Detrait et al., 2005; Beaudin and Stover, 2009; Saxena et al., 2011). The most common form of NTDs is anencephaly and meningomyelocele where the exact cause is still not clearly defined in literature. Certainly, NTDs involve large number of variants across single gene and these variants (genes) differ in population between different ethnic groups. Epidemiological studies reveal that periconceptional folic acid supplementation may reduce

---

*Corresponding author. E-mail: draksaxena1@rediffmail.com.

risk factor up to 75% (MRC, 1991; Botto and Yang, 2000; Van der Put et al., 1995). Meningomyelocele (MMC), the most severe form of spina bifida and possessor is difficult to survive because of dysplastic spinal cord with lack of neural function. In MMC both meninges and the spinal cord protrude through a gap in the vertebral column and the lesion is not covered by the skin. These anomalies can occur at any point along with the developing neural tube, although, lumbosacral lesions are the most common (Hunter et al., 1996). Most of the children of MMC survived after surgical intervention with lifelong disabilities (Detrait et al., 2005).

Dietary factor such as folate, predominantly exists in the form of polyglutamates and hydrolyzed to monoglutamates before absorption by the enzyme folypoly-γ glutamate carboxypeptidase (FGCP) exists in jejunum. The folate monoglutamates are absorbed in proximal part of small intestine by folate carrier (Chandler et al., 1991). Reduced folate carrier (RFCI) is an essential cofactor for the synthesis of purines and pyrimidines synthesis for maintaining the genomic instability (Simonet and Sang, 2005). Recently, a polymorphism of glutamate carboxypeptidase II (H475Y) gene encoding enzyme is responsible for decrease folate level and increased total plasma homocysteine (tHcy) level in NTDs cases (Devlin et al, 2000). The GCP-II gene is localized on chromosome 11p11.2 and the gene product consists of 750 amino acid residues, termed folylpoly - glutamate carboxypetidase (FGCP) which hydrolyze the terminal glutamate residues before absorption. Thereafter, the monoglutamyl folate derivatives are transported through the membrane via the folate transporter as reduced folate carrier. Similarly, RFCI gene is assigned on chromosome 21 (21q22.2-22.3) with number of common variants in the coding sequences and best studied H27R polymorphism of 80G A alleles changing an arginine into histidine (Chango et al., 2000; Marco et al., 2001). RFCI (SLC19A1), a cell surface transmembrane protein and is involved in bidirectional movement of folate across the membrane (Matherly et al., 2007; Hou and Matherly, 2009). In human, the exact causes of neural tube defects are still unknown but it seems to be involvement of mutivariants genes differs between different ethnic groups. RFCI, an essential molecule to carry folate for the developing embryo through placenta during organogenesis and has been associated with risk factor in NTDs. It has been observed that polymorphism of GCP-II (1561CT; H475Y) and RFCI (SLC19A1; 80GA) gene regulate the availability of dietary folate and increase susceptibility towards MMC risk in population (Shang et al., 2008; Pei et al., 2009). The rationale behind the selection of this gene is based on polymorphic variation with other genes, epidemiological studies and diversified biological function (Williams et al., 2002; Zhu et al., 2007; Shang et al., 2008; Pei et al., 2009). Linkage studies showing interesting findings but due to lack of reproducibility and

their association with other genes (Greene et al., 2009; Beaudin and Stover, 2009; Copp and Greene, 2010). Hence, logically hypothesize that these gene variants are associated with development of MMC. However, the present study becomes imperative and curiosity has been generated with the aim to evaluate the frequency of polymorphic genetic variants of GCP-II and RFC I gene to access the "risk factor" in developing NTDs and their clinical association has not been documented earlier in Indian literature.

MATERIALS AND METHODS

The majority of MMC probands and their parents were enrolled after obtaining written consent from the participant's attendant/guardians. The criteria for inclusion of an individual were based on clinically diagnosed MMC. The level of defect was determined by review of image of radiographs and from medical records. In present study those cases having continues chemotherapy or previous history of genetic disorder other than NTDs were excluded. The project was dually approved by Institutional review and ethical committee.

Blood samples (n=100) from proband, mother and their respective controls were collected from the OPD of the Department of Pediatric Surgery and transfer to Human Molecular Cytogenetic Laboratory of Centre of Experimental Medicine and Surgery, Institute of Medical Sciences for genetic studies in EDTA vials and stored at − 20℃ under sterile condition till further analysis. Present study was dually approved by Institute ethical committee and samples were collected after informed written consent from the participant's attendant/guardians.

Genomic DNA was isolated from isolation kit (Bioneer, Korea) for further genetic analysis. Polymerase chain reaction (PCR) was carried out by using RFC-I specific forward 5'AGT GTC ACC TTC GTC CCC TC3' and reverse 5'CTC CCG CGT GAA GTT CTT 3'primers as reported by Chango et al. (2000), while for GCPII forward 5'- CAT TCT GGT AGG AAT TTA GCA-3' and reverse 5'-AAA CAC CAC CTA TGT TTA ACA-3' (Devlin et al., 2000) in total volume of 50 µl containing 50-100 ng of genomic DNA, 20 pmole of each primer, 200 µM of each dNTPs mix with Taq buffer (10 mM Tris HCI pH 8.3, 50 mM KCI), 3.0 mM MgCl$_2$ and 3 unit of Taq polymerase (New England Biolab). PCR product was separated on agarose gel and RFLP analysis was carried out using Hhal & Acc I restriction enzyme for RFC and GCP-II respectively. The guanine (G) changed into adenine (A) at position 80 in RFCI while for GCP-II cytosine change into thymine at position 1561. The amplified products (6 µl) were digested at 37℃ for 3 h in reaction volume of 25 µl containing 1U of Hhal & Acc I restriction enzyme and NEB buffer (2.5 µl) (New England, Biolabs). Digested products were separated on 3% agarose gel stained with Et.Br and DNA fragments were visualized on Gel Doc system (SR Biosystem).

Statistical analysis

Fischer exact two tailed probability test was used to observe the significant differences (p < 0.05) between NTDs cases, mothers and their respective controls. The relative risk factor, the odd ratio (O.R) was calculated at 95% confidence interval (C.I.) for combination of different genotype. The Hardy Weinberg Equilibrium was used to determined individual allele frequency.

RESULTS

GCP II H475Y polymorphism was analyzed on genomic

Table 1. Genotype frequency of GCPII (H475Y) gene polymorphism in NTDs cases and their controls.

Cases/Control of NTDs	Genotype frequency (%)			% Allele frequency	
	CC	CT	TT	C	T
NTD cases	8.00	92.00	0	13.5	11.5
Control child	20.00	75.00	0	11.5	7.5
NTD mothers	6.67	93.30	0	8.0	7.0
Control mothers	33.34	53.30	0	9.0	4.0

Table 2. GCPII (H475Y) genotypes showing odd ratio (O.R) and C.I at 95% in NTD cases and their respective controls between homozygous and heterozygous condition.

GCPII C → T	O.R	(95% C.I)	p-value
NTDs cases vs. control			
CC/CC	0.35	0.067-1.865	0.03
CT/CT	3.833	0.738-19.288	0.124
NTDs mother vs. control mother			
CC/CC	0.143	0.020-1.129	0.169
CT/CT	12.250	1.563-87.603	0.035*

Statistical analysis shows level of significant*(p<0.05) differences using Fisher exact probability test between homozygous and heterozygous condition with respect to controls.

DNA of the patients (NTD cases), mothers and their respective controls. The prevalence of this polymorphism shows three types of genotype variants that is, CC (wild type), CT, and TT (rare type) with highest frequency (92%) of CT between NTD cases and controls as summarized in Table 1. Because the genotype distribution did not differ significantly among NTD cases and their respective controls hence, we combined cases and mothers with their respective controls to increase statistical power for examination of possible interaction of gene polymorphism. The odd ratio was calculated at 95% C.I (0.73-19.28) between cases and controls seems to increase three fold (OR: 3.8) but shows lack of significant (p> 0.12) differences. Interestingly, the significant differences (p = 0.035) were observed between the cases mothers and control in heterozygous condition (Table 2). The individual allele frequency (T allele) was also calculated using Hardy Weinberg equilibrium which reveals highest frequency (11.5 %) in NTD cases when compared with controls (7.5 %).

RFC I (80 GA: R27H) gene polymorphism showing three type of genotypic variants that is, homozygous GG (wild type) and AA (rare type) and heterozygous (GA) condition as documented in Table 3. The highest genotype frequency was observed in homozygous (AA) condition in NTDs cases (20%) when compared with controls (12%), however the highest frequency (48%) was observed in NTDs mother. The individual allele (A) frequency was also calculated in NTDs cases (0.42) and their respective controls (0.28). Statistical analysis was carried out using Fischer exact two tailed probability test shows significant difference between NTD mothers and control mothers in heterozygous condition (p=0.006) with O.R (odd ratio) was 2.37 at 95% C.I (1.32-4.26) as mention in Table 4.

On the basis of severity of disease in preclinically diagnosed NTDs cases and their mothers, the data was further analyzed to determine the genotype frequency of *RFCl* variants considered as an important candidate gene regulate folate metabolism for the development of NTD as documented in Table 5. The rare genotype frequency (50%) was observed in homozygous (AA) state of thoracomeningomyelocele (TMMC) cases. Similarly the highest frequency (60%) was also pragmatic in mother of TMMC cases in heterozygous condition.

DISCUSSION

According to MRC (1991) report, folic acid supplementation to the mother having previous history of having NTDs may reduce the incidence up to 70% for developing risk of congenital malformations associated with central nervous system in population, hence, maternal folate act as modifier for NTD (Botto et al., 2005).

Present study suggested that the significant genotypic variants of GCPII (H475Y) in heterozygous condition in NTDs mother act as risk factor for developing NTDs because such (GCPII) variants are associated with

Table 3. Distribution of *RFCI* genotype and their allele frequency between NTDs cases and their respective controls.

Groups	Genotype frequency (%)			Allele frequency (%)	
	GG	GA	AA	G	A
NTD cases	36.00	44.00	20.00	0.58	0.42
Control	52.00	32.00	12.00	0.68	0.28
NTD mothers	36.00	48.00	16.00	0.60	0.40
Control mothers	64.00	28.00	8.00	0.78	0.22

Table 4. *RFC I* genotypes showing odd ratio (O.R) and C.I. at 95% in NTD cases and their respective controls between homozygous and heterozygous conditions.

RFCI 80 G → A	O.R	(95% C.I)	p-value
Cases vs. control			
GG/GG	0.52	0.30-0.91	0.03*
GA/GA	1.67	0.94-2.97	0.11
AA/AA	1.83	0.85-3.94	0.18
Case mother vs. control mother			
GG/GG	0.32	0.18-0.56	0.00*
GA/GA	2.37	1.32- 4.26	0.006**
AA/AA	2.19	0.91- 5.26	0.13

Statistical analysis showing significant*, highly**(p<0.05) differences between cases and controls in homozygous and heterozygous condition after using Fischer exact probability test.

Table 5. RFCI genotypes showing allele frequency (%) in clinically diagnosed NTD cases.

Mutation 80 G → A	Genotype (%) in NTD case			Genotype (%) in NTD mother		
	LMMC	TMMC	Open NTD	LMMC	TMMC	Open NTD
GG	41.7	37.5	75.0	38.5	25.0	33.5
GA	33.0	12.5	00.0	30.7	60.0	50.0
AA	33.4	50.0	25.0	38.5	25.0	16.7

LMMC: *Lumbosacral myelomeningocele*, TMMC: *Thoraco myelomeningocele*.

lowering the folate level and higher homocysteine level as reported earlier by Devlin et al. (2000). Alternative mechanism is also quite possible that GCPII polymorphism may also influence FGCP (folylpoly- γ-glutamate carboxypeptidase) activity through altered post translational processing and /or activity due to altered configuration of the catalytic region of the enzyme. The intestinal FGCP cleaves glutamate residues from FGCP play an important regulatory role in intestinal absorption from dietary FGCP (Halsted, 1990). Hence, the polymorphic variants in the present study affecting the activity of FGCP would predictably decrease the intestinal absorption and consequently decrease folate level and increase homocysteine level in the body. Our findings also suggests that GCPII allele variance may also affects folate absorption resulting in lower serum folate levels and increasing incidence of NTDs in eastern region of the population. Our data agree with those of Brancaccio et al. (2001) findings between control and NTD cases for H475Y of GCPII polymorphism decreases risk for NTD with an OR of 0.35 (0.06-1.86). In the present study we are unable to observed homozygosity (TT rare type) in NTDs cases may be due to small sample size as also observed by earlier study of Afman et al. (2003).

De novo synthesis of folate does not take place in mammals and required efficient reduced folate carrier (*RFCI*) for the cell proliferation and tissue regeneration (Matherly and Goldman, 2003). *RFCI* act as an essential cofactor for the synthesis of purines and pyrimidines synthesis for maintaining the genomic instability (Simonett and Sang, 2005).

RFCI is an important carrier of folate and across the

placenta easily during the embryonic development (Anthony, 1992). The transport of folate molecules in blood cells occur either through a carrier or as receptor mediated mechanism of 5-methyltetrahydrofolate to maintain adequate intracellular concentration of folate in cytoplasm and prevent the incidence of NTD. Hence, the study of polymorphic variation of *RFCI* gene becomes an imperative for determining "risk factor" in the present case cohort study because of earlier information elucidate the controversial reports regarding distribution of frequency between wild and rare type alleles. Interestingly, present study shows significant (p=0.006) difference with 2.3 fold increase of odd ratio at 95% C.I. 1.32 - 4.26 confirming risk for NTDs in heterozygote (GA) state (Table 3), similar findings are also reported by De Marco (2003) and Relton et al. (2003). Although, 80GG genotype in combination with low red blood cell folate levels was associated with a 4.6-fold increase risk in NTDs (Morin et al., 2003). Earlier studies are evident that homocysteine and folate status of the mother has direct impact on NTDs outcome, hence required to further evaluate whether the maternal genotype has a direct impact on development of NTD risk (Kirke, 1993; Molloy et al., 1998). In fact, it is possible that maternal genotype could play an etiopathogenic role in NTDs either due to inadequate supply of folate to the embryo or accumulation of homocysteine with increased concentrations may disrupts the process of neural tube closure. We have observed that the variation in prevalence of TMMC cases were highest among other clinical sub groups may be either due to epigenetic factor or patients belongs to heterogeneous ethnic groups. Environmental factors contributing significant role in the development of NTDs if mothers might have exposed to teratogens acts as carrier mediated interruption transport of folic acid into the cells from out sources as evident from our study. Our findings from homogenous samples reveal significant variation of the G/A allele between subgroups (NTD case, mother and controls). Several earlier studies of *RFCI* 80G allele are contradictory with evidence that either guanine or adenine is associated with increased risk in NTD affected cases and their mothers (Botto and Yang, 2000; Anthony, 1992; Kirke, 1993).

Conclusion

On the basis of earlier linkage studies have yielded some interesting findings but due to lack of reproducibility and their association with certain genes required are evaluation of polymorphic variation of GCPII and RFCI as candidate gene (Greene et al., 2009; Beaudin and Stover, 2009; Copp and Greene, 2010). To elucidate the etiology of NTDs whether the genotypic variants of GCPII influences the function of FGCP enzyme and other studies are also required based on controlled dietary folate supplement levels in individuals having different

genotypes. The variable frequency of genotype in heterozygous state for both the gene GCPII H475Y and RFCI 80GA (475 CT/80 GA) in NTD mothers may play a significant role in regulation of folate metabolism either together or independently to maintain the adequate supply and absorption of folate through dietary supplement as discuss earlier. Similarly, the phenotypic heterogeneity in NTDs may also help to explain mixed response of *RFCI* gene polymorphism with different degree of severity in NTDs due to an increase frequency of "A" allele may act as an independent risk factor for "Birth Defects". Moreover, the present study is small and having interesting findings reporting first time in eastern region has not been reported earlier. However, further study is required to confirm the hypothesis that the variations in dietary intakes of folate are influenced by genetic variants of H475Y GCPII and 80GA RFCI polymorphism. Therefore, larger group of populations are necessary to investigate this associations further to reestablish the functional significance in different clinically defined NTDs. Such study are still continue to increase more samples of the same ethnic group to make the study significant in Indian population. Although, the author unable to reach in conclusion that how both the genes are interact (linked) to each other and responsible for increasing either as independent or together as risk factor for the development of NTDs.

ACKNOWLEDGEMENTS

Ajit Kumar Saxena thankfully acknowledges the Department of Biotechnology, Ministry of Science and Technology, Government of India, grant no. BT/PR/9022/MED/12/331/2007, for providing financial assistance to carry out this research project. The authors are also thankful to Department of Biotechnology for providing the fellowship to the second author J. Gupta.

REFERENCES

Afman LA, Frans JM, Trijbels , Blom HJ (2003). The H475Y polymorphism in the glutamate carboxy peptidase II gene increase plasma folate without affecting the risk for Neural tube defects in Humans. J. Nutr. 133(1):75-77.

Anthony AC (1992). The biological chemistry of folate receptors. Blood 79:2807-2820.

Beaudin AE, Stover PJ (2009). Insights into metabolic mechanisms underlying folate-responsive neural tube defects: a minireview. Birth Defects Res. A Clin. Mol. Teratol. 85:274-284.

Botto LD, Lisi A, Robert-Gnansia E, Elisabeth Robert-Gnansia, Erickson JD, Vollset SE, Mastroiacovo P,Botting B, Cocchi G, de Vigan C, de Walle H, Feijoo M, Irgens LM, McDonnell B, Merlob P, Ritvanen A, Scarano G, Siffel C,Metneki J, Stoll C, Smithells R, Goujard J (2005). International retrospective cohort study of neural tube defects in relation to folic acid recommendations: Are the recommendations working? Br. Med. J. 330:571.

Botto LD, Yang Q (2000). 5,10- methylenetetrahydrofolareductase gene variants and congenital anomalies: a HuGE review. Am. J. Epidemiol. 151:862-877.

Brancaccio R, Passariello A, Buoninconti A (2001). Multiple genotyping for genes of the folate/methionine metabolism in Italian NTD patients. Abstract book of 3rd international Conference on Homocysteine metabolism (abs).

Cabrera RM, Hill DS, Etheredge AJ, Finnell RH (2004). Investigations into the etiology of neural tube defects. Birth Defects Res C Embryo. Today 72:330-344.

Chandler CJ, Harrison DA, Buf fington CA, Santiago NA, Halsted CH (1991). Functional specificity of jejunal brush-border pteroylpolyglutamate hydrolase in pig. Am. J. Physiol. 260:G865-G872.

Chango A, Emery-Fillon N, de Courcy GP, Lambert D, PWster M, Rosenblatt DS, Nicolas JP (2000). A polymorphism (80G A) in the reduced folate carrier gene and its associations with folate status and homocysteinemia. Mol. Genet. Metab. 70:310-315.

Copp AJ, Greene ND (2010). Genetics and development of neural tube defects. J. Pathol. 220:217-230.

De Marco P, Calevo MG, Moroni A, Arata L, Merello E, Cama A, Finnell RH, Andreussi L, Capra V (2001). Polymorphisms in genes involved in folate metabolism as risk factors for NTDs. Eur. J. Pediatr. Surg. 11:S14-S17.

De Marco PMG, Calevo A, Moroni LA, Merello E, Cama A, Finnell RH (2003). Reduced folate carrier polymorphism (80A G) and neural tube defects. Eur. J. Hum. Genet. pp. 245-252.

Detrait ER, George TM, Etchevers HC (2005). Human neural tube defects: developmental biology, epidemiology, and genetics. Neurotoxicol. Teratol. 27:515-524.

Devlin AM, Ling E, Peerson JM, Fernando S, Clarke R, Smith AD, Halsted CH (2000). Glutamate carboxypeptidase II: a polymorphism associated with lower levels of serum folate and hyperhomocysteinemia. Hum. Mol. Genet. 9:2837-2844.

Finnell RH, Gelineau-Van Waes J, Bennett GD (2000). Genetic basis of susceptibility to environmentally induced neural tube defects. Ann. NY Acad. Sci. 919:261-277.

Greene ND, Stanier P, Copp AJ (2009). Genetics of neural tube defects. Hum. Mol. Genet. 18:R113-R129.

Halsted CH (1990). Intestinal absorption of dietary folates. In Picciano, M.F., Stokstad, E. L.R and Gregory, J. F. (eds),Folic acid metabolism in health and disease. Wiley-Liss, New York, NY, pp. 23-45.

Hou Z, Matherly LH (2009). Oligomeric structure of the human reduced folate carrier: identification of homooligomers and dominant–negative effects on carrier expression and function. J. Biol. Chem. 284:3285-3293.

Hunter AG, Cleveland RH, Blickman JG, Holmes LB (1996). A study of level of lesion, associated malformations and sib occurrence risks in spina bifida. Teratology 54:213-218.

Kirke PN, Molloy AM, Daly LE, Burke H, Weir D G, Scott JM (1993). Maternal plasma folate and vitamin B12 are independent risk factors for neural tube defects. Q. J. Med. 86:703-708.

Matherly LH, Goldman DI (2003). Membrane transport of folates. Vitam. Horm. 66:403-456.

Matherly LH, Hou Z, Deng Y (2007). Human reduced folate carrier: translation of basic biology to cancer etiology and therapy. Cancer Metastasis Rev. 26:111-128.

Molloy AM, Mills JL, Kirke PN, Ramsbottom D, Mcpartlin JM, Burkett, Conley M, Morin I, A.M. Devlin, D. Leclerc, N. Sabbaghian, Halsted CH, Finell R, Rozen R (1998). Low blood folates in NTD pregnancies are only partly explained by thermolabile 5,10-methylenetetrahydrofolate reductase: low folate status alone may be the critical factor. Am. J. Med. Genet. 78:155-159.

MRC (1991). Vitamin Study Research Group: Prevention of neural tube defects: results of the Medical Research Council Vitamin Study. Lancet 338:131-137.

Pei L, Liu J, Zhang Y (2009). Association of reduced folate carrier gene polymorphism and maternal folic acid use with neural tube defects. Am. J. Med. Genet. B Neuropsychiatr. Genet. 150B:874-878.

Relton CL, Wilding CS, Jonas PA, Lynch SA, Tawn EJ, Burn J (2003). Genetic susceptibility to neural tube defect pregnancy varies with offspring phenotype. Clin. Genet. pp. 424-428.

Shang Y, Zhao H, Niu B, Li WI, Zhou R, Zhang T, Xie J (2008). Correlation of polymorphisms of MTHFRS and RFC-1 genes with neural tube defects in China. Birth Defects Res. A Clin. Mol. Teratol. 82:3-7.

Simonett F, Sang WC (2005). Gene Nutrient Interaction in one carbon susceptibility to neural tube defect pregnancy varies with offspring phenotype. Clin. Teratol. 85:274-284.

Van der Put NMJ, Steegers-Theunissen RPM, Frosst P (1995). Mutated Methylenetetrahydrofolate reductase as a risk factor for spina bifida. Lnacet 346:1070-1071.

Williams LJ, Mai CT, Edmonds LD, (2002). Prevalence of spina bifida and anencephaly during the transition to mandatory folic acid fortification in the United States. Teratology 66:33-39.

Zhu H, Cabrera RM, Wlodarczyk BJ, Andreussi L, Cama A, Capra V (2007). Differentially expressed genes in embryonic cardiac tissues of mice lacking Folr1 gene activity. BMC Dev. Biol. 7:128.

2

Challenges in conserving and utilizing plant genetic resources (PGR)

Ogwu, M. C., Osawaru, M. E. and Ahana, C. M.

Plant Conservation Unit, Department of Plant Biology and Biotechnology, Faculty of Life Sciences, University of Benin, Benin City, Edo State, Nigeria.

The problems of food and income security are of global significance and are further compounded by precedential increase in world population resulting in overexploitation of natural resources and by extension plant genetic diversity. Plant genetic resources (PGR) refer to the heritable materials contained within and among plant species of present and potential value. In the recent past, genetic diversity found in landrace, weedy and wild cultivars have been reported to savage animal and plant population diseases, pest and environmental changes. Nevertheless, these resources are lost at alarming rates due to anthropogenic product and by products such as climate change, pollution, genetic erosion, gross mismanagement of these resources and population growth. Hence, the need for conservation and sustainable utilization of these resources. PGR conservation is the management of varietal diversity in plant occasioned by interaction between genes and the environment for actual or potential and present or future use. A complimentary application of *in situ* and *ex situ* conservation technique is recommended for their effective conservation. Efficient survey, collection and documentation is also pertinent. International, national and individual appreciation of the value of this vast genetic diversity would facilitate their sustainable utilization. PGR utilization refers to the use value of these genetic resources. There is need to create avenues through which these can be easily accessed and enact effective policies for their protection especially in their hotspot and regions of high endemism.

Key words: Plant genetic resources (PGR), conservation, utilization, environmental changes, population growth, genetic erosion.

INTRODUCTION

Plant genetic resources (PGR) is the pillar upon which world food security and agriculture depends especially with expanding global population. PGR refers to the heritable materials contained within and among plant species of present and potential, economic, scientific or societal value. They include materials considered of systematic importance and applicable in cytogenetic, phylogenetic, evolutionary biology, physiological, bioche-mical, pathological and ecological research and breeding; encompassing all cultivated crops and those of little to no agricultural value as well as their weedy and wild relatives (Ulukan, 2011). Hammer and Teklu (2008) opined that the genetic adaptation and the rate of evolutionary response of a species to selective forces (changing environments, new pests and diseases and new climatic conditions) depend on inherent levels of genetic diversity present at the time. The importance of PGR is reflected in every facet of human endeavor as it provides the gene pool from which resistant and improved varieties can be engineered. The economic value of increasing crop productivity through the diffusion of improved, modern varieties has been extensively documented, particularly in the context of industrialized agriculture by Alston et al. (2000) and Evenson and Gollin (2003). Costs and benefits for plant genetic resources conserved in gene banks, destined principally for use by commercial farmers have also been estimated by Koo et al. (2004) and Smale and Koo (2003). Thus, PGR can play roles in ensuring income security especially for developing and under developed countries where majority of livelihood is hinged upon these natural resources.

In the past, plant genetic diversity loss was largely caused by natural processes, mainly as a result of climate changes and is still occurring although at a very minimal rate. By contrast, the recent acceleration in the loss of plant diversity is mainly due to the activities of the earth's dominant species. Land clearing, overgrazing, the cutting and burning of forests, the indiscriminate use of fertilizers and pesticides, war and civil strife have all impacted negatively to destroyed natural habitats and the diversity contained therein. Sodhi and Erhlich (2010) supported this by stating that earth dominant species have destroyed, degraded and polluted earth's natural habitat which are key in life support. More so, serious illnesses, water contamination and ecological destruction can be attributed to the drilling of oil which has caused widespread destruction of rainforest and endangered the lives of tens of thousands of people (Hvalkoff, 2001). The diversity of genes within species increases its ability to adapt to adverse environmental conditions. When these varieties or populations of these species are destroyed, the genetic diversity within the species is diminished. In many cases, habitat destruction has narrowed the genetic variability of species lowering the ability to adapt to changed environmental conditions. The greater the variability of the species, the more is the ecosystem stability.

This report is aimed at creating awareness on the threats to PGR and the need for their effective conservation and sustainable utilization.

THREATS TO PGR CONSERVATION AND UTILIZATION

The impact of humans upon biological diversity (biodiversity) has gradually increased with growing industrialization, technology, population, production and consumption rates. Food sovereignty, accessibility and security, landscapes and environmental integrity along with gross mismanagement are contending issues which impact on PGR. Since the era of green revolution, Indus-

trial agriculture and increasing globalization of markets, tastes and cultures, much of this wealth is being lost both on-farm and in genebanks, as increasingly the integrity of these resources is being compromised by genetically modified organisms.

This is further compounded by issues arising from patent rights. The world faces major challenges of population growth, climate change, increasing social and economic instability and a continuing failure to achieve food and income security.

Population growth and Urbanization

As human population break new grounds machinery are set up that modifies the natural environment to his thirst resulting in a strangling pressure on land and other natural resources for food, industries, shelter and agriculture; ultimately leading to habitat destruction and loss of plant genetic resources. For example, Malik and Singh (2006) estimated that the food grain demands by the year of 2020 is anticipated to be around 250 million tonnes, which means an extra 72 million tonnes of food grains are required. This could lead to over exploitation of PGR as witnessed during the Green Revolution. Social disruptions or wars and poverty also pose a constant threat to genetic wipeout as it is associated with heavy reliance directly on natural resource which often leads to overexploitation and destruction of wild PGR.

Pollution

Soil and atmospheric biodiversity including microbial diversity and the diversity of pollinators and predators are also under threat of pollution. Threats to these resources include pollution by genetically modified material and the increasing use of intellectual property rights (IPRs) to claim sole ownership over varieties, breeds and genes, which restricts access for farmers and other food producers. This loss of diversity is accelerating and sliding down the slippery slope of food insecurity that today sends more than 1.2 billion people to bed, hungry.

Habitat loss and modification

Exploration and extraction of natural resources affect and alter the geophysical environment of the areas where they are carried on. An example is the environmental impact of oil exploitation in the Niger Delta region of Nigeria which contribute in no small measure to the destruction of the fragile ecosystem, thus making the region 'one of the world's most severely petroleum impacted ecosystems and one of the five most petroleum-polluted environments in the world' (Niger Delta Natural Resource Damage Assessment and

Restoration Project, 2006). With the exploration of oil; spillage, deforestation, noise pollution and other ecological effects, they are not willing to yield to their demands for adequate attention to their polluted and depreciating environment (Olubisi and Oluduro, 2012).

Climate change

Climate change is having a significant negative impact on the environment and on PGR often leading to perturbations such as drought, flood and disease. Changes in rainfall patterns and extreme weather events are likely to diminish crop yields in many areas. Mores so, rise in sea level, causing loss of coastal land and saline water intrusion, also leads to crop depletion (Pisupati and Warner, 2003). This will impact on the distribution of PGR and most likely alter their physiognomy.

Diseases

The effects of human activities have introduced certain levels of stress to natural resources including PGR. This stress will overtime weaken the immunity of the affected population. More so, PGR are now susceptible to different new diseases absent in the original population. Reduction in gene pool increases vulnerability. Control of fungal diseases by chemicals is expensive and can have negative impacts on natural eco-systems whereas genetically based resistance offers efficient and ecologically sound control (Bhullar et al., 2012).

Alien invasive species (IAS)

IAS are also commonly referred to as invasives, aliens, exotics or nonindigenous species. IAS are species, native to one area or region, that have been introduced into an area outside their natural distribution, either by accident or on purpose, and which have colonized or invaded their new home, threatening biological diversity, ecosystems and habitats, and human well-being. The threat posed to biodiversity by IAS is considered second only to that of habitat loss (CBD, 2005). On small islands, it is now comparable with habitat loss as the lead cause of biodiversity loss (Baillie et al., 2004). Invasive species may out-compete native species, repressing or excluding them and, therefore, fundamentally change the ecosystem. They may indirectly transform the structure and species composition of the ecosystem by changing the way in which nutrients are cycled through the ecosystem (McNeely et al., 2001). Entire ecosystems may be placed at risk through knock-on effects. Given the critical role biodiversity places in the maintenance of essential ecosystem functions, IAS may cause changes

in environmental services, such as flood control and water supply, water assimilation, nutrient recycling, conservation and regeneration of soils. Although not all alien species will become invasive or threaten the environment there is need for a clear policy approach because of its potentially wide-ranging impacts when they do become invasive, and because of the difficulties, including financial costs, in reversing its impacts. Virtually all countries in the world are affected at different degrees by IAS. In 2004, IUCN - the World Conservation Union identified 81 IAS in South Africa, 49 in Mauritius, 44 in Swaziland, 37 in Algeria and Madagascar, 35 in Kenya, 28 in Egypt, 26 in Ghana and Zimbabwe, and 22 in Ethiopia (IUCN/SSC/ISSG, 2004).

Patent rights for the protection of plant varieties

Patents are the strongest form of intellectual property (IP) protection in the sense that they normally allow the right holder to exert the greatest control over the use of patented material by limiting the rights of farmers to sell, or reuse seed they have grown, or other breeders to use. Although there is an imbalance between the IP rights afforded to breeders of modern plant varieties and the rights of farmers who were responsible for supplying the plant genetic resources from which such varieties were mainly derived. Patents on plant varieties are only allowed in the US, Japan and Australia. The number of patents relating to rice issued annually in the US has risen from less than 100 in 1995 to over 600 in 2000. The assignment of IPRs to living things is of relatively recent origin in developed countries. The protection of plant varieties (through plant breeder's rights- PBRs), became widespread in the second half of the 20th Century. Thus systems for the protection of plants derive from the economic structure and circumstances of agriculture prevailed in developed countries in this period. All these are challenges because they consider plant not for what they are but for the value that can be derived thereof. More than 90% of crop varieties have been lost from farmers' fields in the past century due to artificial selection.

Replacement of Traditional varieties with modern ones

In recent years, there has been a loss of traditional conservation practices and other customs. This has been mainly because of the expansion of the use of high yielding species and varieties in commercial agriculture, climatic factors, pests and diseases, inappropriate agrarian policies and development activities and poverty, which increase the migration of indigenous youth (with their knowledge, experience and customs of traditional Andean agriculture). The single most important reason for genetic erosion is the replacement of traditional varieties with modern, high yielding, and genetically uniform ones (Rosendal, 1995). Although gene banks play essential role in conserving and maintaining the varieties, FAO (1998) however reported that widespread genetic erosion is also taking place in some, perhaps even many, genebanks, as a result of poor management, poor maintenance, and scarce financial resources, as well as limited institutional capacities. Based on their mixed experience of the Green Revolution, the farmers were sceptical of GM crops. Contamination by GM maize imported from the USA has been found in a wide area of Oaxaca and Puebla states in Mexico on a large scale. The revealing factor is the presence of the cauliflower mosaic virus, which is used widely in GM crops as a promoter to "switch on" insecticidal properties of genes which have been inserted into them. Monsanto, Syngenta and Aventis all use the same technology. Farmers and consumers are unwilling victims of this pollution. Most importantly local races of crops may well become contaminated through cross-pollination, mixed seed stock, illegal imports of GM seed or contaminated food aid grain being unwittingly used as seed. The Biosafety Protocol should be especially vigilant on releases of GM seeds in Centers of Crop Diversity.

Genetic vulnerability and erosion

Genetic vulnerability results when a widely planted crop is uniformly susceptible to a pest, pathogen or environmental hazard as a result of its genetic constitution, with a potential for widespread crop losses. This phenomena continues to be a significant threat in certain crops and countries (for example hybrid rice in China based on a single male sterile source). A significant example of the impact of genetic vulnerability is the outbreak and continued spread of the Ug99 race of wheat stem rust, to which the large majority of existing varieties is susceptible (Pretorius et al., 2000). Creating and maintaining diversity of crops and their varieties in production systems can help to reduce vulnerability and can be said to impact on ecosystem stability. Manioc - originating in South America - is a major food source for more than 200 million people in 31 African countries. According to Prada (2009), the genetic improvement of crop plants relies on the cultivation of genotypes that possess favourable alleles/genes controlling desirable agronomic trait. This process reduces the levels of genetic diversity. As most of the modern genotypes cultivated today have descended from a relatively small number of landraces, the genes controlling important traits have reduced diversity as compared to the gene pool of landraces and wild relatives (Bhullar et al., 2012).

CONSERVATION OF PLANT GENETIC RESOURCES

It has become increasingly clear during the last few

decades that meeting the food needs of the world's growing population depends, to a large extent, on the conservation and sustainable utilization of the world's remaining plant genetic resources. Conservation of plant genetic resources is the process that actively retains the diversity of the gene pool with a view of actual or potential utilization. Utilization is the human exploitation of that genetic diversity. The aim of conservation is to collect and preserve adaptive gene complexes for present or future use (Hammer and Teklu, 2008). The conservation and use of genetic resources is as old as agriculture itself. For over 12,000 years farmers have conserved seed for future planting, domesticated wild plants, and selected and bred varieties to suit their specific needs and conditions. Over the millennia, hundreds of different plant species have been domesticated and within each species, human and natural selection have combined to produce thousands of different varieties. The conservation of genetic resources enables breeders to find the raw materials needed to develop new varieties and farmers to modify their crops in response to changing environments and markets.

The two main conservation strategies are *ex situ* and *in situ*, and each includes a range of different techniques. The products of conservation activities are primarily conserved germplasm, live and dried plants, cultures and conservation data. To ensure safety, conservation products should be duplicated more than one location (Hammer and Teklu, 2008).

Ex-situ conservation

Ex-situ conservation is defined as the conservation of components of biological diversity outside their natural habitat. This includes field gene banks, tissue culture, green house, cryopreservation, seed gene banks, etc. *ex situ* conservation allows the reintroduction of crops in areas where they have been lost through environmental degradation, replacement or war and the stored materials are readily accessible, can be well documented, characterized and evaluated, and are relatively safe from external threats.

Among the various *ex situ* conservation methods, seed storage is the most convenient for long-term conservation of plant genetic resources. This involves desiccation of seeds to low moisture contents and storage at low temperatures (Hammer and Teklu, 2008). For vegetatively propagated and recalcitrant seed species (seed that quickly lose viability and do not survive desiccation), living plants can be stored in field gene banks and/or botanical gardens. Botanical gardens are recommended for the reproduction of rare species. It guarantees freedom from pest infestation and diseases. However, it is extremely labor and cost intensive. Besides, only a limited amount of genetic variation that can be stored and it is vulnerable to natural and human disasters. Biotechnology has generated new opportunities for genetic resources conservation.

Techniques like *in vitro* culture and cryopreservation have made it possible to collect and conserve genetic resources, especially of species that are difficult to conserve as seeds. Cryo-conservation (storage in extreme deep freeze situations) allows for extremely long storage of many species and is accomplished with liquid nitrogen at -196°C. Nonetheless, it is really expensive to maintain and a constant supply of liquid nitrogen has to be available at all times. DNA and pollen storage also contribute to *ex situ* conservation.

In-situ conservation

In situ involves the setting aside and management of natural reserves, where the species are allowed to remain in their ecosystems within a natural or properly managed ecological continuum. This method of conservation is of significance to the wild relatives of crop plants and a number of other crops, especially tree crops and forest species where there are limitations on the effectiveness of *ex situ* methods of conservation (Hammer and Teklu, 2008). It enables species to be conserved under conditions that allow them to continue to evolve.

In situ conservation comprises two main concepts and/or techniques, which may be distinguished as "genetic reserve conservation" and "on-farm conservation." Both involve the maintenance of genetic diversity in the locations where it is encountered, but the former primarily deal with wild species in natural habitats/-ecosystems and the latter with domesticated species in traditional farming systems. The location, management and monitoring of genetic diversity in natural wild populations within defined areas designated for active, long-term conservation is known as genetic reserve conservation. An example of this technique is the establishment and management of forest reserves especially areas of high species diversity.

On-farm conservation is the sustainable management of genetic diversity of locally developed crop varieties (land races), with associated wild and weedy species or forms, by farmers within traditional agricultural, horticultural or agricultural systems and farmer play a major role in this technique through their selection of plant material which influences the evolutionary process and through their decisions to continue with a certain landrace or not. Plant populations on farms have the capacity to support a greater number of rare alleles and different genotypes. The main drawback is the difficulty in characterizing and evaluating the crop's genetic resources and susceptibility to hazards such as extreme weather conditions, pests and disease. For a successful implementation of on-farm conservation, a better understanding of both crop populations on the farming systems that produce them is needed to create active cooperation

between farmers and conservationists. To adequately conserve the full range of genetic diversity of a target species or gene pool, an application of a range of *ex situ* and *in situ* techniques applied in a complementary manner is recommended (Hammer and Teklu, 2008).

SUSTAINABLE UTILIZATION OF PGR

Plant genetic resources are conserved for use by people as food, medicine, fuel, fodder and building materials. According to Hammer and Teklu (2008), conservation without use has little point; conversely, use without conservation means neglecting the genetic base needed by farmers and breeders alike to increase productivity in the future.

Over the last few decades, awareness of the rich diversity of exotic or wild germplasm has increased. This has lead to a more intensive use of this germplasm in breeding and thereby yields of many crops increased dramatically. Domesticated tomato plants are commonly bred with wild tomatoes of a different species to introduce improved resistance to pathogens, nematodes and fungi. Resistance to at least 32 major tomato diseases has been discovered in wild relatives of the cultivated tomato.

To be of use, material held in genebanks must be well documented. This entails maintaining: passport data, giving location, site characteristics, species, cultivar name, characterization data, recording highly heritable characteristics that can be used as a basis to distinguish one accession from another; and evaluation of data, giving traits such as yield, quality, phenology, growth habit and reactions to pest, disease and abiotic stresses. Access to information is becoming increasingly important and information systems which improve access to data are now been made available. For example The International Crop Information System (ICIS) is a data management model and computer based information system developed by CIMMYT.

Networking is another important way of widening the use of plant genetic resources in which priorities are established and tasks shared. Networks bring together all those with an interest in crop genetic resources, whether it is germplasm collectors, curators, researchers, breeders or other users, and provide a means for identifying the genetic resources within a genepool and for taking collective action to conserve and use them.

Over 150 countries are involved in some form of genetic resources networking, and many of the networks themselves have become world-wide fora for sharing resources, ideas, technologies and information (Hammer and Teklu, 2008). They have become an efficient mechanism for enabling countries to share the responsibilities and costs of training, conservation and technology development, and to promote the establishment of joint conservation strategies based on common interests and goals.

RECOMMENDATION AND CONCLUSION

Future development in the improvement of crops largely hinge on immediate conservation of genetic resources for their effective and sustainable utilization. A vast amount of plant genetic resources are threated, endangered and some have even gone extinct and it is more prominent in recent times, mostly due to genetic erosion and environmental transformation by anthropogenic effect.

In other to meet current global challenges all countries and institution must as a matter of primary obligation discover, collect and conserve valuable and potentially valuable plant genetic resource and utilize it sustainably. To this end the following recommendation are of maximum importance for efficient conservation of PGR:

1. An understanding of the extent and distribution of diversity in species and ecosystems is pertinent and this can be achieved through efficient survey, inventory, appropriate research, field studies and analysis.

2. Sustainable agriculture should be promoted through diversification of crop production and development and commercialization of under-utilized crops and species.

3. On-farm management and improvement of plant genetic resources should be supported and this will require integrated approaches combining the best of traditional knowledge and modern technologies.

4. More natural reserved areas should be created and those existing should be properly managed, financially supported and an effective enforcement of laws should guard them.

5. It is important that this diversity be made more useful and valuable to breeders, farmers, and indigenous and local communities, by providing better and more accessible documentation.

6. The best method of conservation is the use of complementary approach of the different *ex situ* and *in situ* conservation techniques. Since part of the worldwide *ex situ* collections is endangered, priority should be placed on securing and providing financial support for existing collections.

7. Means are needed to identify, increase and share fairly and equitably the benefits derived from the conservation and sustainable use of plant genetic resources.

8. Access to and the sharing of both genetic resources and technologies are essential for meeting world food security and needs of the growing world population, and must be facilitated. Such access to and sharing of technologies with developing countries should be provided and/or facilitated under fair and most favourable terms, including concessional and preferential terms, as mutually agreed to by all parties to the transaction. In the case of technology subject to patents and other intellectual property rights, access and transfer of technology should be provided in terms which recognize and are consistent with the adequate and effective protection of intellectual property rights.

9. A comprehensive information retrieval systems for plant genetic resources need to be constructed and development of monitoring and early warning systems for loss of plant genetic resources would be a plus.

10. Public awareness of the value of plant genetic resources through training, seminar and the media should be promoted. Also an integration of conservation priorities into the educational curricula is encouraged.

The value of PGR to human survival cannot be overemphasized and its conservation and sustainable is within our reach. The challenge is now in our cult to preserve these limited resources and secure the future generation.

REFERENCES

Alston JM, Chan-Kang MC, Marra PP, Wyatt TJ (2000). A meta-analysis of rates of return to agricultural R&D: ex pede herculem? Research Report 113, Intl. Food Policy Res. Institute, Washington DC. 118p.

Baillie JE, Hilton-Taylor C, Stuart SN (2004). 2004 IUCN Red List of Threatened Species. A Global Species Assessment. IUCN- World Conservation Union, Gland. 25p

Bhullar NK, Zhang Z, Wicker T, Keller B (2012). Wheat gene bank accessions as a source of new alleles of the powdery mildew resistance gene *Pm3*: a large scale allele mining project. BMC Plant Biology. 10:88
http://www.biomedcentral.com/1471-2229/10/88

CBD (2005). Invasive Alien Species. Convention on Biological Diversity. http://www.biodiv.org/programmes/cross-cutting/alien/

Evenson RE, Gollin D (2003). Assessing the impact of the Green Revolution, 1960 to 2000. Science. 5620: 758-762.

FAO (1998). The State of the World's Plant Genetic Resources for Food and Agriculture. FAO Rome Italy. 89 p

Hammer K, Teklu Y (2008). Plant Genetic Resources: Selected issues from genetic erosion to genetic engineering. J. Agric. Rural Development in the Trop. Subtropics. 109:15-50.

Hvalkoff S (2001). Outrage in rubber and oil. In: People, Plants and Justice. Zerner, C. (ed). New York. Columbia University Press. 56 pp.

Koo B, Pardey PG, Wright BD (2004). Saving Seeds: The Economics of Conserving Crop Genetic Resources Ex Situ in the Future Harvest Centers of the CGIAR. Wallingford UK: CABI Publishing. 67p.

Malik SS, Singh SP (2006). Role of plant genetic resources in sustainable agriculture. Ind. J. Crop Sci. 1(1-2): 21-28.

McNeely JA, Mooney HA, Neville LE, Schei P, Waage JK (2001). Global Strategy on Invasive Alien Species. IUCN - the World Conservation Union, Gland. 46 pp.

Niger Delta Natural Damage Assessment and Restoration Project. (2006). Report. Phase 1- Scoping Report. Federal Ministry of Environment, Abuja Nigeria Conservation Foundation, Lagos, WWF UK, CEESP-IUCN Commission on Environmental, Economic and Social Policy. 267 p.

Olubisi F, Oluduro OF (2012). Nigeria: In Search of Sustainable Peace in the Niger Delta through the Amnesty Programme. J. Sustainable Devel. 5(7):4-7

Pisupati B, Warner E (2003). Biodiversity and the Millennium Development Goals. IUCN/UNDP. p 8.

Prada D (2009) Molecular population genetics and agronomic alleles in seed banks: searching for a needle in a haystack? J. Expt. Bot. 60:2541-2552.

Pretorius ZA, Singh RP, Wagoire WW, Payne TS (2000). Detection of virulence to wheat stem rust resistance gene *Sr31* in *Puccinia graminis*. f. sp. *tritici* in Uganda. Plant Diseases. 84:203-208

Rosendal GK (1995). 'Genbanker—bevaring av biologisk mangfold' (Genebanks—conservation of biodiversity) In: Nils Christian Stenseth, Kjetil Paulsen, and Rolf Karlsen (eds.), Afrika—natur, samfunn og bistand (Oslo: Ad Notam Gyldendal). pp 375-392.

Smale M, Koo B (2003). Genetic Resources Policies: What is a gene bank worth? *Research at a Glance*. Briefs 7-12. IFPRI, IPGRI, and the Systemwide Genetic Resources Program. p. 34

Sodhi NS Erlich PR (2010). Introduction. In: Conservation Biology for all. Sodhi NS Erhlich PR (eds). Oxford University Press. p. 1.

Ulukan H (2011). Plant genetic resources and breeding: current scenario and future prospects. Intl. J. Agric. Biol. 13:447-454.

3

Amplification and sequencing of *Rosaceae* expressed sequence tags (ESTs) as a resource for functional genomics databases

Haddad El Rabey[1]* and Francesca Barale[2]

[1]Genetic Engineering and Biotechnology Institute, Minufiya University, P. O. Box 79 Sadat City, Egypt.
[2]Faculty of Agriculture, Milan University, Milan, Italy.

A total of 30 successful ESTs (expressed sequenced tags) were amplified and sequenced to be intended as a resource for *Rosaceae* functional genomics data base. 23 EST were isolated from the amplification of Earlygold peach (*Prunus persica*) cultivar DNA, 5 ESTs were isolated from the amplification of Texas almond cultivar and two ESTs were isolated from the amplification of F1 DNA of their hybrid. All the sequences were tested for similarity using blast in the nbci database. Because these sequence data are new, only 13 sequences found similarity (10 belong to Earlygold peach cv., two belong to Texas almond cultivar and one belongs to their F1 hybrid), whereas the other 17 (13 belong to Earlygold peach cv., three belong to Texas almond cultivar and one belongs to their F1 hybrid) found no significant similarity. The resulting database can be used as a resource of data and links related to peach and almond EST sequences.

Key words: Peach, almond, expressed sequence tags, functional genomics.

INTRODUCTION

Rosaceae genome size is small (300 Mb) of about twice that of *Arabidopsis* (Baird et al., 1994). *Rosaceae* members are characterized by a relatively short juvenile period (2 - 3 years) and extensive genetics and genomics resources such as molecular marker maps, interesting mutants and clone library resources (Georgi et al., 2002). In addition, it has been demonstrated that molecular marker tools developed in peach are easily applied to other species in the family (Joobeur et al., 1998 and Zhebentyayeva et al., 2003). Peach (*Prunus persica*) is being developed as a model organism for *Rosaceae*, an economically important family that includes fruits and ornamental plants such as apple, pear, strawberry, cherry, almond and rose. To demonstrate the utility of the integrated and fully annotated database and analysis tools, they described a case study where they anchored *Rosaceae* sequences to the peach physical and genetic map by sequence similarity (Jung et al., 2004). Several marker maps of *Prunus* fruit crops have been published, three of them, using peach (Rajapakse et al., 1995), almond x peach (Foolad et al., 1995) and almond

(Viruel et al., 1995) progenies, were constructed mainly with RFLP markers. Joobeur et al. (1998) found that the Texa x Earlygold map has a level of saturation similar to these maps, and it covers most of the distance of the *Prunus* genome and has a sufficient marker density for use in plant breeding. However, its total distance (491 cM) is clearly shorter when compared to the potato (684 cM), tomato (1276 cM) and rice (1491 cM) maps. This difference may be due either to the small nuclear DNA content of the *Prunus* genome, about two and four times smaller than the rice and tomato genomes, respectively (Arumuganathan and Earle, 1992).

Expressed sequence tags (EST) is considered as a functional genomic resource in plant molecular biology. It is produced by transcriptome (the transcribed portion of the genome) sequencing. EST was analyzed in many plant species that is, in *Arabidopsis* (Spiegelman et al., 2000), in grapes (Scott et al., 2000), *Pinus radiata* and *Pinus taeda* (Cato et al., 2001), sugar beet (Schneider et al., 2002), rice (Jin et al., 2003), in *Ginkgo biloba* Brenner et al. (2005) and in tomato (Labate and Baldo, 2005). Annotations of ESTs include contig assembly, putative function, simple sequence repeats, and anchored position to the peach physical map where applicable. The importance of high-quality fruit and the

---

*Corresponding author. E-mail: elrabey@hotmail.com.

intrinsic difficulties of breeding in a perennial species require the development and application of structural and functional genomic databases for the sustained improvement of rosaceaous fruit crops. Identification and characterization of genes controlling the genetic basis of the traits, and their tagging with molecular markers, permits facilitated introgression of important characters, speeding development of new breeding material combining the best traits formerly isolated in separate varieties (Abbott et al., 2006).

The ESTree db (Lazzari et al., 2004, 2005, 2007, 2008) is an Expressed Sequence Tags (ESTs) database that was developed by the Italian ESTree Interuniversitary Centre as a platform for easy genomics and functional genomics data integration and retrieval. Together with the GDR database (genome database for *Rosaceae*), it represents the most complete online resource for peach EST analysis. The ESTree db sequence analysis is based on a semi-automated Perl pipeline that during its steps feeds the tables of a MySQL database. Queries to the database can be performed via a PHP-based web interface. The ESTree and the GDR databases represent the only existing online resources dedicated to peach EST analysis. The two databases are very similar in terms of entry number (71,540 peach sequences in the ESTree db, 70,939 in the GDR db), but quite different in terms of information and its retrieval. The ESTree db clustering procedure produced a data set of 27,097 unigenes, 4,303 of which were derived from our in-house prepared libraries (Lazzari et al., 2008).

The aim of this study was to amplify, isolate and sequence some peach, almond and their F_1 progeny ESTs for GDR and ESTree databases as functional genomic resources.

MATERIALS AND METHODS

Earlygold peach cultivar DNA, Texas almond cultivar DNA and F_1 of cv Texas (almond) x cv Earlygold (peach) DNA and 60 ESTs primers (30 forward and 30 reverse primers as in Table 1) were obtained from Parco Tecnologico Padano, Lodi, Italy as a partial contribution in the ESTree project.

ESTs amplification

50 µl PCR reaction were prepared as follows: 45 µl PCR super mix (Invitrogen), 1 µl forward primer, 1 µl reverse primer and 3 µl DNA (20 ng/µl).

PCR cycle profile

Primary denature at 95°C for 3 min for one cycle and 35 cycles with the following profile: 95°C for 45 s, 57°C for 1 min and 68°C for 1 min, and a final extension at 68°C for 5 min. Two µl of the PCR product were then visualized on ethidium bromide stained 1.5% agarose gel and photographed (Figure 1).

DNA sequencing

The PCR product was purified, quantified (and adjusted as 10 ng/µl) and automatically sequenced using Applied Biosystem Prism 377 Semiadaptive Version 3.2 DNA sequencer (Perkin-Elmer).

EST sequence analysis

The resulted sequences were aligned to NCBI sequences to look for similarities using BLASTn.

RESULTS

Thirty successful EST sequences were obtained from PCR amplification using 60 EST specific primers (Table 1). Figure 1 shows some amplified ESTs.

The first EST resulted from the amplification of Texas almond cultivar DNA using primer no. A7 that annotates to farnesylated protein (ATFP6) is 659 bp. The resulted sequence of this EST is as follows:

TGATGCTCGGGGTGGGAGTTCAGGCTCAGGAACGC
AAGCAATTCCAGGTATTTTTTTTCCTTCACAAAATCTC
CATATCCAAAAAGGGCTTATAAAAAGTAAATCAAGAA
AATCCAAAGTGATTTTTAAATTTTTTTTTTTTTTGGGGG
CGGAAAAAAAGAATTTTTTTTTACTTGTTTTTTGTTGGA
GGGTTTTTTTTTTTTTATTCTTGGAATTTCCAAAAGGGG
AGGCAAGGGATCCCAACACCCCAAAAGTTGTGGGCT
TTTTTTTTTTTTTTTAAAAAAAAAACTTTTTTTTTATTCTTTT
TAGCTGGGGGGGGGGGGCTCCTCTTTTTTTTTTTTTCTT
TAGCGGGTAGCTCCCCCCAAAACCCCAAAAAAAAAA
CAACGGGGGGGGGTGCCGCCTGTACAAAACCAACTAA
ACAACAATACTTCTTTGCCGACATTTTGGTTTTTTTTT
CCCACGGGGGGAAACCCGGAGAAAAAGGCCCGCTAA
GAACGAGAGAGGGGGGAAGTAAAACCTGGTGAAAGG
GGAAAGAAGGGGACGCCAGTGAGGGCGGAGAAAAG
GCCCCCCATACACCTTCCCCGGCAATTAGGACCCCC
CGGGGGGGGGGGGGGGGGGCGTTGTCCCCCTCCCAG
AGGGGGGAGAAAAGGGGGGGGTTTTCGCGCCGCAAAA
AA.

The NCBI BLAST search of this EST showed no significant similarity.

The second EST resulted from the amplification of Earlygold peach cultivar DNA using primer no. A12 that annotates to phosphorybosyl anthranilate transferase 1 (*Arabidopsis thaliana*) is 241 bp. The resulted sequence of this EST is as follows:

TCCATTACCGTTGGGAGGACTCCTACAATGGCTCTC
GATCATGATGAGTTCAAAACTGATATCTGAGGCATCC
TTAATAAACAATTGGATCTTTCTTCATTGTCTGAAATC
ATTAATCAAAACGGATTAAAAGAATTGATAATTTTTAT
CACAACCTGAAGATGTTAATTATATATGGGGGACCAC
TCACATTGGATATGTCTACCCCTTGATCAAGGGGGCT
GATAGCCTTCCCTGAAAA.

Table 1. The EST primers used in EST amplification in this study. F = Forward and R = Reverse.

Primer	Sequence	Primer	Sequence
A7-F	GCCATGGGTGCTCTAGATCATC	A39-F	ACAAAGTAAGATGGCCCCAGAC
A7-R	TGGCCAGAACTCCACCTTCTTG	A39-R	ACTCAATCTTCCTGCTAGAGGG
A12-F	TCTTGCCAACATCTTTCTGCCC	A41-F	TCGTAGCAGTGATAGCCAGAGC
A12-R	TTCAGGGAAGGCTATCAGAACC	A41-R	TAATGCCGGCTGCTTTGTACCC
A16-F	GCGATAAAGCAAGTCGATGTGG	A42-F	GGCACCCAAAGATTGGTGATGG
A16-R	GTAGCCAGCAAAGAAGTTGGAG	A42-R	TCTTTCCCTCCAAGCAGCCTAG
A20-F	TGTGCGTAACAGGAGCATCTGG	A44-F	CGAGCCGGAACTTATAACACCG
A20-R	TCTGTCTGTGGGTCAGTGGATG	A44-R	GAGTCTACCAGCCAAAGGGTAG
A21-F	TTGATGCTCTCAAAGGCTGCTC	A47-F	GCAACAGCATCAGGCAAATTCC
A21-R	ATCAGTCCAGTGTCGTTCATCC	A47-R	TTCCTGCAGTTGTAGACCCATC
A22-F	AGTGATCCCAAGGATTGGCAGG	A49-F	CAGCTGGAGTAGTTCTTCCAAC
A22-R	GTCCCACACATGAACCTCAACG	A49-R	CTAAGCAACCGAGTTACAGCAG
A23-F	TTTTTTCCCTCAGCAACCCGGG	A50-F	GTCCCAGGTGTTTATGCAATCG
A23-R	GAGGGCTGCACTGAAACATACG	A50-R	TGACGAGTCCTTCAGCATCATC
A24-F	TTCCCCTGGCTATTGAGACTGC	16-F	AGCCTAACCAGGGCTCAAAAGG
A24-R	AGTCGACAGCCAAATCAGCACC	16-R	AGCTGTCATCAGTGCTTGGAGG
A25-F	TGACATCAACTCGGCTGTTGGC	19-F	GGCCTCTGGCAAAATTCAAGGC
A25-R	CCTCAACAATGACACCAGAGCC	19-R	TGGAACACACTTGGCAGTGACC
A27-F	CGGCAAACAGAGATGAAGAGCG	37-F	TGGATCCTGATCACGAGGACAG
A27-R	GCCACAACATCACAGTGGTAGG	37-R	ATAGTTCGGGCCAAAACGGTGC
A32-F	GAGGCTTGGCACAAAGGGTTTC	71-F	TTTGCCATTGCCGACGGAGTTC
A32-R	ACCCGCAATGCACGTGTCAATG	71-R	TTGAACTGGTCCCGTTAGTCCC
A35-F	GCGAATGAAGATCACAGCCAGC	132-F	TTGCTGCACCGGAACTGACTAC
A35-R	CGCATTCCTGTAGTGCTACTCG	132-R	TCTTGTCCGTTGCATCAACGCAG
A36-F	TTCCTCCATTGACCTCTACACC	144-F	AAGCTTAGCTGATATGGCAGCC
A36-R	ACATGAGGTTGGCACTGATGAG	144-R	CAGCACACCAAAATTGTCAGGG
A37-F	CCCTTATGCAACAGATGGACTC	147-F	GTATGATCCTCTGGTGCTTTGG
A37-R	TGACCCAAATGATGGTGGTGAG	147-R	TGCCGAGCACATATCCTTTGAG
A38-F	CTACTGCCGTTTATCAGACGAC	160-F	TCCATTGTCAGCCATTGCAGTG
A38-R	AGCTTCATTGCTCAGCTTTGGC	160-R	TGATTTGGCACAGCAGCTATCC

The NCBI BLAST search of this EST showed no significant similarity.

The third EST resulted from the amplification of Earlygold peach cultivar DNA using primer no. A16 that annotates to isoflavone reductase related protein (*Pyrus communis*) is 322 bp and showed the following sequence:

CGTCCGCCTCAGTTGGCTGACAAGGTACGATCATCG CTGCTATCAAGGAAGCTGGCAATGTTAAGGTAAAAGT TTCGATCTTTTATCTTGGTTTTCTCTGTTTTTGCTTCTT GTGTTCGTGTTTTCTGTGTTGCGATCGTGAATGTGTC GTTGGGTTTCAGAGGTTTTTCCCATCTGAGTTTGGAA ACGACGTGGATCGAGTTCATGCTGTTGAGCCAGCAA AAACTGCATTTGCAACCAAGGCCAAAATTCGCAGAAC GATTGAGGCTGAGGGGATCCCTCACACCTATGTGGC CTCCAACTTCTTTGCTGGCGACAATAT.

The NCBI BLAST search of this EST showed no significant similarity.

The fourth EST resulted from the amplification of Earlygold peach cultivar DNA using primer no. A20 that annotates to cinnamyl-alcohol dehydrogenase, putative (CAD) (*Arabidopsis thaliana*) is 695 and showed the following sequence:

TCTTTTTGGTGTGTAACAGGAGCATCTGGTTTCATAG CATCCGGGCTGGTGAAGCTCTTACTGGAACGAGCTT ATATTGTCAAAAGCAACCGTCCGTGACCCAAGTCAGT GTATTTATATAGATATGCCACTACTATTTAATCTCTTC TCCCTTTTTTCTCTCTCCCGAAAAAAGGTTTTTCTTTG GAGATTTTCATTTTCAACTTCTTGGTTTTGATGACATC AACAGAAATTTCTGGATTTCAATTATTGTTCTCATCCA CGATCATATCGAAATAACACTTTTCTTCAGATCCAGC CTTATCATCATGTGCTAAAAATATGTCAAACTAGATTT TGGACCCAACACCATGAGGAGAGAGGAGAGAGGAG AAGACAGAGTGAATATTAAAGAGGAAAACAATTGTGT CAAACTAGATTGAATGAGAGTTGTATCTCCCTAACAT ATAACTAACTCTCTAACTGTATTCCATAACAATACAAT TAGTGACTAATCTTGTACGGTGAGACCTTTTTTTA

A

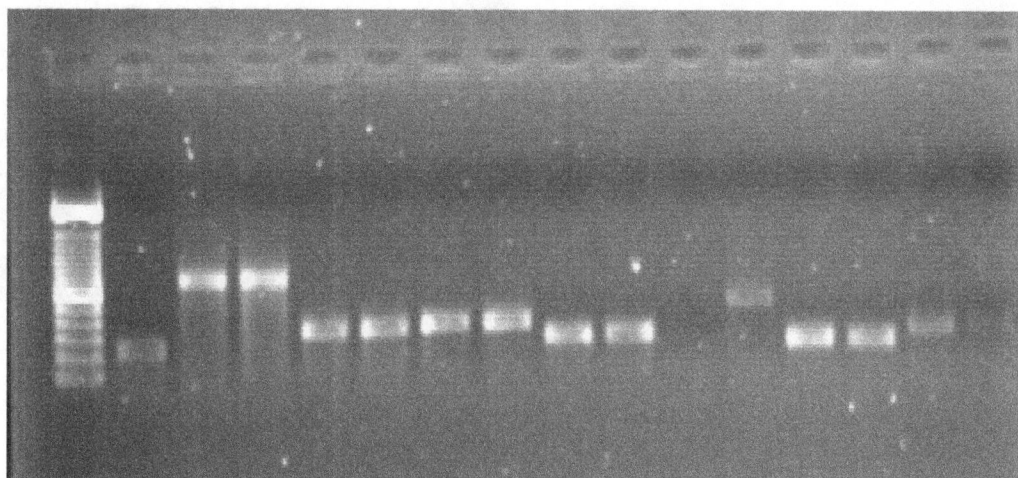

B

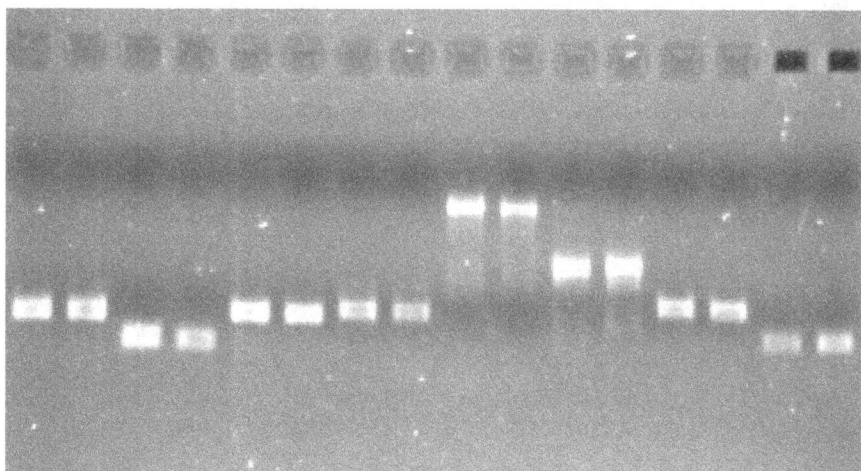

Figure 1. A and B are examples of amplified ESTs in the ESTree project, some of them were sequenced and presented in the current investigation.

AACTAATTACTTTTGTAATTATGCATATGCCACAGATG
ACCCGAAGAAAACAGAACACTTATTGGCACTTGAGG
GGGCAAAAGAAAGGCTCCATTTGTTCAAAGCAGATTT
GTTAGAAGAAGGATCTTTTGATGCTGTTGTGATGGAT
GTGAGGGTGTTTTCATACAAGCTTCC

The NCBI BLAST search of this EST showed no significant similarity.

The fifth EST resulted from the amplification of Earlygold peach cultivar DNA using primer no. A21 that annotates to putative cinnamoyl-CoA reductase (*Oryza sativa*, japonica group) is 676 bp and showed the following sequence:

CCCCCCTTACAAGAGGGGGGGGGGGGCGCCGCCC
GGTTTTTCATATGGAGGGAAGGAGTCATCGACTCCC
CTTAATTTTTGTTGCGGGGGGGGATGATGGGCCTTTG
GAACCATAAGATCACCCCTTCCAAAAAATTGGGGTCT
GAATTTGGGTGGGGAAAAAAGGGGACAACATACAGT
GAAAAATTTTGATTTGTGGCAACAAAGTATTTGGTTT
GTGCCATTTCTCTTTTCAGAAAGAAACTCGCACTTTTT
TCGTTTCTACATTTAAGGAGAGAAAAAAATTGTCTTTT
TTTGCCCGTGTCAGAAATGGGATTTGAACCTCTCTTT
TTTTCTTATCCCAACCACCATAATGTTTTACAATGTAA
TAAGTCCCTTACATCCCTTCCAGAAAAAAATTACGGG
ATTAATATTTGTTGAAGAACACAGTCTTTTAGTCAAGA
AAGCCAATAACGCACAGGGGTTTTATTTCTTTAAAGT
AATAATTAATATTTTTTTTTTCCATTAGGATTTTAAGAT
ATAAATTAAGCTCCGCCACTGTTATCAGCCGACTTTC
TCATGTTTTTGGTAGCCCAATTGGGTTTTGGACCTAA
AAAAAAGATAAAATAAACCACCTTTCTTCCCGAAAAA
AACCTCCACTATTAGGCAAATATTAAATTAAACAAGG
TGTAATACCA

The NCBI BLAST search of this EST showed no significant similarity.

The sixth EST resulted from the amplification of Earlygold peach cultivar DNA using primer no. A22 that annotates to cinnamic acid 4-hydroxylase (*Lithospermum erythrorhizon*) is 161 bp and showed the following sequence:

CCTCTTTGGGGGGGCTCTTTCCCTTCAAAAACATCTCT
CGGCGTTAACTCCTCCGGCGTCCTCTAGAAGGCAGG
GGTGTTTTTGGGCAAATGCATCAAAAGGATTTTTTTC
TCCTCCGGAATGGGGAGCTGCCAGCTTTTCTTTTTTT
ATGTCTTTTGGGGAAAAAAAATAT.

The NCBI BLAST search of this EST showed no significant similarity.

The seventh EST resulted from the amplification of Earlygold peach cultivar DNA using primer no. A23 that annotates to putative cinnamoyl-CoA reductase (*Prunus avium*) is 783 bp and showed the following sequence:

TTTTTTTTTTTTTTTTCTCCCCCGACCGCCCCGGGGTG
GCGTTCCTTCTCGCCACACCGCCACAATAAAAACAAT

ATTTCTTGGGCGAAGCTATTTTCCAACCCTTATGTTC
CCGAACTTGACCCGTAAAAGGGGGTCAGTGAAAGAT
AAAAGCAGTACCAAGGTAGCATTTTGATCTTCTTTAC
AATTTATTTTAGTAAAAATACCTTCTTTTTTTTTCCTTA
TTTTGGGAGTAGGCCTTTCCTTTTTTTTTAGTAAGCCT
TACCAATCCGTTACAATCTCTTCCCCTCAACTCACCC
CGACCTATTTAATTAAATCCAATCCCTTCGACCCTTT
GAAGAGAACACCATTTCAGTCCATTGACCTGGAGCT
GTGTTGAGATGAGGTCGAGCTATATCTTGTGTTGTTT
TGCTTTCCATCATAACGTAGGGGATCAGCAACAAACA
CTTCAGATTTGACGTTGTGAAGCTTATTGTTGTGGAC
ACGAAACCCAACTCCTAAACTTAAGATATGCACATAG
TTAAGCCAAAAAAAATATATATGCACATAGTTATGGAA
AAATATATATATGCACATAGATTCAGTTGCAAAACCA
CTTTCCAGGTGTCTAAGGCCTGGAATTTACCTCGGC
ACCTTGTACTCAGGGTAAAGTTCAGCTACCTTGGCCA
CGAAGTCCCCATAATGTGATATAGCTTCAACGCACAG
GTGTCTACCAGTGGCCGATTTGTTCTCATACACTAAA
ATGTGTGCAAGAGCTACATCTTTAAAATCCACCGGCC
CTCAT

The NCBI BLAST search of this EST showed 19% similarity with *Prunus avium* putative cinnamoyl-CoA reductase (CCR) mRNA.

The 8th EST resulted from the amplification of Texas almond cultivar DNA using primer no. A24 that annotates to ripening-induced protein (*Fragaria vesca*) is 453 bp and showed the following sequence:

TTTTGAACTGGTCCCGTTAGTCCCAAATACATATGCT
CCTGCAATAGCCAAAAATTCATTAGTGACTTGTGAAA
ATTGGTCAAAAACAAAATACTTAGCTGATAACTGAGA
CACTAGGAATGTTACCAAGTGTAGGCATCAAAAGGG
AGAGCACAGCAATTCCAGCAATCTTTAATGGCAACCC
GGTTTTGATCATATCTTGGATTTCAATGTGGCCAGTG
GTAAATCCAACTACATTTGAAGGTGTTCCTGTTGGAA
GCAAGAAAGAAAATTGTGCCCCAATGGCTCCAGGAA
CCATAAGGAGGAGTGGATTTACATGCATGATTTTGGC
TATTTGAATTAGAAGAGGCACAACCAGCGTGGTGGT
GGAGTTGTTTGATGTAAACTCAGTGATGGTGCTACTT
ATGAGACAGACGGCAGGCGCAATGGCAAAATATGGA
ACTGCCCTCCCCA

The NCBI BLAST search of this EST showed 98% similarity with *Prunus persica* (peach) BAC clone 82I18, complete sequence, 67% similarity with *Vitis vinifera* contig VV78X109361.11, whole genome shotgun sequence and 66% with PREDICTED: *Vitis vinifera* hypothetical protein LOC100255398 (LOC100255398), mRNA.

The 9th EST resulted from the amplification of Texas almond cultivar DNA using primer no. A25 that annotates to ripening regulated protein DDTFR10 (*Lycopersicon esculentum*) is 272 bp and showed the following sequence:

TGGGAAGGATGCTCATTCGTTTCATGCAGCTCTTTCA

AAGCCTCCATCATCAGAATTTGTGAACGTGTCTCGGT
GGTACAACCATATCACTGCACTTCTTAGGATTTCGTA
AGTTCTTATAACTTGTGATGGGTTTCTTTTATCTTTAT
TATTAGATTATGATGGATTTCTTCAAAGCTCATATATT
TGGGTTAAATCTTTGACTTATTTGATTATCTCTGGTGG
TGTAGGGGTGTTTCTGGACAAGGCTCTGGTGTCTTT
GTTGAAGAAAA

The NCBI BLAST search of this EST showed no
significant similarity.

The 10th EST resulted from the amplification of
Earlygold peach cultivar DNA using primer no. A27 that
annotates to diphenol oxidase is 312 bp and showed the
following sequence:

TTTGGATAAAACGCCCGGGATTTGCGTAGTGATTTTG
ATGCAGCCGTTCCGTCGGCGAAATCAGGAAGAATCG
GAGTGGGAATGGTTGAGGGGGTGGTGCTGCCGTTG
TATTGAAGAATGGCAGAGGTGGTGCTGTTGTTGAAT
GCAACATCCCCATCAACAAAAGGATGGGAAGCTACG
TGATAGTGGCTGGGAGACTGGTTTGCAACTACCAAA
ATGTCCATGGTTTGGCCTGGAGTTATCATGAGGTAG
GAGGTGGTTATAGGTTTTATGTATGCACCATCTTGAG
CTACCACTGTGATGTTGTGGCAA

The NCBI BLAST search of this EST showed no
significant similarity.

The 11th EST resulted from the amplification of
Earlygold peach cultivar DNA using primer no. A32 that
annotates to aldehyde dehydrogenase (NAD+) (*Nicotiana
tabacum*) is 342 bp and showed the following sequence:

TTGGCGAGAAACATAACCATATTTACCTACAATCTGA
AACAGGATTATGATTGATCTAAACATGGTTGGCTCTT
GGGGGAGGATGACATGCCAAAGCACAGGATGAGATT
TTTGGTCCAGTGCAGTCCATCTTGAAATACAAGTGAG
CAATAAAGCTTTCTTCTCTAAACCTGTTGGTATCCAAT
CCCTTTTGTTAGAATTAACATTAACATTATGGCTGATT
GCAGGGACCTTGATGAGGTGGTAAGAAGGGCAAATA
CTACGCGATACGGGCTTGCTGCAGGGGTCTTCACAC
AAAACATAGATACTGGAAACACATTGACACGTGGATT
GGGGGGTAAA

The NCBI BLAST search of this EST showed 18%
similarity with both *Solanum lycopersicum* cDNA, clone:
LEFL1042BA06, HTC in leaf and *Lycopersicon
esculentum* clone 132363R, mRNA sequence.

The 12th EST resulted from the amplification of
Earlygold peach cultivar DNA using primer no. A35 that
annotates to mevalonate disphosphate decarboxylase
(*Hevea brasiliensis*) is 240 bp and showed the following
sequence:

CCTTTCCATTTTTTTCATTTTGGTTGGAATGGGGGGT
GCGGGGTCAGGGAATTTTTTAAGTCCTGATAAGGTC
AAAGTTATTATGTTTGCTACTATATGGACCCCATGGG
AATGGATTTCACTATGGGGGTTTTTAAGTTCCTCTTG

GAAGTGGGGTAATAAATCCCTTTTCCCCTGGATTCCA
TTATAAAAATACAAGTATTGGAGTTGGGGAAGATAAT
TCTCTAAATTTGTTGTTCT

The NCBI BLAST search of this EST showed no
significant similarity.

The 13th EST resulted from the amplification of Texas
almond cultivar DNA using primer no. A36 that annotates
to mevalonate kinase (*Hevea brasiliensis*) is 694 bp and
showed the following sequence:

TTTTCCTTCCCATGGACCTCTACACCTATGTCTCTCTT
CGCTTTCCCCACTCCTTCTGGGTACGCTCCTATCTCT
CTCTGGCTATTTGGGGTTTCTTCTTCTTCTTCTTCTTC
TTCTTCTTTTTTCTTCCATCTATCCGATTTATTTATTTG
GTTTGGTCTGGTTTTTGGTGCCTTTGGGATTGGTTTT
GATTTGGGTTATGGTTGCGATGTCATTCGGATACAAT
TTTGAGTTCTTTTTACCGTGTCTCTCCTTGTATCATTA
TTGTATTTTCTCTATTTGGGTTCATGGGTTTTTCTTAA
ACTTTGTGATTTTACCTATGCAAAAGTTTGTGTCTTTT
TGTGAGCTTAACATTTGCCTATTGGGTTGCTTGAAAT
CATATGATTAAAAGAAACCCTTTTGATATGATGTATAA
ATCTATTTCGTGGATTACTCTGCTAGCCTAGTTGCAG
GGTTTAAGTGATATTAGAAATCTGATTGATTGCTTTCA
GTAATGTGTTTTCAAGAACGATAAATTTTTTGATCTGG
AAACTAATGGGATTCATATGCTCAAGTGATATTGAAT
GAATTTCTCACTGCTGTTTCCTGCTTTTTTCATAGACA
ATGATGATGCACTAAGACTCCAGCTCAAGGATGTTG
GATTAGAGTTTTCATGGCCAATTGGTAGAATAAAGAA
AGCCCTTTCCAGACAAAT

The NCBI BLAST search of this EST showed no
significant similarity.

The 14th EST resulted from the amplification of Texas
almond cultivar DNA using primer no. A37 that annotates
to lipoxygenase (*Nicotiana attenuate*) is 198 bp and
showed the following sequence:

GCATTGTGAATTGGTCCGAACCTATGTCAATTACTAC
TATCCTGATGCAAGTGCGGTTAATTTTGATACTGAAC
TGCAGGCCTGGTACAATGAGTCAATCAATTTAGGCC
ATGCTGATCTTCGCCATGCTAGCTGGTGGCCTAAAC
TCTCTACTCCAGATGATCTCACATCCATTCTCACCAC
CATCATTTGGGTCAA

The NCBI BLAST search of this EST showed 98%
similarity with *Prunus persica* lipoxygenase 1 mRNA.

The 15th EST resulted from the amplification of
Earlygold peach cultivar DNA using primer no. A38 that
annotates to lipoxygenase is 285 bp and showed the
following sequence:

TGAACATGATGGCTCTAAACAAAAAAAGGGAATCATG
ATCATTTCTCCCACCACAAGGCCTAAGGAAAAGTGTC
TTGTTTTACAGCTTGTTAGTACTGAAACTGAAGCAGG
TAAAGCTATACAACTACTATTATCGTATTATTATTATTT
TATCAATATTAATTTTATTTTTTTATTTTACAATATTTCT

CTTTGGTGCCATATCTTTCTTCTTTGACTTTTCTTCTT
AATGTCGCATTGAAACAAAAGAGTCCACGAAGCCAA
AGCTGAGCAATGAAGCTTAAC

The NCBI BLAST search of this EST showed 40% similarity with *Prunus persica* lipoxygenase 1 mRNA.

The 16th EST resulted from the amplification of Earlygold peach cultivar DNA using primer no. A39 that annotates to acetyltranferase-like protein (*Arabidopsis thaliana*) is 276 bp and showed the following sequence:

TATCTTTGGTGAGGCTTCTAGTCGAGACATCATCTAT
AACTCCTCATTCCTCGAGAACCCTACAGCTCTCTGTT
TTGGATCAGATGGTTCTTAGTCACGTTTACTTCCCAA
CGCTTCTCTTCTATTCCGGAACAATAATATTACTGGTT
CAGGAGGTGGAGCTACTTCTACAGACATGGCGGCCA
TGAGGATGGAGAAAAATTATTGTGTCATCTAATTGGG
TCATTAGCTAAAATCTCTTCACTTCTACCCCCTCGCA
GAAAAATTAAGTGAAAA

The NCBI BLAST search of this EST showed no significant similarity.

The 17th EST resulted from the amplification of F$_1$ of cv *Texas* (almond) x cv *Earlygold* (peach) DNA using primer no. A41 that annotates to nicotinate phosphoribosyltransferase-like protein (*Medicago truncatula*) is 194 bp and showed the following sequence:

CTCTTTTTTTATACTCAATTTCCACATTGGGGGTTATG
TTCCCTCAAGTGTTTTCGGGTTTGAATTAAAATTATAA
TACTTTATCTGAGAGGGTGGGGTCCCCAGCCCCTCC
GCCTCCTAAAAGGATTGGGGTTTCCAAATTAAATTTT
TTTTGTCTATGCCTTCATATGGGTATCCTGGGAAAGG
GAAAAAAA

The NCBI BLAST search of this EST showed no significant similarity.

The 18th EST resulted from the amplification of Earlygold peach cultivar DNA using primer no. A42 that annotates to serine O-acetyltransferase 1 (*Glycine max*) is 159 bp and showed the following sequence:

TGGCCAGGACTTGTATTTTGGGGACATCAAAATTGGT
GAAGGGGCAAAGATTGGGGCTTGTTCTGTGGTTCTA
AAGGAAGTGCCTCCAAGGACTACTGCAGTTGGGAAC
CCAGCTAGGCTGCTTGGAGGGAAAGAACCACCCCC
CTTTTTGGGCCACAA

The NCBI BLAST search of this EST showed no significant similarity.

The 19th EST resulted from the amplification of Earlygold peach cultivar DNA using primer no. A44 that annotates to alcohol acyl transferase (*Pyrus communis*) is 159 bp and showed the following sequence:

CCCGAGAAAATAGGACGCCAGGACGGCCTTCGGTTT
CTTTTCAGTCATCATATCTTATAAAAACAATCCTTCAA

TGAAAGGAAACGACGCCGTTATGGTGATCAAGGAAG
CATTGAGTAGAGCACTAGTGGATTACTACCCTTTGGC
TGGGAGACTCAG

The NCBI BLAST search of this EST showed 94% similarity with *Prunus armeniaca* alcohol acyl transferase (AAT) mRNA.

The 20th EST resulted from the amplification of Earlygold peach cultivar DNA using primer no. A47 that annotates to 3-ketoacyl-CoA thiolase B; acetyl-CoA C-acyltransferase (*Mangifera indica*) is 211 bp and showed the following sequence:

GGTAAAAACAGAAAATTTGGTCTCCTGGGTGACTGTC
TTTCATTTTTCTGTAATTAACTATTTGTATGCATAGATT
GTGGATCCGAAAACTGGAGAGGAGAGGCCTGTTACA
ATTTCTGTGGATGATGGGATCCGGCCAAATGCAAAC
ATGAATGATTTGGCAAAGCTGAAGCCTGCGTTTAAAG
CAGATGGGTCTACAACTGCAAGAAAA

The NCBI BLAST search of this EST showed no significant similarity.

The 21st EST resulted from the amplification of Earlygold peach cultivar DNA using primer no. A49 that annotates to N-myristoyl transferase (*Arabidopsis thaliana*) is 183 bp and showed the following sequence:

ACCAGTTACTGGCATAGGGTCTTTGACCCAAAGAAG
CTTATTGATGTTGGGTTTTCTAGGCTTGGTGCCAGGA
TGACTATGAGCCGAACCATAAAACTGTACAAGTTACC
AGATTCACCAGCTACTCCTGGATTCAGGAAAATGGAA
CTTCGTGATGTCCCTGCTGTAACTCGGTTGCTTAGA

The NCBI BLAST search of this EST showed 98% similarity with predicted: *Vitis vinifera* hypothetical protein LOC100256549 (LOC100256549), mRNA and 88% similarity with *Vitis vinifera* contig VV78X105607.9, whole genome shotgun sequence.

The 22nd EST resulted from the amplification of Earlygold peach cultivar DNA using primer no. A50 that annotates to lipoamide dehydrogenase (*Pisum sativum*) is 237 bp and showed the following sequence:

CGGTAAGGGGGAACTTTCAACCTCGGTATTCAACAC
CAAGGGACTTCACTTGTTCCTCGGTCTTTCCAACAAA
TGCAACTTCAGGGGGGGGTATAGACAACCCCAGGGG
CCAAGTCATAGTCCACATGCCCAACCTTACCAGCAA
GGGACTCCACGCATGCAACCCCATCCTCTTCTGCCT
TGTGGGCTAACATAGGTCCAGGAATAACGTCCCCGA
TTGGATAAACACCTGGGAAAA

The NCBI BLAST search of this EST showed 93% similarity with the following three sequences; *Populus trichocarpa* precursor of dehydrogenase dihydrolipoamide dehydrogenase 1 (LPD1), mRNA, *Populus trichocarpa* x *Populus deltoides* clone WS01314_P09 unknown mRNA and *Populus tremuloides* mitochondrial lipoamide dehydrogenase (LPD1) mRNA; nuclear gene for mitochondrial product.

The 23rd EST resulted from the amplification of F$_1$ of cv *Texas* (almond) x cv *Earlygold* (peach) DNA using primer no. 16 that annotates to pectinesterase 2 precursor is 343 bp and showed the following sequence:

GGGGCCAATTTAAAGGGCTGAAGGCTAGGGAGTAC
GGAGCTGTCAAGGACTGCTTGGAGGAGATGGGTGAT
ACCGTGGACAGGCTCAGCAAATCAGTCCAGGAGCTA
AAGAACATGGGCAAATCCAAGGGCCAGGATTTCGTG
TGGCACATGAGCAATGTGGAGACTTGGGTTAGTGCT
GCTTTGACTGATGACAATACTTGCCTTGATGGGTTCT
CTGGCAAGGCCTTGGATGGCAAAATCAAGGCCTCAA
TCAGAGCTCAGGTGCTTAATGTTGCACAGTGCACTA
GCAATGCTTTGGCCTTGTGCAACAGGTTTGCCTCCA
AGCACTGATGACAGCTTAA

The NCBI BLAST search of this EST showed 46% similarity with *Populus trichocarpa* predicted protein, mRNA and 17% similarity with *Populus trichocarpa* predicted protein, mRNA.

The 24th EST resulted from the amplification of Earlygold peach cultivar DNA using primer no. 19 that annotates to anthocyanidin synthase (*Prunus persica*) is 353 bp and showed the following sequence:

ATGGGGGGGGGGGACTACTTCTTCCACCTTTGTGTAC
CCTGAGGACAAGCGTGACTTGTCCATTTGGCCTCAA
CACCTGCTGATTACATGGAAGTACTCAAGGGAAAAAA
AAAAAATTCTTATTATGCGTGTTGATTGGAAGATTTGT
TGAAAATCTACCAAAAAAAACAACCTCTTTAACAACC
GAATAATGGGGGAGGGGGGTTTTTTCATAAGAAAAA
AATGAGGGTTATCAACAATTTTGTTTTTTTGACGCTCA
CCATTCTTGAGAAAAAAAGGTAACCTTTTTTGGGGGG
GGCCCCGAGAAGTAGGGAGTGGGGGGGGGGGTAAAA
CAAGTGGGCGGGGGGGGGAATTTT

The NCBI BLAST search of this EST showed 49% similarity with the following three sequences; *Prunus persica* leucoanthocyanidin dioxygenase (LDOX) gene, complete cds, *Prunus persica* leucoanthocyanidin dioxygenase (LDOX) gene, LDOX-1 allele, complete cds, *Prunus cerasifera* anthocyanidin synthase (ANS) gene, partial cds and 30% similarity with *Prunus persica* leucoanthocyanidin dioxygenase (LDOX) gene, LDOX-2 allele, complete cds.

The 25th EST resulted from the amplification of Earlygold peach cultivar DNA using primer no. 37 that annotates to catalase (*Prunus persica*) is 214 bp and showed the following sequence:

CGGCCTTGAAGATTTTCTTTCCCCTGCAGCCAGTTG
GCCGTTTGGGTCTGAATAAAAACATCGATAACTTCTT
TGCAGAGAATGAACAACTTGCGTTTAACCCTGCCCAT
GTTGTCCCTGGTGTCTACTATTCAGATGATAAGATGC
TCCAAACTCGAATCTTCGCCTATTCTGATACTCAGAG
GCACCGTTTTGGCCCCGGAAAAAAAAAAAAA

The NCBI BLAST search of this EST showed 92% similarity with the following four sequences; *Prunus persica* mRNA for catalase 1, partial, *Prunus persica* mRNA for catalase (cat2 gene), *Prunus avium* catalase (cat2) mRNA, complete cds, *Prunus persica* mRNA for catalase (cat1 gene), 91% similarity with *Prunus persica* mRNA for catalase (cat1 gene), 88% similarity with *Zantedeschia aethiopica* catalase 1 (cat1) mRNA, complete cds and 80% similarity with *Zantedeschia aethiopica* catalase 1 (cat1) mRNA.

The 26th EST resulted from the amplification of Earlygold peach cultivar DNA using primer no. 71 that annotates to putative sodium-dicarboxylate cotransporter protein (*Arabidopsis thaliana*) is 445 bp and showed the following sequence:

GGGGAAGGCAGTTCCATATTTTGCCATTGCGCCTGT
CGACTGTCCCATAAGTAGCACCATCACTGAGTTTACC
TCTTACAACTCCACCACCACGCTGGTTGGGCCTCTT
CTAATTCAAATAACCAAAATCATGCATGTAAATCCACT
CCTCCTTATGGTTCCTGGAGCCATTGGGGCAGAATT
TTCTTTCTTGGTTCCACAGGAACACCTTCAAATGGAT
TGGATTTACCACTGGCCACATTGAAATCCAAAATATG
AAAAAAACCGGGGTTGCCATTAAAAAATTGCAGAATTGCT
GTGGTCTCCCTTTTGATGCCACACTTGGAACATTCCT
AATGGCTCAGTTTCAACTAATATTTTTTTATTTAACATT
TTCACAAGTCACTAATGAATTTTTGGCTATTGCGGAA
CAATGTTTTGGGACTAACGGGACCAATTCAAAAAATA
A

The NCBI BLAST search of this EST showed 97% similarity with *Prunus persica* (peach) BAC clone 82l18, complete sequence.

The 27th EST resulted from the amplification of Earlygold peach cultivar DNA using primer no. 132 that annotates to cysteine protease (*Anthurium andraeanum*) is 661 bp and showed the following sequence:

AGGTTCTTCTGGGCCTCAAGGGTAGGGAAATTCTAT
CATTCGTTTCATTTCAATATCTACTTGGTGTTTGACAC
TGAAAACCAACATTATAGTGTACTTCGCAAGTTGCAA
ACTAATACTTAACCTAGTGTAGCAAGTTATGAAGCAT
ATTAGAAAAGGGAAAAAAGTCGTCAAGTGTATACAGA
ATCATTTCATTTACCTCTCAAGGTAGATACATAGAATC
AATAGAGCCGGCCGTTAAACTAATAACTAATGTCTCT
CATATTTTCATTACACAGAATGCGGGAGATTTCTGGG
GAGTATCTGCAAAGAAGAGAAAAATGGCCAAGCACA
AGCTCCCATGGACTAAAGTAGAGCAAACAGAGAAGA
CATACCACCCACTGCAGTGGAAGATGAACCGGTTTG
CTGCAATGCGCTGAAAAAGGGAAAGGAAGTTATTGT
TTCTGTGAAGTTTGAAGAAGGTTGTCACTATTGGTAT
TTAAATTTATGTAAAACAGGGATTTGTTCCTCCTGGTT
TGTTTTTTTGGTTTTTTAACAAATGTTCTTGGGGAGCA
TTGGTTTCTTAAACATATGGGGGGGAACGAAAAACCTG
TAAAAGAAAAAATATTTTGAGAAAAGATGATTTTTTCT
CCTGCCGTTGGATGCAGCGGAGAAAAAAAAAA

The NCBI BLAST search of this EST showed no

significant similarity.

The 28th EST resulted from the amplification of Earlygold peach cultivar DNA using primer no. 144 that annotates to Loring-Oro-Red haven is 777 bp and showed the following sequence:

TCCCTTTTTCTATCTTTGTGGTTCATGTTATCAGCCTG
TGCGCTTTTCTCAAAAACATGGATTGAAGTTTCTGGA
TCCTCTGCCAAGGATGTTGCCAAGCAGCTCAAGGTA
TTAAATCTGAATATATCAGTTATGTTTTCTCTAATATA
CCAAAGAAAGTGTTTATATTTTTGTTGGGATCTCTTTC
TTGCAAATGAGGGGGTTTGTGTGGGTAGATACCTAC
CCGTTATATGTATTTTCTAATATATTTCTTATGATTTGT
TTCACTTGATGCTTTCGAGTTTAATGACCAGATCATC
CTTTTTTTATTTTCATACCGGAAGCGCTATTTTTATAA
TAAAGGGGAGTTTGAACAATTCTTAAGTCAGAACTTT
ATGGATTATACTCAGAGATAAGCATCATGCTAACAAA
ATCCCTTTATAATTTGCAAGAACAACAAATGGTGATG
CCTGGTCATCGTGAATCAAACTTGCAAAAGGAGTTGA
ACCGCTACATTCCCACAGCTGCTGCTTTTGGAGGCA
TGTGCATCGGAGCACTGACAGTGTTGGCCGATTTCT
TGGGCGCAATTGGTTCAGGAACATGAATTCTGCTTG
CAGTGACAATCATCTATCAGTACTTTTAGACATTCGA
GAAAGAAAGAGCTTGGAGCTCGATTTCTCTATGATTC
CATGCAATACTGTTGCCAGAGATGGTGGGTCTCACC
GACGATTTCTGGTGTGATCGAGCAATTTGTTGCCAGT
AATGTTTGGTCTCCCTGACAATTTTGGTGGTGCTGAG

The NCBI BLAST search of this EST showed 27% similarity with both *Solanum lycopersicum* chromosome 2 clones C02SLe0011K05, complete sequence and *Solanum lycopersicum* cDNA, clone: FC03DH06, HTC in fruit.

The 29th EST resulted from the amplification of Earlygold peach cultivar DNA using primer no. 147 that annotates to glucose acyltransferase (*Lycopersicon pennellii*) is 632 bp and showed the following sequence:

AGAAAAACAGGAGCCGTCCCTTTCGTTCGTCCCGAT
ATTATACCTCATAAAATCTGCATTTTTAGATGAATTAT
TATATTTTCTAATTTTATGAAATTTTATAATCAGAAAAT
CTTTTTAAAAGAAAGGGAAGGCTCTGTTTTCCCTGTT
TTTTTTGTTCTTTTTTGGGTTAGGCAGGGCGCCAAAG
CACACAAGTAAACCTAATGAGCAACTAACCGAGCCC
AAGAGCCTCAAAACCTATCAAAAGCCATACAAAGTTG
ATTAGCGTTAGCTTTCTAAGTATATTCTGTGTATTTTT
CACTAATACACAGAATATAATTGTAACTCTGCACATTA
AAGTTCCAACTGTTTAACACATACTTTGTTGTAGGTC
CCCTATCCTTTGACTATGCACATTCCATTGGCAACAA
ACCAAAATTAAAATTGAATCCATATTCCTGGACAAAG
GTAGGTTTTCGTGTTTGTGTCTTTGTATGTGATCTTTA
AAACTATTCAGGTGCGTATGACATTGAGAAGGAAAT
ATCCCAAAACTTTCAGGTTGCCCACATAATATTTTTTT
AGACGCACCAGGGGGCACAGGGATTCTCCTTATGCG
AAAAAAATTGGGAAAGGATATAGCAATTCTCATGAC

The NCBI BLAST search of this EST showed no significant similarity.

The 30th EST resulted from the amplification of Earlygold peach cultivar DNA using primer no. 160 that annotates to fantasia-Bolero-Red haven is 305 bp and showed the following sequence:

GTGCAAGCCCTACCTCTTTATTACGCTGTAACTCGAG
ACGAATTGCTAGGGATGGCGGGAGAGGTATTTGGTA
ATGTTCAATCAGGTGTATTGCGGGTTCGAGTAAATCA
CACCTACCCATTGTCTCAAGCAGCACAGGCACACGA
AGACCTTGAGAATAGGAAAACATCTGGATCTGTTGTG
CTTATCCCCTAAGGCAAATATGAGTTTGGTCTGTTCA
TTTTAAAGGACTCGTGGGGGGTGTGTGAAAATGAAT
AAGGAACGTTTGCCTTCCTGTTCCGGATAGCTGCTG
TGCCCAATCAAAA

The NCBI BLAST search of this EST showed no significant similarity.

DISCUSSION

EST localization derived from candidate genes for a specific function in genetic maps. Therefore, peach and almond functional genomics has been very important due to the effort to improve peach fruit properties such as flesh softening, ethylene metabolism, aroma productions, nutraceutical, etc. However, the genomic databases of peach and almond and their progeny can play a significant role in the gene discovery and the genetic understanding of related species (Lazzari et al., 2005).

In the current study, the thirty ESTs were sequenced for the first time and so far, they were not submitted to the databases. Consequently, 17 ESTs (13 belong to Earlygold peach cv., three belong to Texas almond cv. and one belongs to their F[1] hybrid) -and resulted from the amplification using the following 17 primers; A7, A12, A16, A20, A21, A22, A25, A27, A35, A36, A39, A41, A42, A47, 132, 147, 160- out of the thirty ESTs showed no significant similarity when they were subjected to the BLAST search in the NCBI web site for sequence alignment. On the other hand, thirteen sequences (that resulted from the amplification using the following 13 primers; A23, A24, A32, A37, A38, A44, A49, A50, 16, 19, 37, 71, 144) had different levels of similarities ranges from 19 - 99%. Yhis results are consistent with (Lazzari et al., 2005). The distribution of these 13 ESTs is as follows; 10 belong to Earlygold peach cv., two belong to Texas almond cultivar and one belongs to their F[1] hybrid.

The sequenced EST of the current study will be submitted to databases to be used as a resource for ESTree (Lazzari et al., 2004, 2005, 2008) and genome database for *Rosaceae* (GDR) (Jung et al., 2004). The ESTree database pipeline was used in EST analyses for related projects, with different input datasets; data flow was maintained through the entire process, but allowing the preparation of dataset-specific outputs. The contig assembly process was kept apart from the putative SNP detection procedure, allowing the two processes to be carried out independently. In some cases, different features were added and easily integrated in the procedure; that is blast analysis versus species specific genomic sequences (Lazzari et al., 2004). On the other hand, Lazzari et al. (2008) introduced version VI of ESTree database. This ESTree database offers a broad overview on peach gene expression. They reported that EST provides systematic sampling of the transcribed portion of the genome, provides "sequence tags" allowing unique identification of genes, provides experimental evidence for the positions of exons, provides regions coding for potentially new proteins and provides clones for DNA microarrays. On the other hand EST has some limitaions; some cDNA are over-represented and rediscovered many times before a weakly expressed gene can be identified, partial representation due to tissue-specific and developmental regulation of gene expression and non-overlapping reads from the 5' end are scored as independent genes (Lazzari et al., 2008).

SUMMARY

A total of 30 successful ESTs (expressed sequence tags) were amplified and sequenced to be a resource for *Rosaceae* functional genomics data base. Twenty-three EST were isolated from the amplification of Earlygold peach (*Prunus persica*) cultivar DNA, 5 ESTs were isolated from the amplification of Texas almond cultivar and two ESTs were isolated from the amplification of F_1 DNA of their hybrid. All the sequences were tested for similarity using BLAST in the NCBI (National Center for Biotechnology Information) database. Because these sequence data are new, only 13 sequences found similarity (10 belong to Earlygold peach cv., two belong to Texas almond cultivar and one belongs to their F_1 hybrid), whereas the other 17 (13 belong to Earlygold peach cv., three belong to Texas almond cultivar and one belongs to their F_1 hybrid) showed no significant similarity. The resulting database will be used as a resource of data and links related to peach and almond EST sequence databases.

ACKNOWLEDGEMENT

The authors would like to thank Prof. Salamini from Parco Tecnologico Padano, Lodi, Italy for dedicating materials of this study.

REFERENCES

Abbott AG, Zebentenvayeva T, Georgi L, Garay L, Horn R, Jung S, Main D, Lalli JD, Decroocq V, Badenes ML, Baird WV, Reighard GL (2006). The *Rosaceae* genome database: A tool for improving apricot genetics and agriculture. ISHS Acta Horticulturae International Symposium on Apricot Breeding and Culture 717: 13.

Arumuganathan K, Earle ED (1992). Nuclear DNA content of some important plant species. Plant Mol. Biol. Rep. 9: 208-218.

Baird WV, Estager AS, Wells J (1994). Estimating nuclear DNA content in peach and related diploid species using laser flow cytometry and DNA hybridization. J. Am. Soc. Hort. Sci. 119: 1312-1316.

Brenner ED, Katari MS, Stevenson DW, Rudd SA, Douglas AW, Moss WN, Twigg RW, Runko SJ, Stellari GM, McCombie WR, Coruzzi GM (2005). EST analysis in *Ginkgo biloba*: an assessment of conserved developmental regulators and gymnosperm specific genes. BMC Genomics 15(6): 143.

Cato SA, Gardner RC, Kent J. Richardson TE (2001). A rapid PCR-based method for genetically mapping ESTs. Theor. Appl. Genet. 102: 296-306.

Foolad MR, Arulsekar S, Becerra V, Bliss FA (1995). A genetic map of *Prunus* based on an interspecific cross between peach and almond. Theor. Appl. Genet. 91: 262-269.

Georgi L, Wang Y, Yvergniaux D, Ormsbee T, Inigo M, Reighard G, Abbott G (2002). Construction of a BAC library and its application to the identification of simple sequence repeats in peach [*Prunus persica* (L.) Batsch]. Theor Appl. Genet. 105: 1151-1158.

Jin Q, Waters D, Cordeiro GM, Robert Henry RJ. Reinke RF (2003). A single nucleotide polymorphism (SNP) marker linked to the fragrance gene in rice (*Oryza sativa* L.). Plant Sci. 165: 359-364.

Joobeur T, Viruel MA, de Vicente MC, Jauregui B, Ballester J, Dettori MT, Verde I, Truco MJ, Messeguer R, Batlle I, Quarta R, Dirlwanger E, Arus P (1998). Construction of a saturated linkage map for *Prunus* using an almond X peach F_2 progeny. Theor. Appl. Genet. 97: 1034-1041.

Jung S, Jesudurai C, Staton M, Du Z, Ficklin S, Cho I, Abbott A, Tomkins J, Main D (2004). GDR (Genome Database for *Rosaceae*): integrated web resources for *Rosaceae* genomics and genetics research. BMC Bioinformatics. 9 (5):130.

Labate JA , Baldo AM (2005). Tomato SNP discovery by EST mining and resequencing. Molecular Breeding. 16: 343-349.

Lazzari B, Caprera A, Cosentino C, Stella A, Milanesi L, Viotti A (2007). ESTuber db: An online database for *Tuber borchii* EST sequences. BMC Bioinformatics 8: 13.

Lazzari B, Caprera A, Vecchietti A, Merelli I, Barale F, Milanesi L, Stella A, Pozzi C (2008). Version VI of the ESTree db: an improved tool for peach transcriptome analysis. BMC Bioinformatics. 9 (2): S9doi:10.1186/1471-2105-9-S2-S9.

Lazzari B, Caprera A, Vecchietti A, Stella A, Milanesi L, Pozzi C (2005). ESTree db: a Tool for Peach Functional Genomics. Italian Society of Bioinformatics (BITS): Annual Meeting.

Lazzari B, Caprera A, Milanesi L, Stella A, Bianchi F, Vecchietti A, Cosentino C, Viotti A, Pozzi C (2004). ESTree DB and ESTuber DB: a fully automated procedure for EST sequence analysis and database management. Proceedings of the XLVIII Italian Soc of Agric Genet. SIFV-SIGA Joint Meeting: Lecce.

Rajapakse S, Beltho LE, He G, Estager AE, Scorza R, Verde I, Ballard RE, Baird VW, Callahan A, Monet R, Abbott AG (1995). Genetic mapping in peach using morphological, RFLP and RAPD markers. Theor. Appl. Genet. 90: 503-510.

Schneider K, Schäfer-Pregl R, Borchardt DC, Salamini F (2002). Mapping QTLs for sucrose content, yield and quality in a sugar beet population fingerprinted by EST-related markers. Theor. Appl. Genet. 104: 1107-1113.

Scott KD, Eggler P, Seaton G, Rossetto M, Ablett EM, Lee LS, Henry RJ (2000). Analysis of SSRs derived from grape ESTs. Theor. Appl. Genet. 100: 723-726.

Spiegelman JI, Mindrinos MN, Fankhauser C, Richards D, Lutes J, Chory J, Oefner PJ (2000). Cloning of the Arabidopsis RSF1 Gene by Using a Mapping Strategy Based on High-Density DNA Arrays and Denaturing High-Performance Liquid Chromatography. The Plant Cell. 12: 2485-2498.

Viruel MA, Messeguer R, de Vicente MC, Garcia-Mas J, Puigdomenech P, Vargas FJ, Arus P (1995). A linkage map with RFLP and isozyme markers for almond. Theor. Appl. Genet. 91: 964-971.

Zhebentyayeva N, Reighard L, Gorina M, Abbott G (2003). Simple sequence repeat (SSR) analysis for assessment of genetic variability in apricot germplasm. Theor. Appl. Genet. 106: 435-44.

Four novel mutations detected in the exon 1 of MBL2 gene associated with rheumatic heart disease in South Indian patients

Radha Saraswathy[1]*, V. G. Abilash[1], G. Manivannan[1], Alex George[1] and K. Thirumal Babu[2]

[1]Division of Biomolecules and Genetics, School of Biosciences and Technology, VIT University, Vellore 632014, Tamilnadu, India.
[2]Heart line Medical and Research Centre, 72, Thennamaram Street, Vellore-1, India.

The aim of this study was to determine the genetic variations associated with the mannose-binding lectin 2 (MBL2) gene in rheumatic heart disease (RHD) patients in the Vellore region, South India. This study included 50 patients with RHD and equal number of age and sex matched healthy controls. The genomic DNA was extracted from peripheral blood, to find out the genetic variations if any in MBL2 gene. The exon 1 of MBL2 gene was amplified by polymerase chain reaction (PCR) and then screened with Single Strand Conformation Polymorphism (SSCP) analysis. DNA sequencing was carried out in ABI PRISM® 3730 DNA analyzer. The sequence data were edited as required using the sequence analysing software and sequences were aligned using Autoassembler version 2.0 software. Four novel mutations in four RHD patients in exon 1 of MBL2 gene were observed, (1) 46 G/A (Heteroplasmic mutation) (2) 47 G (deletion), 3) 67 G → A (serine to phenyl alanine) and (4) 96 G (insertion). This is the first report of these novel mutations detected in exon 1 of MBL2 gene of RHD patients in South India. The clinical importance of the study is understanding the genomic nature of every population may show variation in its degree of susceptibility to any environmental insults.

Key words: Rheumatic heart disease, RHD, MBL2 mutation, mannan-binding lectin.

INTRODUCTION

Cardiovascular diseases are the known major and growing contributors to mortality and morbidity in South Asia (Nishtar, 2002). RHD continues to be an important problem in our region, Vellore, South India, though a decline in some other countries has been reported (Jose and Gomathi, 2003; Lalchandani et al., 2000; Krishnaswami et al., 1991). Death rate of coronary heart disease in India rose from 1.17 million in 1990 to 1.59 million in 2000 and is expected to rise to 2.03 million in 2010 (Ghaffar et al., 2004). Almost half of the adult disease burden in South Asia is attributable to non-communicable diseases and the environmental factors that are major determinants. However other factors such as sedentary lifestyles, extreme poverty and inadequate

health systems also contribute to the disease.

It is well known that Rheumatic fever (RF), which results from a nonsuppurative sequela of pharyngitis caused by group A streptococcus (GAS) in untreated genetically susceptible hosts, displays a wide spectrum of clinical manifestations including carditis, arthritis, chorea, subcutaneous nodules and erythema marginatum (Bisno, 2000). Recent reports from the developing world have documented Rheumatic Fever incidence rates as high as 206/100 000 and RHD prevalence rates as high as 18.6/1000. The high frequency of RHD in developing world necessitates aggressive prevention and control measures.

GAS displayed strong binding to mannose-binding lectin (MBL) (Neth et al., 2000) encoded by the MBL2 gene in the chromosome region 10q11.1-q21. It plays a major role in innate immunity due to its ability to opsonize pathogens, to enhance their phagocytosis and to activate

---

*Corresponding author. E-mail: r_saraswathy@yahoo.com.

the complement cascade via the lectin pathway (Jack et al., 2001). Inherited insufficiency of MBL that impairs the innate immune function and enhances susceptibility to infection has been shown to be essentially due to three structural variants in exon 1 of the MBL2 gene at codons 52 (C to T), 54 (G to A) and 57 (G to A), corresponding, respectively, to changes of arginine to cysteine (Arg52Cys), glycine to aspartic acid (Gly54Asp) and glycine to glutamic acid (Gly57Glu) in the protein. Any of the three amino acid changes disrupts the collagen helix of the MBL molecule and homozygosity or compound heterozygosity for any of the three alleles results in MBL deficiency (Sumiya et al., 1991).

The worldwide high prevalence of these variants in MBL suggests an apparently ambivalent role for MBL2 in a number of pathogenetic and homeostatic processes (Garred et al., 1994; Ezekowitz, 1998; Saevarsdottir et al., 2004).

Further the knowledge of MBL2 expression may help to understand the pathogenetic mechanisms involved in cardiovascular disease (Best et al., 2009).

It is known that the structure of human populations is relevant in various epidemiological contexts. As a result of variation in frequencies of both genetic and non-genetic risk factors, rates of disease and of such pheno-types as adverse drug response vary across populations (Rosenberg et al., 2002). It was thought it is reasonable to check for the type and nature of MBL2 mutations if any associated with RHD that may provide different type of useful information in a new rare geographical population such as one of South Indian population, Vellore and its surrounding area. As expected four new mutations in exon 1 of MBL2 gene presenting in RHD patients were detected. All those mutations are assigned under four types of novel MBL2 mutation observed for the first time showing the importance of the population genomes in mutation analysis. It is advisable to investigate their usefulness both in basic and clinical studies.

MBL2 gene polymorphism are well established and hence in this study it was tempted to explore the possibility of further genetic variations in the MBL2 gene in rheumatic heart disease (RHD) patients in the Vellore region, South India and this study was undertaken.

MATERIALS AND METHODS

Patients and controls

This study included 50 patients with rheumatic heart disease and equal number of healthy controls. The clinical details of RHD patients are presented in the Table 1.

Both patients and controls provided written consent. Clinically suspected RHD patients were carried out with routine blood investigations, Electrocardiogram (ECG) and Chest X-ray. Confirmation of RHD was done by 2 D colour Doppler (ESAOTE MYLAB 5O X Vision, USA). Patients with at least one deficit valve involvement were included. Cases already underwent surgery or interventional procedures or on drugs were not included in this study. Out of 200 cases referred, 50 were selected for this study.

MBL genotyping

Blood samples were drawn in EDTA vacutainers from the patients. The molecular studies were carried out in all the cases with equal number of control samples.

DNA extraction, quantification and PCR analysis

9 ml of intravenous blood was sampled from all the patients and control by using EDTA coated vaccutainer. The genomic DNA was extracted from peripheral blood by using modified method of Lotery et al. (2000) and standardized at Biomedical Genetics Research Lab at VIT University. Qualitative analysis of DNA was carried out by 0.8% Agarose Gel Electrophoresis and quantification of DNA by using Biophotometer (Eppendorf). Dilutions of DNA were made up to 10 ng/µl concentration by using Tris Ethylene diamine tetra acetic acid (TE) buffer, pH-8.0. The 10 ng/ul of concentration of DNA solution was checked on 0.8% agarose gel. The exon 1 of MBL2 gene was amplified with a pair of primers derived from the published sequence (NCBI accession no. NG_008196 (F -5'-AGGCAGCCAGGCTACTATCA- 3', R (5' TTTGGGGTTGGATGGAAATA -3'). To confirm the amplification of PCR product of exon1 of MBL2 gene, the PCR products were checked by electrophoresis in a 2% agarose gel containing ethidium bromide (0.5 mg/ml) and the bands visualized under UV illumination.

Single strand conformation polymorphism (SSCP) analysis

SSCP was performed by the modified protocol (Orita et al., 1989) that was standardized in Biomedical Genetics Research Laboratory in VIT University, Vellore. Samples were denatured at 95°C for 5 min and immediately placed on ice for 1 min. 4 µl of PCR product was loaded on 6% polyacrylamide gel. The electrophoresis set up was run at 80 v for 5 h, silver stained and photographed under gel documentation system (Lark innovative, India).

DNA sequencing and analysis

It was carried out using Dideoxy chain termination method (Sanger and Coulson, 1975). The Sequencing was carried out in ABI PRISM® 3730 DNA Analyzer. The sequence data were edited as required using Sequencing Analysis Software™ (Applied Biosystems, USA) and sequences were aligned using Autoassembler version 2.0 software (Applied Biosystems, USA) for identification of mutations/polymorphisms.

RESULTS

The patients (n = 50) undertaken for this study were diagnosed by the clinician and their clinical features are presented in Table 1. In this study, the patients were affected in the first decade and 22 out of 50 patients (44%) showed consanguinity.

The SSCP analysis of the PCR products of the exon 1 of MBL2 gene in RHD patients is presented in Figure 1a.

Out of 50 samples analysed, only 5 (10%) samples showed polymorphism by Single Strand Conformation Polymorphism analysis (SSCP) in MBL2 gene exon1 (RHD 13-1, RHD 19-1, RHD30-1, RHD 31-1, RHD 46-1) and presented in Figure 1a.

Table 1. Clinical and general characteristics of RHD patients

Sl. No.	Individuals studied (case no)	Age at reporting	Sex	Consanguinity	Clinical features (valve defects)
1	RHD-1-1	19	F	Yes	MS/MR
2	RHD-2-1	45	M	Yes	MS
3	RHD-3-1	30	F	Yes	MS
4	RHD-4-1	32	F	Yes	MS
5	RHD-5-1	23	F	No	MS/MR
6	RHD-6-1	11	M	Yes	MR
7	RHD-7-1	21	F	No	MS
8	RHD-8-1	28	F	No	MS
9	RHD-9-1	31	F	No	MS
10	RHD-10-1	30	F	No	MS
11	RHD-11-1	28	F	Yes	MS
12	RHD-12-1	37	F	No	MS/MR
13	RHD-13-1	35	M	No	MS/MR
14	RHD-14-1	50	F	No	MS
15	RHD-15-1	35	M	Yes	MS
16	RHD-16-1	23	F	No	MS
17	RHD-17-1	26	M	No	MR
18	RHD-18-1	48	F	No	MS
19	RHD-19-1	50	F	Yes	MS
20	RHD-20-1	11	F	No	MR with MVP
21	RHD-21-1	34	M	No	MS/MR
22	RHD-22-1	50	F	No	MS/MR
23	RHD-23-1	30	F	Yes	MS/MR
24	RHD-24-1	13	M	Yes	MS/MR
25	RHD-25-1	35	F	No	MS/MR
26	RHD-26-1	32	F	No	MS/MR
27	RHD-27-1	11	M	Yes	MS/MR
28	RHD-28-1	49	M	Yes	MS
29	RHD-29-1	10	M	Yes	MS/MR
30	RHD-30-1	14	F	Yes	MS
31	RHD-31-1	35	F	No	MS/MR.
32	RHD-32-1	15	M	Yes	MS
33	RHD-33-1	39	M	No	MS/MR
34	RHD-34-1	32	F	No	MS/MR
35	RHD-35-1	45	F	Yes	MS/MR
36	RHD-36-1	43	F	Yes	MS
37	RHD-37-1	42	M	No	MS/MR
38	RHD-38-1	40	F	No	MS
39	RHD-39-1	36	M	Yes	MS/MR
40	RHD-40-1	13	M	No	MS
41	RHD-41-1	28	F	No	MS
42	RHD-42-1	26	F	No	MR
43	RHD-43-1	41	M	Yes	MS
44	RHD-44-1	40	F	No	MR
45	RHD-45-1	38	F	Yes	MR
46	RHD-46-1	18	M	Yes	MS
47	RHD-47-1	17	M	No	MR
48	RHD-48-1	50	F	Yes	MS
49	RHD-49-1	19	F	No	MS/MR
50	RHD-50-1	16	M	No	MR

MS, Mitral stenosis; MR, Mitral regurgitation; MVP, Mitral valve prolapsed.

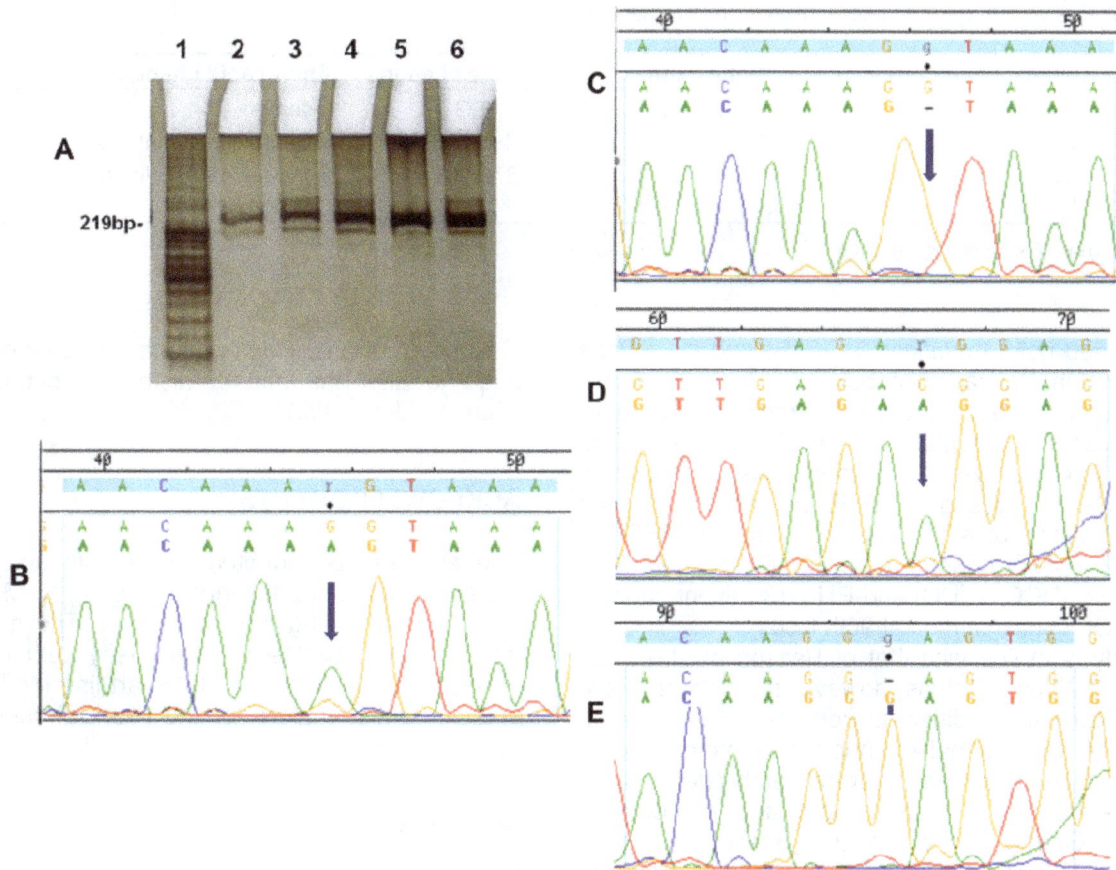

Figure 1. A; SSCP analysis of the PCR products of the exon 1 of MBL2 gene in RHD samples. 1→10bp DNA ladder; 2→RHD 30-1; 3→RHD 31-1; 4→RHD 19-1; 5→RHD 13-1; 6→RHD 46-1. B, C, D and E; Novel mutations observed in exon 1 of MBL2 gene. The reference sequence is aligned on the top of the sequences. The arrow indicates the position of the mutation.

In this study, sequencing of MBL2 gene exon1 for the RHD samples revealed four types novel mutations in 4 RHD patients. The patient (RHD-46) showed G→A heteroplasmic mutation in 46th locus of MBL2 gene, RHD-19 showed guanine nucleotide deletion at 47th locus, one patient (RHD-30) showed nucleotide change from 67 G→A, serine to phenylalanine (UCC→UUC) and RHD-13 patient showed Guanine nucleotide insertion at 96th locus. They are presented in Figures 1 (b, c, d and e) and Table 2.

DISCUSSION

All the children with RHD were affected in the first decade as been similarly observed by Steffensen et al. (2000). One of the common features of RHD cases was mitral stenosis as reported by Schafranski et al. (2004). Mutation screening of exon 1 of MBL2 gene was done by PCR-SSCP analysis for all the samples.

MBL is considered as a modifying molecule in the physiopathology of RF and RHD, presenting a dual role,

on one side conferring protection against initial infection by GAS and on the other side provoking inflammatory response in the chronic stage of the disease (Messias et al., 2006). Many factors are certainly involved in the course of RF/RHD, where MBL genotypes as well as MBL protein levels may also play a role.

MBL, an important component of the complement system, plays a key role in innate immunity. As the main role of innate immunity is to restrict the multiplication of infectious agents, deficiency in one of the genes involved in innate immunity may delay or impair the clearance of the pathogens and persistence of the pathogens may trigger the immune system response. Indeed, subjects who are homozygous or compound heterozygous for defective MBL2 alleles or who have low serum MBL suffer from recurrent bacterial and viral infections as children (Turner and Hamvas, 2000) and a subset of them have enhanced risk for autoimmune disorders (Thiel et al., 2006). There is increasing evidence that MBL is also involved in the modulation of disease severity in both infectious and autoimmune diseases (Garred et al., 1997; Garred et al., 2000; Tabona et al., 1995). Steffensen et al.

Table 2. The types of MBL gene mutation observed in RHD patients.

Case no.	Clinical characterestics	Gene/exon	Nucleotide changes	Amino acid changes	Remarks
RHD 13	MS/MR	MBL2, exon 1	96G	Insertion	Novel
RHD 19	MS	MBL2, exon 1	46 G/A	Heteroplasmic mutation	Novel
RHD 30	MS	MBL2, exon 1	67G→A	Serine→Phenyl alanine	Novel
RHD 46	MS	MBL2, exon 1	47 G	Deletion	Novel

MS, Mitral stenosis; MR, Mitral regurgitation.

(2000) reported point mutations in 52, 54 and 57 codons of the exon1 in Caucasian population. However in this study, sequencing of MBL2 gene exon1 for the RHD samples revealed novel mutations in 4 RHD patients. The patient (RHD-46) showed G→A heteroplasmic mutation in 46th locus of MBL2 gene, RHD-19 showed guanine nucleotide deletion at 47th locus, one patient; (RHD-30) showed nucleotide change from 67 G→A, serine to phenylalanine (UCC→UUC) and RHD-13 patient showed nucleotide Guanine insertion at 96th locus.

This study is in line with that of Reason et al. (2006) that had mitral valve lesions; however the patients were Brazilian Caucasians different from our South Indian patients. The worldwide high population prevalence of these variants suggests an apparently ambivalent role of MBL2 in a number of pathogenetic and homeostatic processes (Garred et al., 1994; Ezekowitz, 1998; Saevarsdottir et al., 2004). Further understanding of MBL2 expression may improve our ability to understand and disrupt the pathogenetic mechanisms involved in cardiovascular disease (Best et al., 2009).

Rosenberg et al. (2002) feels that "The structure of human populations is relevant in various epidemiological contexts. As a result of variation in frequencies of both genetic and non-genetic risk factors, rates of disease and of such phenotypes as adverse drug response vary across populations. As expected in this analysis, four novel mutations in exon 1 of MBL2 gene confer a greater susceptibility for developing RHD. These are all novel MBL2 exon 1 mutation observed for the first time in our new geographical area Tamil Nadu in South Indian population. Further, information about a patient's population of origin might provide health care practitioners with information about risk when direct causes of disease are unknown.

Conclusion

Certainly, molecular genetics approach has to be performed for proper medical management and genetic counseling for the RHD patients. MBL replacement therapy might be warranted in the future, MBL deficiency leads to increased susceptibility to disease, MBL replacement could be used to increase resistance to that disease. In an acute infection MBL therapy might, by enhancing the immune response, speed the resolution of disease in MBL-deficient patients. MBL therapy could be used to alter the natural history of chronic diseases (Summerfiled, 2003).

ACKNOWLEDGMENTS

The authors are indebted to patients and their family members and extremely grateful to Dr. K. M. Marimuthu, for critically reading the manuscript. The authors would also like to thank the VIT University authorities for providing all the facilities and to Heart line Medical and Research Centre, Vellore for the financial support to carry out this work.

REFERENCES

Best LG, Ferrell RE, Decroo S, North KE, Maccluer JW, Zhang Y, Lee ET, Howard BV, Umans J, Palmieri V, Garred P (2009). Genetic and other factors determining mannose binding lectin levels in American Indians: the Strong Heart Study. BMC. Med. Genet., 10: 5.

Bisno AL (2000). Nonsuppurative poststreptococcal sequelae: rheumatic fever and glomerulonephritis. In Mandell GL, Bennett JE, Dolin R (eds.) Principles and practice of infectious diseases. Churchill Livingstone, New York, NY, pp.1799-1810.

Ezekowitz RA (1998). Genetic heterogeneity of mannose-binding proteins: the Jekyll and Hyde of innate immunity? Am. J. Hum. Genet., 62: 6-9.

Garred P, Harboe M, Oettinger T, Koch C, Svejgaard A (1994). Dual role of mannan-binding protein in infections: another case of heterosis?. Eur. J. Immunogenet., 21: 125-31.

Garred P, Madsen HO, Balslev U, Hofmann B, Pedersen C, Gerstoft J, Svejgaard A (1997). Susceptibility to HIV infection and progression of AIDS in relation to variant alleles of mannose-binding lectin, Lancet. 349: 236-240.

Garred P, Madsen HO, Marquart H, Hansen TM, Sorensen SF, Petersen J, Volck B, Svejgaard A, Graudal NA, Rudd PM, Dwek RA, Sim RB, Andersen V (2000). Two edged role of mannose binding lectin in rheumatoid arthritis: a cross sectional study. J. Rheumatol., 27: 26-34.

Ghaffar A, Reddy KS, Singhi M (2004). Burden of non-communicable diseases in South Asia. BMJ., 328: 807-810.

Jack DL, Klein NJ, Turner MW (2001). Mannose-binding lectin: targeting the microbial world for complement attack and opsonophagocytosis. Immunol. Rev., 180: 86-99.

Jose VJ, Gomathi M (2003). Declining prevalence of rheumatic heart disease in rural school children in India: 2001-2002. Indian Heart J., 55: 158-160.

Krishnaswami S, Joseph G, Richard J (1991). Demands on tertiary care for cardiovascular diseases in India: analysis of data for 1960-89. Bull. World Health Organ., 69: 325-330.

Lalchandani A, Kumar HRP, Alam SM (2000). Prevalence of rheumatic fever and rheumatic heart disease in rural and urban school children of district Kanpur. Indian Heart J., 52: 672.

Lotery AJ, Namperumalsamy P, Jacobson SG, Weleber RG, Fishman GA, Musarella MA, Hoyt CS, Héon E, Levin A, Jan J, Lam B, Carr RE, Franklin A, Radha S, Andorf JL, Sheffield VC, Stone EM (2000). Mutation Analysis of 3 Genes in Patients with Leber Congenital Amaurosis. Arch. Ophthalmol., 118: 538-543.

Messias RIJ, Schafranski MD, Jensenius JC, Steffensen R (2006). The association between mannose-binding lectin gene polymorphism and rheumatic heart disease. Hum. Immunol., 67:991-998.

Neth O, Jack DL, Dodds AW, Holzel H, Klein NJ, Turner MW (2000). Mannose-binding lectin binds to a range of clinically relevant microorganisms and promotes complement deposition. Infect. Immun., 68: 688-693.

Nishtar S (2002). Prevention of coronary heart disease in south Asia. Lancet., 360: 1015-1018.

Orita M, Iwahana H, Kanazawa H, Hayashi K, Sekiya T (1989). Detection of polymorphism of human DNA by gel electrophoresis as single-strand conformation polymorphisms. Proc. Natl. Acad. Sci., USA 86: 2766-2770.

Rosenberg NA, Pritchard JK, Weber JL, Cann HM, Kidd KK, Zhivotovsky LA, Feldman MW (2002). Genetic structure of human populations. Science., 298: 2381-2385.

Saevarsdottir S, Vikingsdottir T, Valdimarsson H (2004). The potential role of mannan-binding lectin in the clearance of self-components including immune complexes. Scand. J. Immunol., 60: 23-29.

Sanger F, Coulson AR (1975). A rapid method for determining sequences in DNA by primed synthesis with DNA polymerase. J. Mol. Biol., 94: 441-448.

Schafranski MD, Stier A, Nisihara R MESSIAS-REASON IJT (2004). Significantly increased levels of mannose-binding lectin (MBL) in rheumatic heart disease: a beneficial role for MBL deficiency, Clin. Exp. Immunol., 138: 521-525.

Steffensen R, Thiel S, Varming K Jersild C, Jensenius JC (2000). Detection of structural gene mutations and promoter polymorphisms in the mannan-binding lectin (MBL) gene by polymerase chain with sequence-specific primers. J. Immunol. Methods., 241: 33-42.

Sumiya M, Super M, Tabona P, Levinsky RJ, Arai T, Turner MW, Summerfield JA (1991). Molecular basis of opsonic defect in immunodeficient children. Lancet., 337: 1569-1570.

Summerfield JA (2003). Clinical potential of mannose binding lectin-replacement therapy. Biochem. Soc. Trans., 31: 770-773.

Tabona P, Mellor A, Summerfield JA (1995). Mannose binding protein is involved in first-line host defence: evidence from transgenic mice, Immunology., 85: 153-159.

Thiel S, Frederiksen PD, Jensenius JC (2006). Clinical manifestations of mannan-binding lectin deficiency. Mol. Immunol., 43: 86-96.

Turner MW, Hamvas RM (2000). Mannose-binding lectin: structure, function, genetics and disease associations. Rev. Immunogenet., 2: 305-322.

5

Healthy carrier parents in partial 7 and 9 chromosome trisomy in two pediatric patients: Report cases at the Hospital para el Niño Poblano, Mexico

J. M. Aparicio-Rodríguez[1,9]*, M. L. Hurtado-Hernández[2,8], M. Barrientos-Perez[3],
S. I. Assia-Robles[4], N. C. Gil-Orduña[5,9], R. Zamudio-Meneses[6], J. S. Rodríguez-Peralta[7],
S. M. Brieke-Walter[5], F. Almanza-Flores[4] and C. Silva-Xilotl[8]

[1]Department of Genétics, Hospital para el Nino Poblano, Mexico.
[2]Department of Cytogenetics, Hospital para el Nino Poblano, Mexico.
[3]Department of Endocrinology, Hospital para el Nino Poblano, Mexico.
[4]Department of Pediatrics, Hospital para el Nino Poblano, Mexico.
[5]Department of Estomatology, Hospital para el Nino Poblano, Mexico.
[6]Department of Cardiology, Hospital para el Nino Poblano, Mexico.
[7]Department of Neurosurgery, Hospital para el Nino Poblano, Mexico.
[8]Department of Perinatology, Hospital de la Mujer S.S.A. Puebla, Mexico.
[9]Department of Estomatology, Benemérita Universidad Autónoma de Puebla, Mexico.

Two pediatric cases are reported in this study, a one year and five months old male patient with partial trisomy for the long arm of chromosome 9 due to chromosome segregation error in the father 46 XY, del (9q-)/46 XY dup (10q+), and a seven years old male patient with partial trisomy for the long arm of chromosome 7 due to chromosome segregation error in the mother 46 XX, ins (10; 7) (q21; q23q35) are described. The major abnormalities in both cases are reported and compared, where both syndromes have defects including central nervous system disorders as mental retardation, hypotonia, craniofacial anomalies, micrognatia, poor feeding and xyphoescoliosis.

Key words: Central nervous system, partial trisomy, translocation, chromosome, healthy carriers.

INTRODUCTION

Aneuploidy refers to abnormal number of chromosomes. An extra or missing chromosome is considered a common cause of genetic disorders. Some malignant cells due to cancer also have different numbers of chromosomes reported by Sen (2000). Aneuploidy occurs during cell division when the chromosomes do not separate properly between the two cells. Chromosome abnormalities occur approximately in 1 of 160 live births, there are some extra chromosomes rising to differentiate a trisomy known as complete or partial, as in this study, however the most common trisomies are 21, 18 and 13 (Driscoll and Gross, 2009). It is important to remember that different species have different numbers of normal

chromosomes and thus the term "aneuploidy" refers to the chromosome number being different for that species. In humans the normal chromosome number is 46, including sexual chromosomes. Most embryos cannot survive with a missing or extra autosome and are spontaneously aborted. The most frequent aneuploidy in humans is trisomy 16, although fetuses affected with the full version of this chromosome abnormality do not survive to term, it has been reported by Griffiths et al. (2000), that human with mosaic cells might survive.

Trisomy of the short arm of chromosome 9 as has been reported to be a clinical syndrome (Rethor et al., 1970, 1973; Rethor and Lafourcade, 1974). However, the clinical syndrome associated with trisomy of the whole of chromosome 9 has not been fully characterized. Although Juberg et al. (1970), Feingold and Atkins (1973) reported several cases as Haslam et al. (1973), Bowen et al. (1974)

---

*Corresponding author. E-mail: jmapar@prodigy.net.mx.

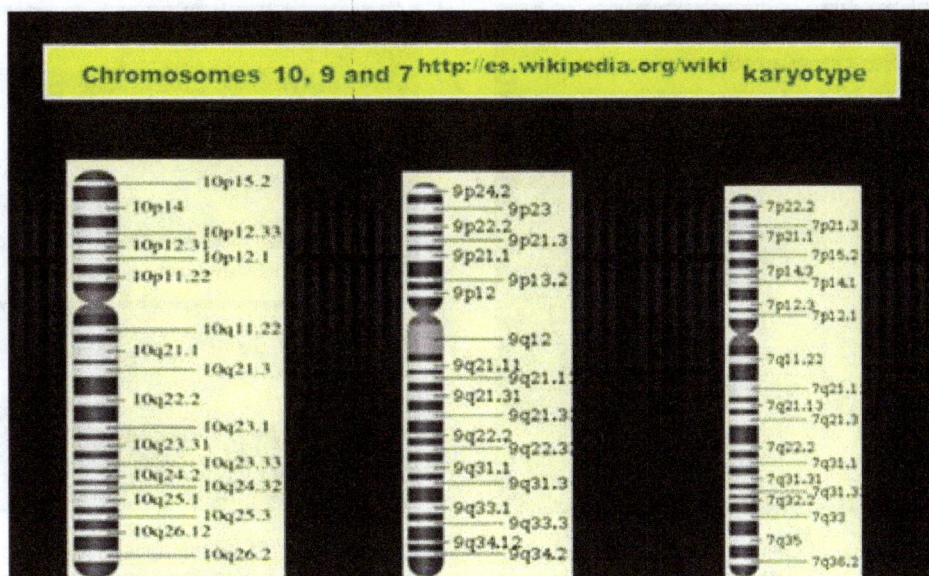

Figure 1. Chromosomes 10, 9 and 7. They show both short and large arms and centromeres. (http://es.wikipedia.org/wiki/Cromosoma_10).

reported mosaisism resulting from meiotic disjunction of a pattern translocation (9;10) (p22;q32).

Direct intra-arm intrachromosomal duplications, such as found in case 2 in this study were an insertion on chromosome 10 is part of chromosome 7, are rare events. Unequal interchange in meiosis between homologous chromosomes and unequal intrachange between chromatids of one chromosome by an inversion or insertion has been reported by Couzin et al. (1986), For Abosco et al. (1988) and Abuelo et al. (1988).

A translocation event can be postulated as possible mechanisms for a chromosomal insertion, either intra or inter chromosomal. As observed by de Arce et al. (1982) occurred with an estimated frequency of 1/5000 newborns and are usually ascertained through healthy carriers as both cases in this study, having unbalanced offspring or being investigated for reproductive wastage Abuelo et al. (1988), as seen in both recombination at chromosome 10, Figure 1 t(7q+) in case 1 and t(9q+) in case 2.

MATERIALS AND METHODS

Blood was taken from both families, in parents and patients to establish a peripheral lymphocyte culture for chromosome analysis. Chromosome results showed a 9q+ in case 1 (Figures 3 and 4) and 7q+ in case 2 (Figures 9 and 10) in each of the 20 complete cells from each patient, who were analyzed. This finding prompted a cytogenetic study of parental lymphocytes.

Case 1

Based on 20 complete cells, the father of patient 1 was determined to be 46, XY del (9q-) dup (10q+). The father was diagnosed to be a balanced translocation of all the studied cells. Most of the long arm of one chromosome 9 was obviously translocated to translocate to the long arm of a 10 (Figure 2).

Case 2

Based on 20 complete cells, the mother of the patient 2 was determined to be 46, XX, ins (10, 7) (q21; q23q35). The mother was diagnosed to have a balanced translocation of all the studied cells, where part of the distal portion of long arm of one chromosome 7 was obviously translocated to the middle part of the long arm of chromosome 10 (Figure 8).

RESULTS AND DISCUSSION

The terms "partial monosomy" and "partial trisomy" are used to describe an imbalance of genetic material caused by loss or gain of part of a chromosome. In particular, these terms would be used in the situation of an unbalanced translocation, where an individual carries a derivative chromosome formed through the breakage and fusion of two different chromosomes. In this situation, the individual would have three copies of part of one chromosome (two normal copies and the portion that exists on the derivative chromosome) and only one copy of part of the other chromosome involved in the derivative chromosome, where aneuploidy have been performed as pre-natal screening (Driscoll and Gross, 2009).

In this study two unrelated patients with partial trisomy in different chromosomes 7 and 9 were studied. Neither dolicocephaly nor severe respiratory distress as reported in other cases of some trisomies seems to be constant features in both cases in this study. The pattern of abnormalities typically associated with the some syndromes

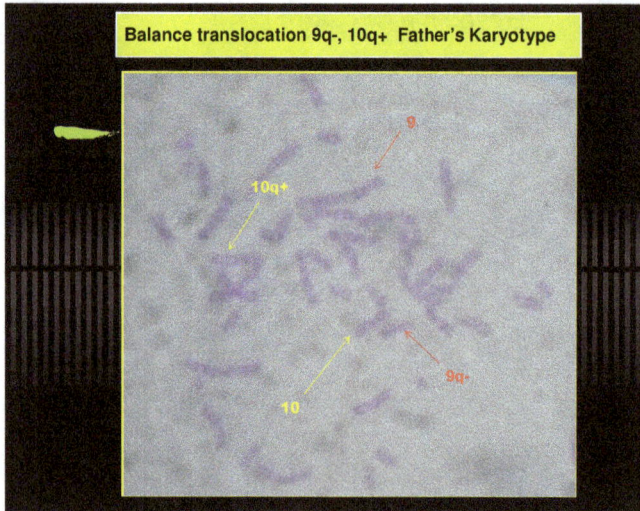

Figure 2. Case 1: The father of patient 1 was determined to be 46, XY del (9q-) dup (10q+). The father was diagnosed to be a balanced translocation of all the studied cells. Most of the long arm of one chromosome 9 was obviously translocated to the long arm of a 10.

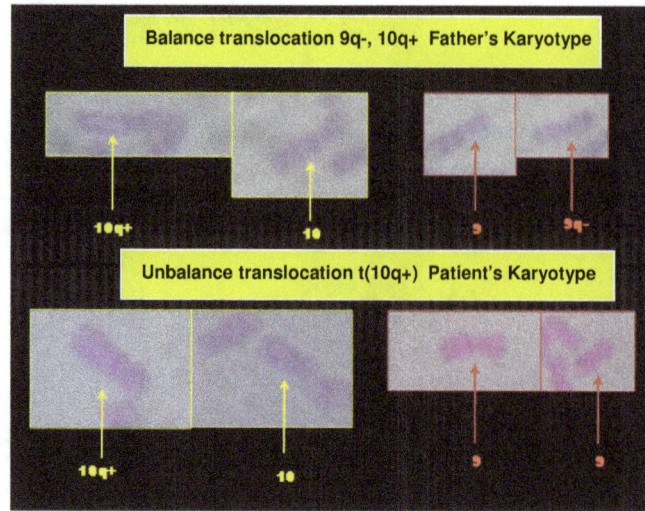

Figure 4. Case 1: In the upper section the father's chromosomes can be seen, where the 10q chromosome has a part of the 9 chromosome, 46XY del(9q-) dup(10q+). In the lower section the patient´s chromosome analysis showed 10q+ due to partial 9 trisomy, where part of the long arm of chromosome 9 was translocated.

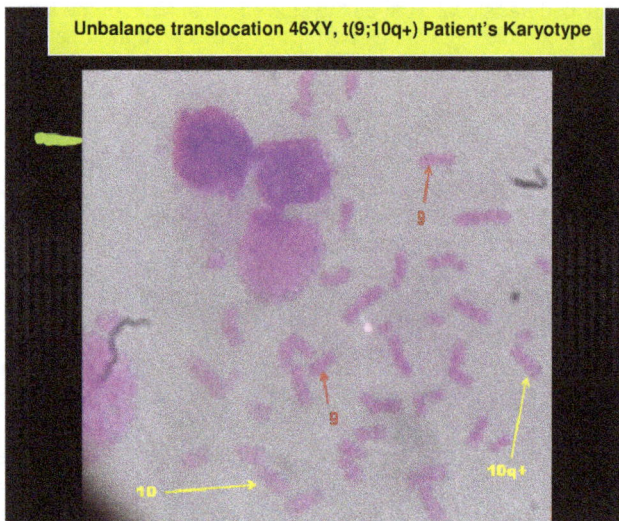

Figure 3. Case 1: Peripheral lymphocyte culture for chromosome analysis. Chromosome results showed 10q+ due to partial 9 trisomy, where part of the long arm of chromosome 9 was translocated.

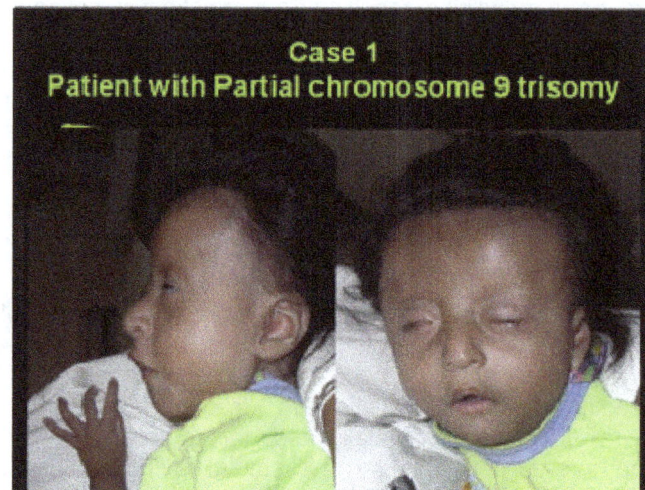

Figure 5. Case 1: Plagiocephaly, facial hypoplasia, micrognatia, hypertelorisim, broad nasal bridge. Short stature and congenital cardiovascular malformation (Interventricular communication).

syndromes associated to trisomy as 7, 9 or 10 (Figure 1), as reported by Couzin et al. (1986), Bowen et al. (1974), Grosse et al. (1975), Hustinx et al. (1974) and Schleiermacher et al. (1974), appears to be: mental retardation, third percentile values for height, weight and head circumference, a bulky forehead, broad nasal bridge, club feet and hypotonia of skeleton muscles. In the cases reported to date, either the patients have had the entire long arm of chromosome 7 or 9, or just one proximal band

less than this amount (Figures 4 and 10). Perhaps some of these abnormalities will not develop in cases yet to be reported in which a smaller amount of the some arms are translocated. In relation to case 1, this male patient presented a partial 9 trisomy (Figure 3), (the father was healthy carrier), with facial hypoplasia, micrognatia, hypertelorisim, broad nasal bridge (Figure 5), short stature and congenital cardiovascular malformation (interventricular communication), small penis normal testicles, camptodactily, muscular hipotonia. The tomography showed infratentorial asimetry and right hipoplasic

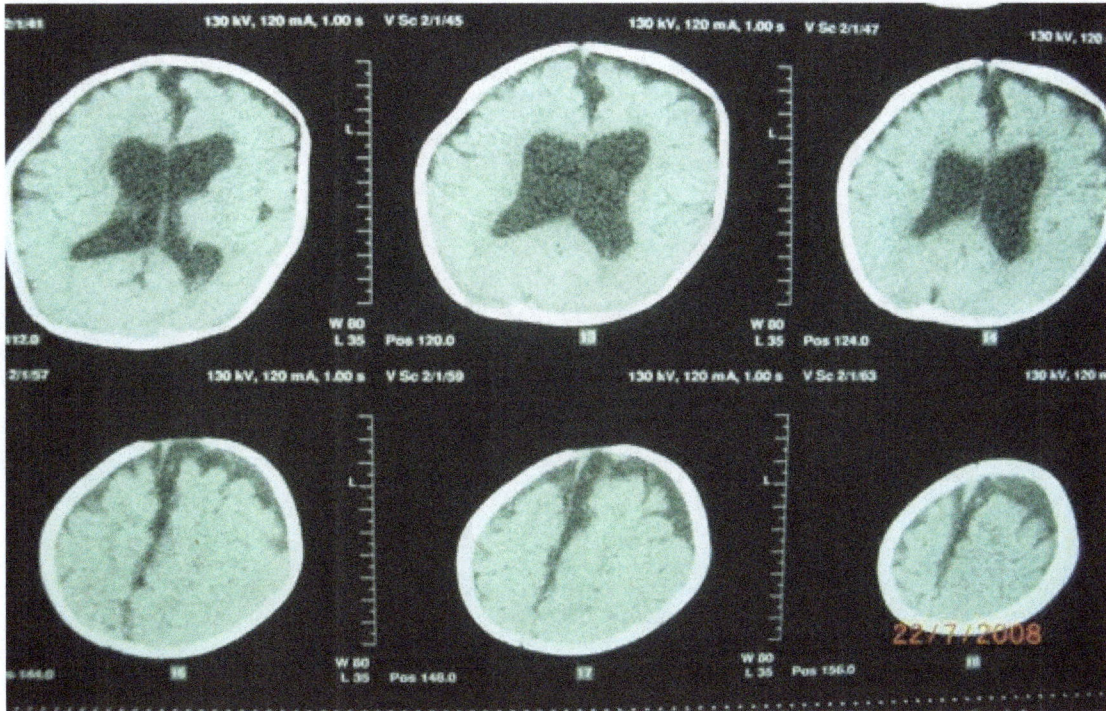

Figure 6. Case 1: Tomography shows infratentorial asymmetry and right hipoplasic temporal brain with a wide basal ganglia and subcortical atrophy and ventriculomegaly and agenesis of the corpus callosum.

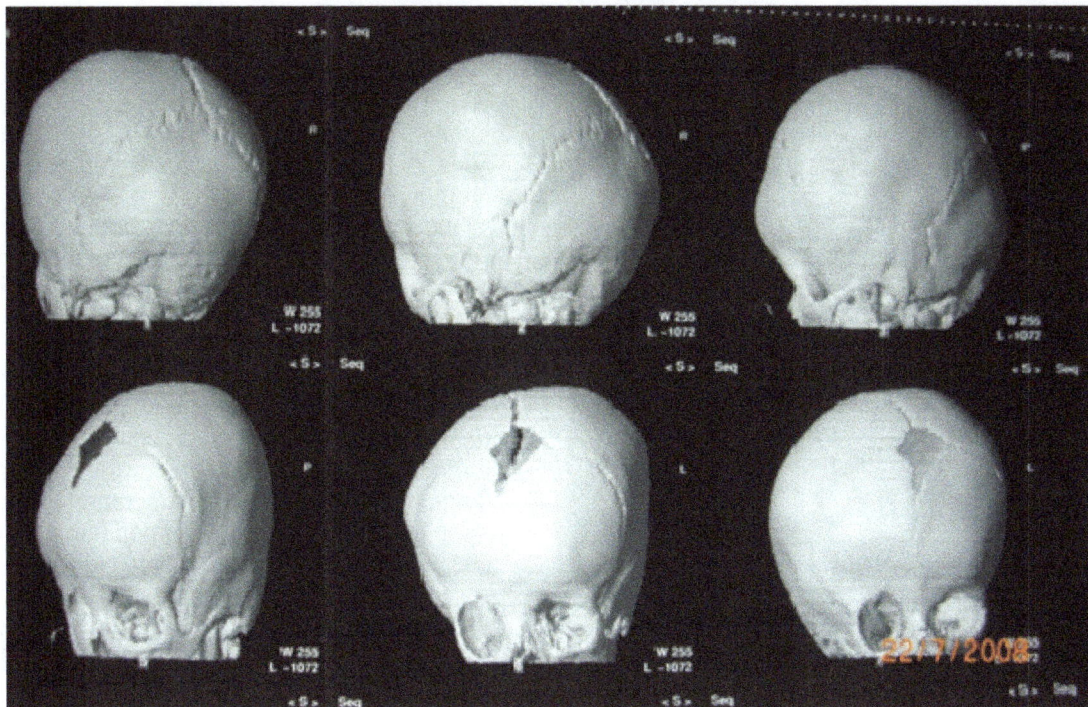

Figure 7. Case 1: An asymmetric plagiocephaly head was confirmed with 3rd dimension tomography due to a craniosynostosis and cranial asymmetry.

temporal brain with a wide basal ganglia and subcortical atrophy (Figure 6.) An asymmetric plagiocephaly head was confirmed with 3rd dimension tomography due to craniosynostosis Figure 7.

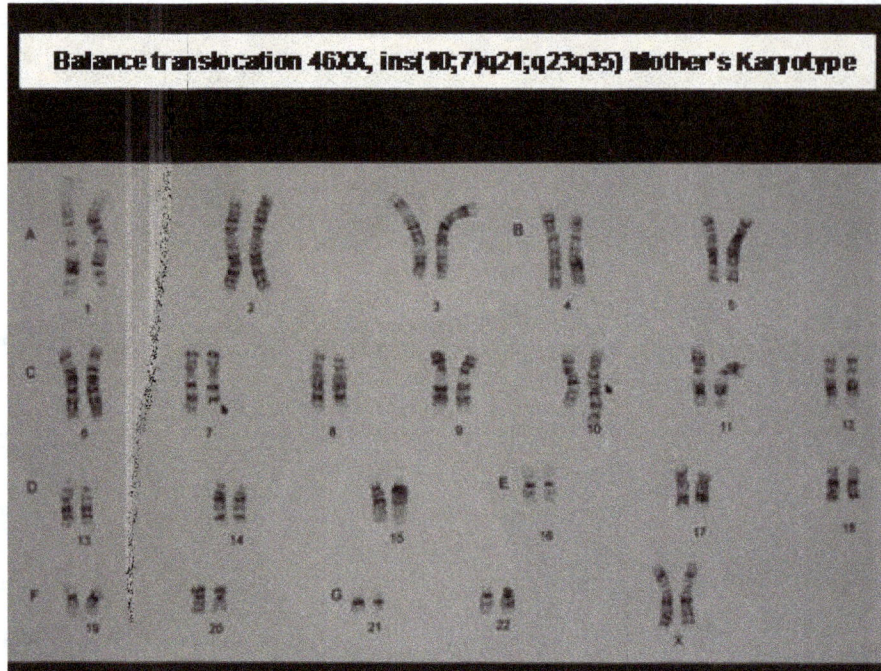

Figure 8. Case 1: Mother of the patient 2 was determined to be 46, XX, ins (10, 7) (q21; q23q35). The mother was diagnosed to have a balanced translocation of all the studied cells where part of the distal portion of long arm of one chromosome 7 was obviously translocated to the middle part of the long arm of chromosome 10.

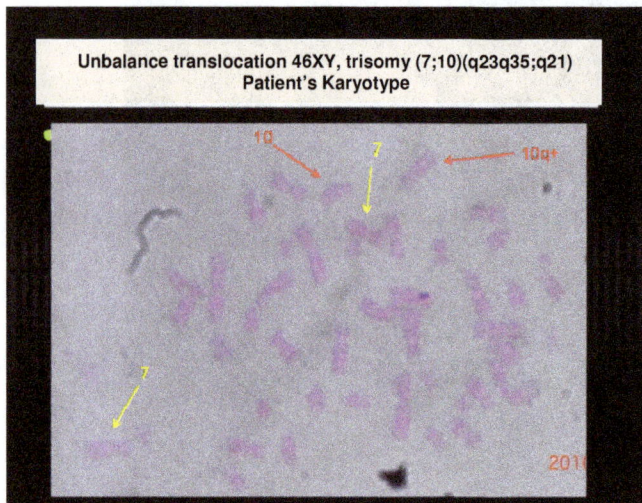

Figure 9. Case 2. Peripheral lymphocyte culture for chromosome analysis. Chromosome result showed 10q+ due to partial 7 trisomy, where part of the long arm of chromosome 7 was translocated and inserted in the middle of the long arm of chromosome 10.

The karyotype analysis confirmed the partial 9 trisomy where the mother has normal karyotype 46 XX and the father is a healthy carrier with a chromosome formula; 46 XY del (9q-)/46 XY dup (10q+) (Figure 2). It was confirmed that the use of drugs by the father that might

be responsible for the chromosome aberration. However, a number of cases involving spontaneous trisomy for varying amounts of chromosome have been recorded a trisomy 9p syndrome, (Rethor et al., 1970, 1973). In most cases this results from a parental translocation involving chromosome 9. The trisomy arises either from 3:1 meiotic disjunction (Lindenbaum and Bobrow, 1975) in cases with 47 chromosomes (tertiary trisomy) or inheritance of an unbalanced form of the translocation in cases with 46 chromosomes. In tertiary trisomic cases there is also trisomy forth telomeric region of the other chromosome involved in the translocation. This additional imbalance cannot be ignored it probably has minimal phenotypic effect since such telomeric regions are most probably heterochromatic, Ford EH, 1973. Clinical features common to case 2 and some reported cases including region 7q22-q31 are developmental delay, frontal bossing, strabismus, epicanthic folds, low set and large ears and small nose. Our patient showed, in addition, dysplasia of the left kidney, subluxable hips and hearing loss, hypotonia, thin upper lip (Figure 11), non severe asymmetry and general hipoplasic brain with a wide basal ganglia, subcortical atrophy (Figure 12), skeletal abnormalities as xyphoescoliosis (Figure 13), osseous hands malformations as campotactily and clinodactyly similar to those features discussed by Forabosco et al. (1988).

Both 7 and 9 partial trisomy have been reported with different clinical features as in trisomy number 10

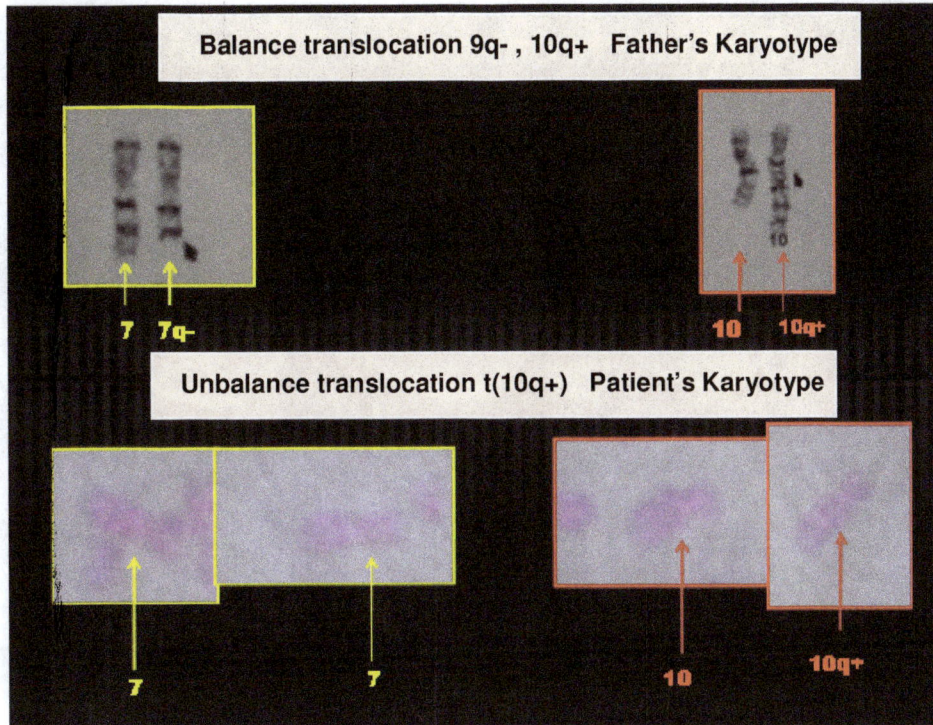

Figure 10. Case 2: In the upper section it can be seen the mother's chromosomes where the 10q+ chromosome has inserted a part of the 7 chromosome, 46, XX, ins (10, 7) (q21; q23q35). In the lower section the patient´s chromosome analysis showed 10q+ due to partial 9 trisomy, where part of the long arm of chromosome 9 was translocated and inserted.

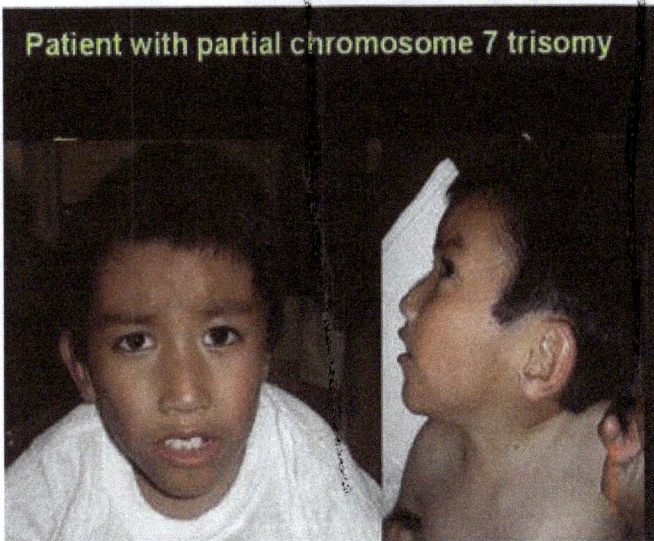

Figure 11. Case 2. The second patient shows a developmental delay, frontal bossing, strabismus, epicanthic folds, small nose, subluxable hips, low set and large ears, hearing loss, hypotonia and thin upper lip.

(F.E.D.E.R.http://www.enfermedadesraras.org/es/default. htm, E.U.R.O.R.D.I.S. http://www.eurordis.org and Fryns

JP in 1979), with central nervous system defects, hypotonia, physical and mental retardation and malnutrition, hypotelorism, maxi-llary dysplasia and congenital cardiopathy. Most of the cases are reported as "de novo" by International genetic and rare diseases associations such as (F.E.D.E.R.http://www.enfermedades-raras.org/es/default.htm, E.U.R.O.R.D.I.S. http://www.eurordis.org). Definitely, the karyotype is important for the final diagnosis in both the patients and their parents to confirm possible healthy carriers.

Although partial 7 chromosome in case 2 (Figure 10) was considered to be due to an expontaneous mutation in the mother and inherited to her son. The family in case 2 (Figure 4) had strong predisponent etiological factors, since drugs were positive for the father who was carrier of this 7 partial trisomy, whether the mother drugs negative, was normal 46 XX. In relation to both cases the genetic counseling was performed to explain the future risk in their progeny, where 25% might be healthy with no translocation, 25% healthy carrier as the balanced trans-located parents (9-;10+) and ins(7-;10+) respectively. 25% of the progeny will have a partial trisomy of the 7 and/or 9 chromosomes and 25% with a deleted 7q- and/or 9q-. Both patients were also evaluated at the department of physical rehabilitation therapy.

It was observed that a great variability of clinical features

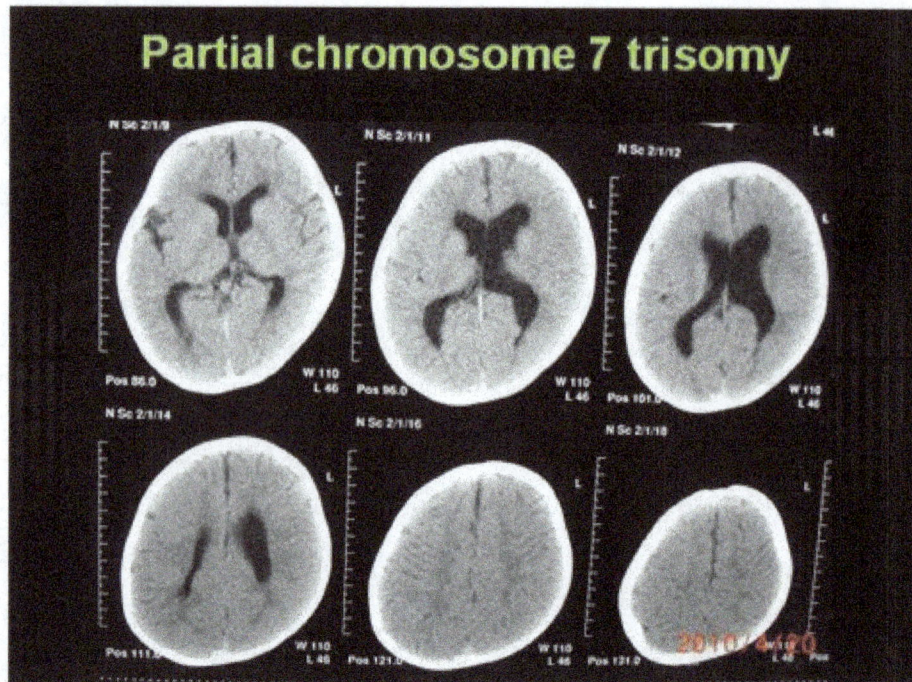

Figure 12. Case 2: Tomography shows non severe asymmetry and general hipoplasic brain with a wide basal ganglia and subcortical atrophy.

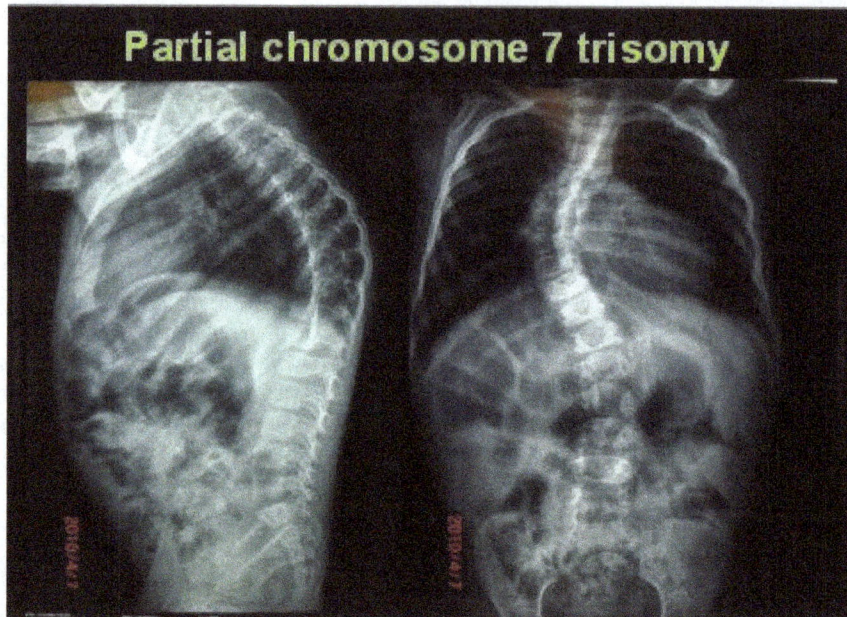

Figure 13. Case 2: The patient X-rays shows skeletal abnormalities as severe xyphoescoliosis.

as craniofacial, neurological and organic alterations can be associated to partial trysomies and how important it is nowadays to have a healthy life free of drugs to have a better opportunity for normal pregnancy and healthier progeny. Although it will always be a risk that any parent might have deoxyribonucleic acid (DNA) or chromosome mutation as it was in the second patient.

An early chromosome diagnosis should be considered

Healthy carrier parents in partial 7 and 9 chromosome trisomy in two pediatric patients: Report cases at the Hospital...

37

important and it could be more relevant whether the chromosome study should be part of prenuptial analysis, to detect healthy carriers of a great variety of chromosome aberrations. This will be important to ensure a better life quality for future children in a family.

ACKNOWLEDGMENTS

The authors would like to thank the authorities of the Hospital Para El Nino Poblano and the Director of the Faculty of Estomatology from the Autonomous University of Puebla, Mtro. Jorge A. Albicker Rivero, for their unconditional support.

REFERENCES

Abuelo DN, Barsel-Bowers G, Richardson A (1988). Insertional translocations: report of two new families and review of the literature. Am. Med. Genet. 31: 319-29.

Bowen P, Ying KL, Chung GSH (1974). Trisomy 9 mosaicism in newborn infant with multiple malformations. J. Pediat. 85: 95-9.

Couzin DA, Haites N, Watt JL, Johnston AW (1986). Partial trisomy 7 (q32-*qter) syndrome in two children. Med. Genet. 23: 461-5.

Chromosomes 7,9,10. http://es *wikipedia.org/wiki/Chromosome_7, 9, 10_(human).* (2010). Chromosome 7,9,10 (human) - Wikipedia, the free encyclopedia.

de Arce MA, Law E, Martin L, Masterton JG (1982). A case of inverted insertion assessed by R and G banding. Med. Genet. 19: 148-51.

Driscoll DA, Gross S (2009). Clinical practice. Prenatal screening for aneuploidy. New Engl. J. Med., 360(24): 2556-62.

European Organization for Rare Disorders (EURORDIS). Plateforme Maladies Rares 102, Rue Didot. Localidad: 75014 Paris eurordis@eurordis.org/ http://www.eurordis.org

Federación Española de Asociaciones de Enfermedades Raras (FEDER) Domicilio: c/ Enrique Marco Dorta, 6 local. Localidad: 41018 Sevilla Correo-e: f.e.d.e.r@teleline.es/ info@enfermedades-raras.org. http://www.enfermedades-raras.org/es/default.htm.

Feingold M, Atkins L (1973). A case of trisomy 9. J. reed. Genet. 10: 184-187.

Fryns JP (1979). Partial trisomy of the distal portion of the long arm of chromosome number 10 (10q24-10qter): A clinical entity. Acta Paediatr. Belg. 32: 141-143.

Forabosco A, Baroncini A, Dalpra L (1988). The phenotype of partial dup (7q) reconsidered: a report of five new cases. Clin. Genet. 34:48-59.

Ford EH (1973). Human chromosomes, London-New York: Academic Press. p. 34.

Griffiths AJF, Miller JH, Suzuki DT, Lewontin RC, Gelbart WM (2000). Chromosome Mutation II: Changes in Chromosome Number. An Introduction to Genetic Analysis (7th ed.). New York: W. H. Freeman.

Grosse K, Schwanitz G, Singer H, Wieezorek V (1975). Partial trisomy 10p. Humangenetik 29: 141-144.

Hustinx T, ter Haar B, Scheres J, Rutten F (1974). Trisomy for the short arm of chromosomeNo. 10. Clin. Genet. 6: 408-415.

Haslam RHA, Broske SP, Moore CM, Thomas GH, Neill CA (1973). Trisomy 9 mosaicismwith multiple congenital anomalies. J. reed. Genet. 10: 180-183.

Juberg PC, Gilbert EF, Salisbury RS (1970). Trisomy C in an infant with polycystic kidneysand other malformations. J. Pediat. 76: 598-603.

Lindenbaum RH, Bobrow M (1975). Reciprocal translocations in man. 3:1 meiotic disjunction resulting in 47-- or 45 chromosome offspring. J. reed. Genet. 12: 29-43

Rethor MO, Ferrand J, Dutrillaux Lejeune J (1970). Trisomie 9p par t(4; 9) (q34; q21)mat.Ann. G~n6t. 17: 157-161.

Rethor MO, Hoehn H, Rott HD, Couturier J, Dutrillaux, Lejcune J (1973). Analyse de la trisomie 9p par denaturation menage. Apropos d'un nouveau cas. Humangenetik 18: 129-138.

Rethor MO, Lafourcade J (1974). Trisomie du bras court du chromosome 9: Syndrome +9p. Journes parisiennes de pediatrie, pp. 379-390.

Schleiermacher E, Schliebitz U, Steffens C (1974). Brother and sister with trisomy 10p: A new syndrome. Humangenetik 23: 163-172.

Sen S (2000). Aneuploidy and cancer. Curr. Opinion Oncol. 12(1): 82-8.

6

Chromosome complement and C-banding patterns in 6 species of grasshoppers

Pooja Chadha* and Anupam Mehta

Department of Zoology, Guru Nanak Dev University, Amritsar, Punjab, Pin Code: 143005, India.

Chromosomes with detailed karyotypic information (nature, number, size, relative length, length of X-chromosome, nature of X-chromosome) and C-banding patterns of six species of grasshoppers belonging to sub-families- Tryxalinae, Oedipodinae and Catantopinae are discussed. The karyotypes comprises of acrocentric chromosomes with complement number 2n=23 (male). Constitutive heterochromatin distribution was found at centomeric, interstitial, terminal sites along with thick and thin bands among all the species except in *Acrida turrita,* which possessed only centromeric C-bands. The number and location of C-bands in Acridids exhibit both intra – and interspecific variations. In the present communication the chromosome complement and C-banding patterns are analyzed for further differences between congeneric species and among genera belonging to the same sub-family.

Key words: Acrididae, Orthoptera, chromosome complement, C-banding.

INTRODUCTION

Orthoptera has been considered as a classical material for karyological investigations. The size and number of their chromosomes are such that both qualitative and quantitative studies on chromosomal anomalies can be detected easily (Turkoglu and Koca, 2002). The karyotype is found to have a cytotaxonomic value. Acridoid group is known for its karyotypic uniformity or conservatism (Aswathanarayana and Ashwath, 2006). The introduction of C-banding technique offers a simple mean of defining constitutively heterochromatic regions.

C-banding technique have made it easier to assess the changes in constitutive heterochromatin and have revealed the existence of remarkable degree of C-band variations within species (King and John, 1980; Lopez-Fernandez and Gosalvez, 1981). Many studies on the C-bands have been done in Acridoids which are known to possess high level of chromosomal variations. The number and location of C-bands in Acridids exhibit both intra- and interspecific variations (Yadav and Yadav, 1993). The present communication deals with the chromosome complement and the distribution of constitutive heterochromatin, along with the differences between congeneric species and among genera belonging to same sub-family are discussed.

MATERIALS AND METHODS

The males of 6 species Acrididae were collected in and around Guru Nanak Dev University campus, Amritsar (Punjab). The testes were excised following standard Colchicine-Hypotonic-Cell suspension-Flame dry technique (Yadav and Yadav, 1983). The flame dried slides were treated for C-banding following method of Sumner (1990) with slight modifications. The chromosomes were classified after Levan et al. (1964) and appropriate karyotypes were prepared.

RESULTS

The perusal of Table 1 revealed that, among the total six species studied; 4 species belong to sub-family Tryxalinae and these are *Acrida turrita* Linn., *Acrida exaltata* Walk., *Phlaeoba infumata* Brunn., *Phlaeoba antennata* Brunn. and one species *Oedaleus abruptus* Thunb. belong to sub-family Oedipodinae and *Oxya velox* (Fabr) belong to sub-family Catantopinae. The diploid chromosome number was found to be 23 and

*Corresponding author. E-mail: poojachadha77@yahoo.co.in.

Table 1. Nature of chromosomes and morphometric characters in 6 species studied.

S.No.	Species	Male 2n number	Chromosome size			*Range of relative length of autosomes	Length of X- chromosomes	Nature of chromosomes m sm a			Nature of X- chromosome
			L	M	S			m	sm	a	
1	Sub-family: Tryxalinae Acrida turrita	23	3	6	2	19.9 ± 0.39 144.2 ± 0.48	199.0 ± 2.26	____		All	Largest
2	Acrida exaltata	23	3	6	2	35.1 ± 0.36 128.9 ± 1.32	148.4 ± 0.42	____		All	Largest
3	Phlaeoba infumata	23	3	6	2	38.0 ± 0.72 128.2 ± 0.46	141.0 ± 1.18	____		All	Largest
4	Phlaeoba antennata Sub-family: Oedipodinae	23	3	6	2	23.2 ± 0.32 139.0 ± 1.02	151.1 ± 1.05	____		All	Largest
5	Oedaleus abruptus	23	3	6	2	22.7 ± 0.38 113.6 ± 1.02	136.3 ± 1.27	____		All	Largest
6	Sub-family: Catantopinae Oxya velox	23	3	6	2	26.3 ± 0.6 149.0 ± 2.52	175.4 ± 1.52	____		All	Largest

sex-determining mechanism was found to be XO: XX type, among all the species investigated. Table 1 shows the various morphometric characters of all the species investigated. It was ascertained that, the chromosome morphology is acrocentric for all the species. X-chromosome is the marker, as it is the largest element among all the species studied. Figure 1 shows the constructed karyotypes of all the species respectively.

Table 2 and Figure 2 are showing the position of constitutive heterochromatin among male grasshoppers studied. It was observed that, the centromeric bands of constitutive heterochromatin were found among all the species under study.

The interstitial bands were seen in A. exaltata, P. infumata, P. antennata only. Terminal bands were seen in all the species investigated except in A. turrita. Thick and thin bands were seen in all the species of hoppers except in A. turrita.

DISCUSSION

The karyology of every species is unique in itself and provides an identity to species (Channaveerappa and Ranganath, 1997). The short horned grasshoppers are characterized bypossessing acrocentric chromosomes. Due to great cytogenetic uniformity, the short horned hoppers are considered as an example of "Karyotypic conservatism" (Aswathanarayana and Ashwath, 2006). In the present study, 6 Acridids have been investigated among which 4 belong to sub-family Tryxalinae and one each belonging to sub-family Oedipodinae and Catantopinae, respectively. It is revealed that, hoppers belonging to family Acrididae have 23 chromosome number. The sex-determining mechanism is found to be XO/XX type among all the studied species. Yadav and Yadav (1986) reported similar results in relation to chromosome number and sex-mechanism among Haryana population of

a

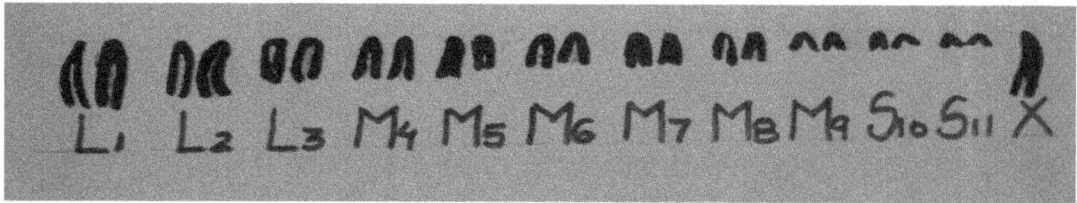

b

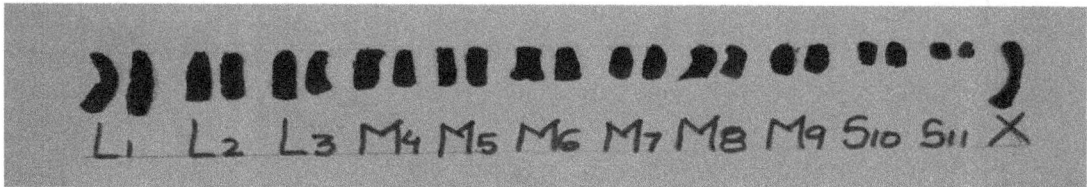

c

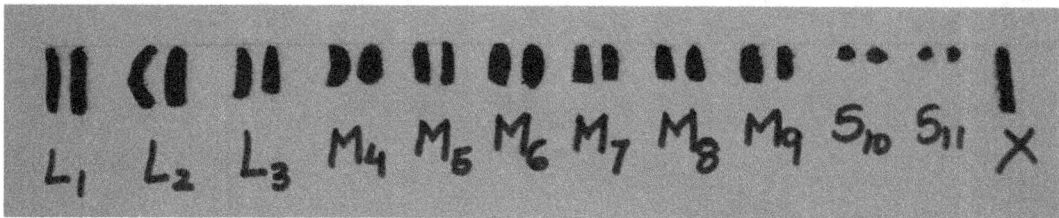

d

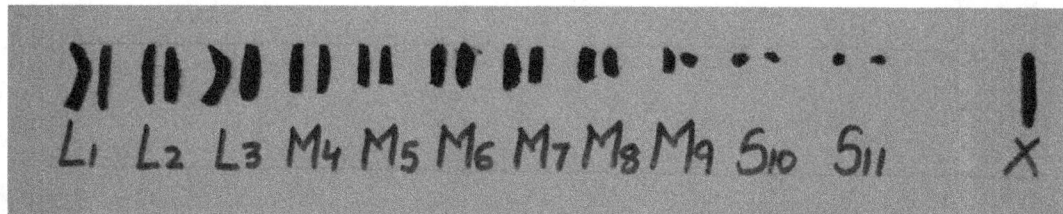

e

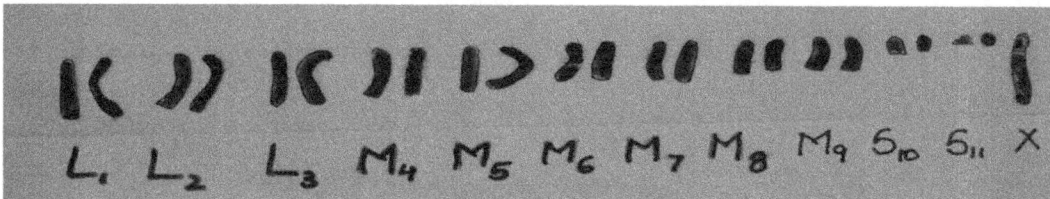

f

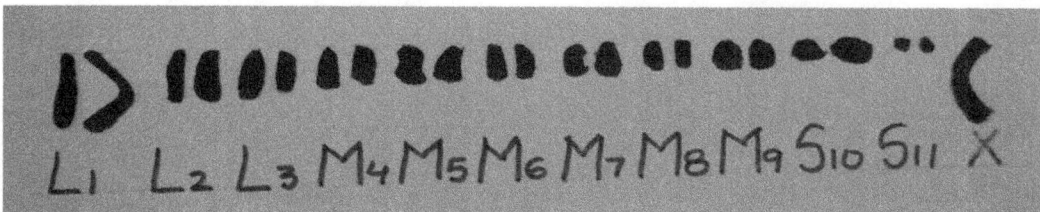

Figure 1. Karyotypes of (a) *A. turrita* (b) *A. exaltata* (c) *P. infumata* (d) *P. anteenata* (e) *O. abruptus* (f) *O. velox.*

Table 2. Showing the position of C-heterochromatin in male grasshopper species under present investigation.

Species Super-family: Acridoidea	C-banding sites				
	Centromeric	Interstitial	Terminal	Thick band	Thin band
Sub family: Tryxalinae Acrida turrita	1-11,X	——	——	1,4,6,7	——
Acrida exaltata	1-11,X	1,X	1,2,4	1,2,3,4,5,6,7,8,9,X	3,4
Phlaeoba infumata	1-11,X	3,5,7,X	1,6,7	1,2,3,5,6,7,X	1,4
Phlaeoba antennata	1-11,X	2	4,5	1,2,3,4,X	1,3,6
Sub family: Oedipodinae Oedaleus abruptus	1-11,X	——	4,5	3,5,6,8,9,X	1,2,5,6,7
Subfamily: Catantopinae Oxya velox	1-11,X	——	1,5,X	1,2,3,4,6,7,X	2,3,5,7,8

Acridoideans. While studying the chromosomes of 11 species of grasshoppers from Simla (H.P), Sharma and Gautam (2002) also revealed similar results. So, the short horned grasshoppers of different regions are showing cytogenetic uniformity, regarding chromosome number and sex-determining mechanism.

During the present investigation, the chromosomes are found to be acrocentric in nature. Up to six metacentrics through fusions have been reported in Tryxalines *Myrmeleotettix maculates* (John and Hewitt, 1966) and *Stauroderus scalaris* (John and Hewitt, 1968). Meanwhile, Aswathanarayana and Ashwath (2006) observed a series of structural changes involving 6th, 7th and 9th pair, exhibiting hetero and homomorphism in *Gastrimargus africanus* orientalis. Mayya et al. (2004) reported short arms in chromosomes of *Aiolopus thalessimus* tumulus and *Acrotylus humbertianus*. Whereas, no such change have been reported in present study. The X-chromosome is found to be largest of all the other chromosomes among the 6 species investigated. But in the population of *O. velox* from Himachal Pradesh, the length of X-chromosome was found to be between L1 and L2 (Sharma and Gautam, 2002). Mayya et al. (2004) also reported the X-chromosome to be largest in all the species except in *A. thalessimus* tumulus and *Spathosternum prasiniferum*. Yoshimura et al. (2005) explored 4 species of Oxya (*Oxya hyla* intricate, *Oxya japonica* japonica, *Oxya chinensis* formosana and *Oxya yezoensis*) and reported that X- chromosome was the longest of medium sized pairs of chromosomes in all the Oxya sp. But in our studies, X- chromosome is the largest and marker in *O. velox*. The C-band represents the constitutive heterochromatin in the homologous chromosome of a karyotype. This type of DNA consist of short repeated polynucleotide sequences. C-bands exhibit centromeric, interstitial and terminal sites. It is a variant state of chromatin and said to be genetically inert (Yunis and Yasminieh, 1971). C-banding pattern in various species of grasshoppers provide important clues that have occurred during the course of evolution. Many studies have shown a remarkable degree of C-band variation.

In the present study, C-bands are also found to be present in three locations, that is centromeric, interstitial and terminal. But their extent is found to vary among the species studied. All the nine species of Acridids studied possess centromeric C-bands. Yadav and Yadav (1993) and Mayya et al. (2004) reported the prevalence of centromeric bands in short horned grasshoppers as a very common feature. Kumaraswamy and Rajasekarasetty (1976) reported centromeric C-bands in *A. turrita*. Aswathanarayana and Ashwath (2006) also revealed centromeric bands in *A. turrita*. According to Yadav and Yadav (1993), restriction of C-heterochromatin to centromeric regions is considered to facilitate whole arm translocation. C-bands found within the centre of the body of chromosome is termed as interstitial C-band. In the present study, interstitial C-bands are seen in all the species of grasshoppers exceptin *A. turrita*, *O. abruptus* and *O. velox*. The interstitial C-bands were also encountered in 10 species of their study by Yadav and Yadav (1993). Mayya et al. (2004) revealed the presence of interstitial C-bands at different locations on chromosomes among Acridoid species. These interstitial bands are found to be inactivated centromere in some species of *Hieroglyphus nigrorepletus* (Yadav and Yadav, 1993). These interstitial C-bands might have an effect on the expression of the flanking euchromatic segment (Aswathanarayana and Ashwath, 2006). Terminal bands were exhibited by all the 6 species except *A. turrita*. Similarly, the absence of terminal bands were also reported in *A. turrita* by Yadav

a

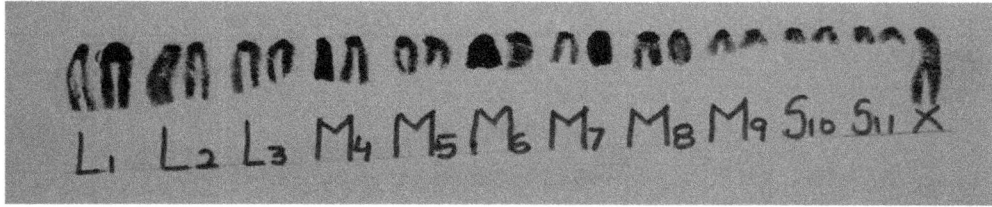

b

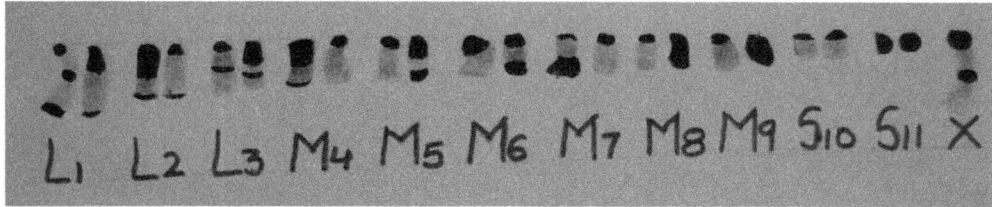

c

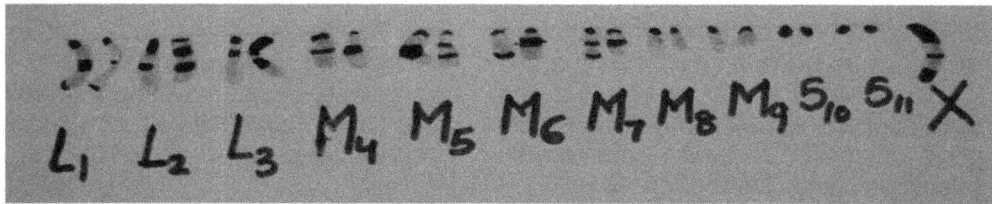

d

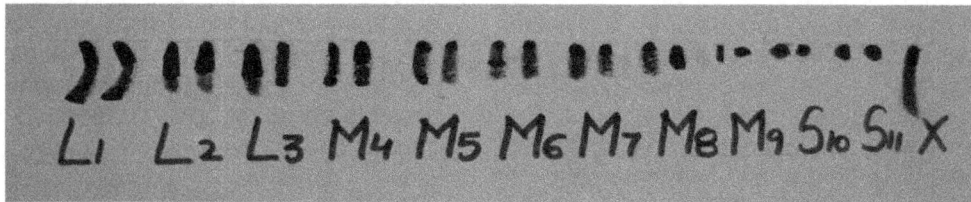

e

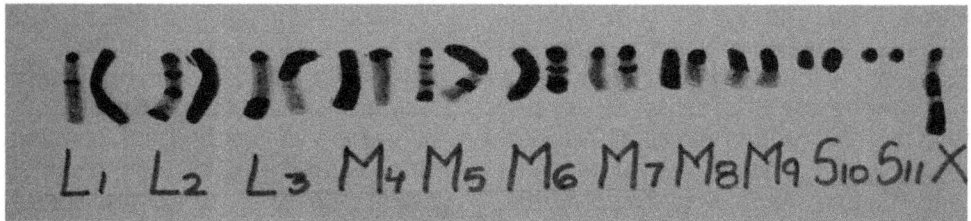

f

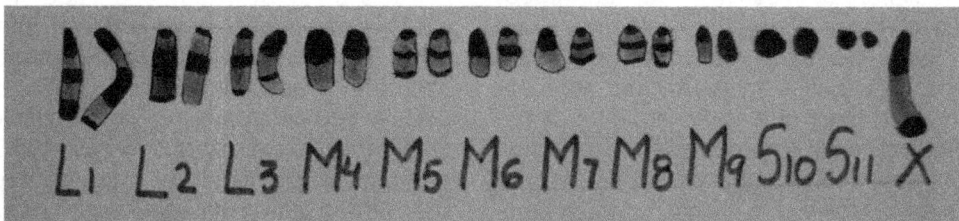

Figure 2. C- Banding karyotypes of (a) *A. turrita* (b) *A. exaltata* (c) *P. infumata* (d) *P. antennata* (e) *O. abruptus* (f) *O. velox*.

and Yadav (1993). On the other hand, our studies revealed the presence of centromeric, interstitial, terminal bands along with thick and thin bands.

The comparison of interspecific C-banding patterns of the same sub-family has no clear correlation. The species from the same genus have not shown uniformity in their C-banding patterns (John and King, 1977; Santos and Giraldez, 1982) which has attributed to dynamic nature of heterochromatin (Yadav and Yadav, 1993). Same situation has been seen in present study where genus coming under different sub-families show some C-banding similarities (*P. antennata, O. abruptus*). The species from the same sub-family differ in their C-band distribution (*A. turrita, A. exaltata*). Likely, such comparisons are such that, one cannot be sure that chromosomes of similar relative lengths are necessarily homologous in all genomes (King and John, 1980). Perhaps, the only exceptions are the X and the megameric chromosomes which presumably have a common origin within the Acrididae (White, 1973). The immediate tendency for C-heterochromatin to vary in grasshoppers has been considered by many reports (Santos et al., 1983; Yadav and Yadav, 1983; present report). The pattern and distribution of C-heterochromatin distribution varies among *Acridoid taxa*, especially karyologically conservative ones. These variations are to be governed by some hidden mechanism of change, other than gross chromosome rearrangements operating in the process of speciation.

ACKNOWLEDGEMENTS

Our sincere thanks to Head, Deptt of Zoology for providing laboratory facilities and a heartful thanks to Prof. A. S. Yadav , Kurukshetra University for identification of grasshoppers.

REFERENCES

Aswathanarayana NV, Ashwath SK (2006). Structural polymorphism and C-banding pattern in a few Acridid grasshoppers. Cytologia, 71(3): 223-228.

Channaveerappa H, Ranganath H (1997). Karyology of few species of south Indian acridids. II Male germ line karyotypic instability in Gastrimargus. J. Biosci., 22(3): 367-374.

John B, Hewitt GM (1966). Karyotype stability and DNA variability in the Acrididae. Chromosoma (Berl.), 20: 155-172.

John B, Hewitt GM (1968). Patterns and pathways of chromosome evolution within the Orthoptera. Chromosoma (Berl.), 25: 40-47.

John B, King M (1977). Heterochromatin variations in *Cryptobothrus chrysophorus* II Patterns of C-banding. Chromosoma, 65: 59-79.

Kumaraswamy KR, Rajasekarasetty MR (1976a). Pattern of C-Banding in *Acrida turrita* (Acrididae:Orthoptera). Curr. Sci., 45(21): 762-763.

King M, John B (1980). Regularities and restrictions governing C-band variation in Acrid grasshoppers. Chromosoma, 76: 123-150.

Levan A, Fredga K, Sandberg AA (1964). Nomenclature for centromeric position of chromosomes. Hereditas, 52: 201-220.

Lopez-Fernandez C, Gosalvez J (1981). Differential staining of a heterochromatic zone in Arcyptera fusca. Experentia, 37: 240.

Mayya S, Sreepada KS, Hegde MJ (2004). Non-banded and C-banded Karyotypes of ten species of short- horned grasshoppers (Acrididae) from south India. Cytologia, 69(2): 167-174.

Santos JL, Giraldez R (1982). C-heterochromatin polymorphism and variation in chiasma localization in *Euchorthippus pulvinatus* gallicus (Acrididae: Orthoptera). Chromosoma, 85: 507-518.

Santos JL, Arana P, Giraldez R (1983). Chromosomal C-banding pattern in Spanish Acridoidea. Genetica, 61: 65-74.

Sumner AT (1990). Chromosome banding. U. Hymen ed. London, Boston, Sydney, Wellington, pp. 39-104.

Sharma T, Gautam DC (2002). Karyotypic studies of eleven species of grasshoppers from north-western Himalayas. Nucleus, 45(1-2): 27-35.

Turkoglu S, Koca S (2002). Karyotype, C- and G- band patterns and DNA content of Callimenus (= Bradyporus) macrogaster macrogaster. J. Insect Sci., 2(24): 1-4.

White MJD (1973). Animal cytology and evolution. Third edition, Cambridge University Press.

Yunis JJ, Yasminieh WG (1971). Heterochromatin, Sat- DNA and cell function. Science, 174: 1200-1209.

Yadav JS, Yadav AS (1983). Analysis of hopper chromosomes with banding techniques. I. *Phlaeoba infumata* (Brunn.) (Tryxalinae: Acrididae) and *Atractomorpha crenulata* (F.) (Pyrgomorphinae: Acrididae). Biologia, 29(1): 47-53.

Yadav JS, Yadav AS (1986). Chromosome number and sex-determining mechanisms in thirty species of Indian Orthoptera. Folia Biol., 34(3): 277-284.

Yadav JS, Yadav AS (1993). Distribution of C-heterochromatin in seventeen species of grasshoppers (Acrididae: Orthoptera). Nucleus, 36(1-2): 51-56.

Yoshimura A, Obara Y, Ando Y, Kayano H (2005). Comparative karyotype analysis of grasshoppers in genus Oxya (Orthoptera : Catantopidae) by differential staining techniques. Cytologia, 70(1): 109-117.

Use of simple sequence repeat (SSR) markers to establish genetic relationships among cassava cultivars released by different research groups in Ghanaian

Peter Twumasi[1] , Eric Warren Acquah[1,2], Marian D. Quain[2] and Elizabeth Y. Parkes[2]

[1]Department of Biochemistry and Biotechnology, Kwame Nkrumah University of Science and Technology, Kumasi, Ghana.
[2]Crops Research Institute (CRI) of Council for Scientific and Industrial Research (CRI-CSIR)-Fumesua, Ghana, P.O. Box 3785, Kumasi, Ghana.

Cassava (*Manihot esculenta*) is an important staple crop widely cultivated in Ghana. The crop also has diverse industrial applications including starch, beer and alcohol productions. Knowledge about the state of the Ghanaian cassava genetic diversity and population structure is paramount in breeding programmes aimed at cultivar improvements or breeding of new cultivars for specific purposes. This study focused on the use of 36 simple sequence repeats (SSRs) to produce SSR allelic polymorphisms for estimation of inter- and intra-population genetic diversity among Ghanaian cassava cultivars from five Ghanaian released and local cassava populations consisting of 11 released and two local cultivars. The results show high diversity among the studied cultivars with an average of seven (7) alleles per locus. Polymorphic loci varied from 68.6 to 100% with an average of 88.58%. A strong genetic diversity was observed within populations (HS =0.552) and therefore suggesting a low rate of inter-population gene flow among the individuals constituting the populations. This high genetic variability among the cultivars provides valuable genetic resource to support any future breeding programmes aimed at establishing new cassava varieties for domestic and industrial purposes.

Key words: Cassava, DNA fingerprinting, genetic diversity, simple sequence repeat (SSR).

INTRODUCTION

Cassava (*Manihot esculenta* Crantz), a member of the family Euphorbiaceae containing 28 wild spp. (Raghu et al., 2007), is one of the most important food crops of sub-Saharan Africa and grown throughout the tropics including

Table 1. Five cassava populations involved in the study: "*CRI*", "*Professor Safo Kantanka*", "*Professor J.P Tettey*", "*SARI*" released materials and Local cultivars.

Cassava population	Breeder	Constituting cultivars
Population 1	Crops Research Released (CRI) materials	*Agbelifia, Essam Bankye, Afisiafi, Doku* Duade, *Bankye Hemaa*
Population 2	Professor Safo Kantaka released materials	*IFAD* and *Nkabom*
Population 3	Professor J.P Tettey released materials	*UCC* and *Bankye Botan*
Population 4	Savanna Agriculture Research Institute (SARI) released materials	*Filindiakoh* and *Nyerikobga*
Population 5	Local cultivars	*Akosua Tumtum* and *Debor*

Asia and Latin America. Cassava plays a famine prevention role wherever it is cultivated widely because it provides a stable base to the food production system (Jarvis et al., 2012; Romanoff and Lynam, 1992). In Ghana, cassava is produced in all but two regions in the northernmost part of the country and according to Nweke et al. (1999), in Collaborative Study of Cassava in Africa (COSCA), villages that did not experience the famine of 1983 in Ghana were those that cultivated cassava as the most important and dominant staple crop. Among the many cassava products in Ghana include fufu, gari, agbelima, agbelikaklo,yakeyake and kokonte (Wareing et al., 2001).

Cassava is assumed to have evolved from inter-specific hybridization among its wild species. The crop is strongly an out crossing monoecious species but suffers from inbreeding depression, making it difficult to develop appropriate stocks for classical improvement of genetic methods (Fregene et al., 1997). The issue of inbreeding has led to the use of crossing blocks for inter mating superior individuals so that inbreeding could be minimized (Falahati-Anbaran et al., 2006).

Cassava is a diploid plant with 2n=36 chromosomes and a DNA content of 1.67 pg per cell nucleus (Prochnik et al., 2012; Awoleye et al., 1994). This is equivalent to 772 Mbp per haploid genome, and occupying the lower end of the genome size range for higher plants (Bennett et al., 1992). This relatively small genome size of cassava, favors development of saturated genetic map and molecular tag which may contribute to understanding of the inheritance of many important genetic traits despite the heterozygous nature of cassava (Fregene et al., 1997). Considerable amounts of genetic variations in cassava germplasm have been reported by many laboratories (Raghu et al., 2007; Acquah et al., 2011; Kawuki et al., 2013).

In recent years, there has been increasing interest in the use of DNA-based markers which is acclaimed to be more reliable marker for genetic diversity studies in contrast to the classical morphological markers. Morphological descriptors are highly subjective, environmentally influenced and of low polymorphism. However, due to the lack of molecular genetic expertise and high cost of establishing modern biotechnolical laboratories, many researchers continue to rely upon morphological descryptions for genetic diversity studies (Falahati-Anbaran et al., 2006). Simple Sequence Repeat (SSR) has been reliably used to quantify genetic diversity and examine popu-lation differentiation in a number of agricultural crops (Morgante et al., 1994; Maughan et al., 1995; Raghu et al., 2007). SSRs are co-dominant and tend to have multiple alleles per locus so that individuals can be identified as homozygotes or heterozygotes. An additional benefit of SSR marker is its ability to detect variations in allele frequency at many unlinked loci which are abundant in plants and the technique is easily adaptable to automation (Donini et al., 1998).

The objective of this study was to use SSR allelic polymorphisms produced by 36 SSRs to estimate inter- and intra-population genetic diversity among eleven (11) released and two (2) local Ghanaian cassava cultivars sampled from five cassava populations derived from local breeders.

MATERIALS AND METHODS

Plant materials

Five Ghanaian cassava populations involved in this study included 11 released and two local cultivars (Table 1). They were collected from farms of the CSIR-Crop Research Institute (CRI) at Fumesua in the Ashanti Region. Fresh stem cuttings 20-30 cm with 5-8 nodes were obtained from disease-free matured cassava plants. They were planted in loamy soil for a period of five weeks in plastic pots at the CSIR-Crops Research Institute screen house. The growing

Table 2. SSR primers and thermocycler programmes used in the study.

SSR Locus	Forward primer	Reverse primer	Thermocycler programme
SSRY 4	ATAGAGCAGAAGTGCAGGCG	CTAACGCACACGACTACGGA	MicroBC1
SSRY 9	ACAATTCATCATGAGTCATCAACT	CCGTTATTGTTCCTGGTCCT	MicroBC1
SSRY 12	AACTGTCAAACCATTCTACTTGC	GCCAGCAAGGTTTGCTACAT	MicroBC1
SSRY 19	TGTAAGGCATTCCAAGAATTATCA	TCTCCTGTGAAAAGTGCATGA	MicroBC1
SSRY 20	CATGGACTTCCTACAAATATGAAT	TGATGGAAAGTGGTTATGTCCTT	MicroBC1
SSRY 21	CCTGCCACAATATTGAAATGG	CAACAATTGGACTAAGCAGCA	MicroBC1
SSRY 34	TTCCAGACCTGTTCCACCAT	ATTGCAGGGATTATTGCTCG	MicroBC1
SSRY 38	GGCTGTTCGTGATCCTTATTAAC	GTAGTTGAGAAAACTTTGCATGAG	MicroBC1/ NewBC1
SSRY 51	AGGTTGGATGCTTGAAGGAA	GGATGCAGGAGTGCTCAACT	MicroBC1/Yucadiv (Yu)
SSRY 52	GCCAGCAAGGTTTGCTACAT	AACTGTCAAACCATTCTACTTGA	MicroBC1/Yu
SSRY 59	GCAATGCAGTGAACCATCTTT	CGTTTGTCCTTTCTGATGTTC	MicroBC1
SSRY 63	TCCAGAATCATCTACCTTGGCA	AAGACAATCATTTTGTGCTCCA	MicroBC1/Yu
SSRY 64	CGACAAGTCGTATATGTAGTATTCAG	GCAGAGGTGGCTAACGAGAC	MicroBC1/Yu
SSRY 69	AGATCTCAGTCGATACCCAAG	ACATCCGTTGCAGGCATTA	NewBC1(Ne)
SSRY 82	TGTGACAATTTTCAGATAGCTTCA	CACCATCGGCATTAAACTTG	MicroBC1/Yu
SSRY 100	ATCCTTGCCTGACATTTTGC	TTCGCAGAGTCCAATTGTTG	NewBC1
SSRY 102	TTGGCTGCTTTCACTAATGC	TTGAACACGTTGAACAACCA	NewBC1
SSRY 103	TGAGAAGGAAACTGCTTGCAC	CAGCAAGACCATCACCAGTTT	NewBC1
SSRY 105	CAAACATCTGCACTTTTGGC	TCGAGTGGCTTCTGGTCTTC	NewBC1
SSRY 106	GGAAACTGCTTGCACCAAAGA	CAGGCAAGACCATCACCAGTTT	NewBC1
SSRY 108	ACCCTATGATGTCCAAAGGC	CATGCCACATAGTTCGTGCT	MiccroBC1/Yu
SSRY 110	TTGAGTGGTGAATGCGAAAG	AGTGCCACCTTGAAAGAGCA	NewBC1
SSRY 135	CCAGAAACTGAAATGCATCG	AACATGTGCGACAGTGATTG	Yucdiv
SSRY 147	GTACATCACCACCAACGGGC	AGAGCGGTGGGCGAAGAGC	Yucadiv
SSRY 148	GGCTTCATCATGGAAAAACC	CAATGCTTTACGGAAGAGCC	Yucadiv
SSRY 151	AGTGGAAATAAGCCATGTGATG	CCCATAATTGATGCCAGGTT	NewBC1
SSRY 155	CGTTGATAAAGTGGAAAGAGCA	ACTCCACTCCCGATGCTCGC	Yucadiv
SSRY 161	AAGGAACACCTCTCCTAGAATCA	CCAGCTGTATGTTGAGTGAGC	Yucadiv
SSRY 164	TCAAACAAGAATTAGCAGAACTGG	TGAGATTTCGTAATATTCATTTCACTT	NewBC1
SSRY 169	ACAGCTTAAAAACTGCAGCC	AACGTAGGCCCTAACTAACCC	Yucadiv
SSRY 171	ACTGTGCCAAAATAGCCAAATAGT	TCATGAGTGTGGGATGTTTTTATG	NewBC1
SSRY 177	ACCACAAACATAGGCACGAG	CACCCAATTCACCAATTACCA	Yucadiv
SSRY 179	CAGGCTCAGGTGAAGTAAAGG	GCGAAAGTAAGTCTACAACTTTTCTAA	MicroBC1
SSRY 180	CCTTGGCAGAGATGAATTAGAG	GGGGCATTCTACATGATCAATAA	MicroBC1
SSRY 181	GGTAGATCTGGATGGAGGAGG	CAATCGAAACCGACGATACA	Yucadiv
SSRY 182	GGAATTCTTTGCTTATGATGCC	TTCCTTTACAATTCTGGACGC	Yucadiv

conditions used were 30ºC day temperature, 24ºC night temperature, 12 h day light, and 55% relative humidity. Plants were watered daily and the experiment repeated in the following year with similar growth conditions.

DNA isolation and polymerase chain reaction

Genomic DNA was isolated from cassava leaves according to Egnin et al. (1998) isolation protocol adopted by the Council for Scientific and Industrial Research- Crops Research Institute (CSIR-CRI) molecular laboratory with some modifications. DNA quality

was determined on 0.8% agarose gel stained with ethidium bromide. 10 μl PCR reaction mixtures [1.0 μl of buffer (10X), 0.9 μl of MgCl₂ (25 mM), 0.4 μl of dNTPs (10 mM), 0.25 μl of both forward and reverse primer (10 μM), 0.125 μl Taq polymerase (5 U), 1.0 μl of genomic DNA template (10 ng/ μl) all together with 6.075 μl of nuclease-free PCR water] were prepared. Amplification was performed for 30 cycles in MyCycler thermal cycler (Bio-Rad Laboratories Inc.) with heated lid to reduce evaporation. The 36 SSR markers used were in three cycling programmes namely: Yucadiv, MicroBC1 and NewBC1 (Table 2).

The following were the cycling profiles: Yucadiv [95°C for 2 min (initial denaturation), 30 cycles of the following steps: 94°C for 30 s

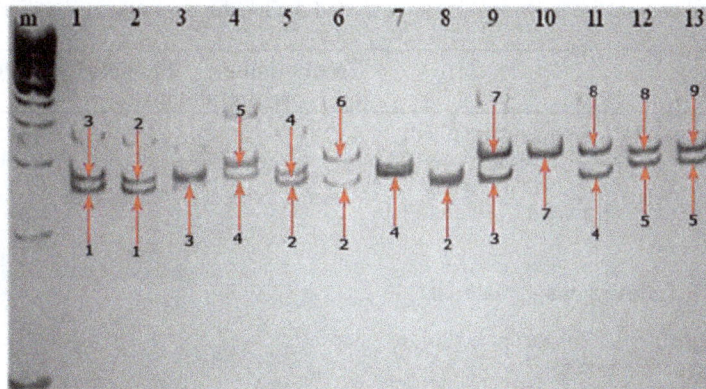

Figure 1. 6% Polyacrylamide gel electrogram showing silver stained PCR amplified allelic fragments of locus SSRY 59 for 11 released and two local Ghanaian cassava cultivars. M, 100 bp marker; 1, *UCC*; 2, *IFAD*; 3, *Agbelefia*; 4, *Nyerikobga*; 5, *Nkabom*; 6, *Esaam Bankye*; 7, *Akosua Tumtum*; 8, *Debor*, 9, *Filindiakoh*; 10, *Afisiafi*; 11, *Doku Duade*; 12, *Bankye Hemaa*;13, *Bankye Botan*.

(denaturation), 55°C for 1 min (annealing), 72°C for 1 min (extension) then 72 C for 5 min (final extension) and storage at 4°C]; MicroBC1 [94°C for 2 min (initial denaturation), 30 cycles of the following steps: 94°C for 1 min (denaturation), 55°C for 1 min (annealing), 72°C for 1 min (extension) then 72°C for 5 min (final extension) and storage at 4°C]; NewBC1 [95°C for 2 min (initial denaturation), 30 cycles of the following steps: 94°C for 30 s (denaturation), 55°C for 30 s (annealing), 72°C for 1 min (extension) then 72°C for 5 min (final extension), 25 cycles of the following steps; 94°C for 30 s, 65°C for 30 s (with annealing decreasing subsequently by 1°C/ Cycle), 72°C for 1 min then 94°C for 30 s (final denaturation), 55°C for 30 s (annealing), 72°C for 1 min (final extension) and storage at 4°C].

Gel electrophoresis and band scoring

After the amplification reactions, 8 µl of the amplified DNA fragments were separated on 6% PAGE gel [12 % acrylamide solution (19:1), TBE (12X) 0.4% APS, TEMED and filtered autoclaved distilled water (FADW)] at 200 V for 35 min in TBE (1X) using a mini-protean 3 cell electrophoretic apparatus. A 100 bp DNA marker (0.05 µg/µl, 25 µg) (Invitrogen) was used as a standard and the DNA fragments were visualized by silver nitrate staining (Figure 1).

The SSR amplified bands were scored as diploid (each individual cultivars score at a locus consisted of two digits) and codominant (each separate allele at a locus was given a score that relates to its size) by visual inspection and "climbing ladder" approach for both F-Statistics (FSTAT) and TOOLS FOR Population Genetic Analysis (TFPGA). For example, in Figure 1 Lane 1, the second band (1) corresponds to an allele seen on a chromosome while the first band (3) corresponds to the alternative allele, with a higher bp than the second allele, located at that same locus. For TFPGA analysis, cultivar 8 and 10 (*Debor* and *Afisiafi*) for example were scored as 0202 and 0707 respectively meaning the cultivars were identified at locus SSRY 59 as homozygous for alleles 2 and 7 respectively (Figure 1). On the other hand, cultivar 6 (*Essam bankye*) was represented as heterozygous for alleles 2 and 6 and this was scored as 0206. However, in FSTAT analysis the scoring 0202 and 0206 would be recorded as 22 and 26 respectively.

Statistical analysis and genetic diversity determination

The bands score data were analyzed for descriptive statistics parameters [number of alleles per locus, allelic richness, allelic frequency, percent polymorphic loci, Hardy-Weinberg equilibrium, gene diversity per locus per population and Nei's (1978) estimation of heterozygosities], F-statistics and cluster analysis, and Nei's (1978) genetic distance. With the exception of Hardy-Weinberg equilibrium, cluster analysis, Nei's (1978) genetic distance and F-statistics which were analyzed using TFPGA version 1.3 (Miller, 1997) all other analysis were carried out with FSTAT version 9.3.2 (Goudet, 1995).

RESULTS

DNA isolation, PCR and gel electrophoresis

High quality genomic DNA assessed on 0.7% agarose gel were successfully isolated from all the 13 cassava cultivars. A total of 35 out of the 36 primers, representing 97.2%, produced clear and scorable bands in electrophoretic gels (Figure 1). Amplification at locus SSRY 177 failed in all the triplicate reactions. *Afisiafi* generated no products at locus SSRY 38, 106 and 164. Similarly, at locus SSRY 179 and 180 no PCR products were observed for *Agbelifia* cultivar. Again there were no successful PCR products in *Nkabom* cultivar at locus SSRY 52. For the 13 cultivars analyzed, all the 35 microsatellite loci were found to be polymorphic.

Genetic diversity determination

The number of alleles observed per locus, at all the 35 SSR loci analyzed, varied from three (3) to eleven (11) and averaging seven (7) alleles per locus. The number of

Table 3. The number of alleles and percentage polymorphic loci recorded among the cassava populations under study

Cassava Population	Breeders	Total number of allele per population	Percentage of polymorphic loci per population
Population 1	Crops Research Released (CRI) materials	163	100
Population 2	Professor Safo Kantaka released materials	86	80.0
Population 3	Professor J.P Tettey released materials	99	100
Population 4	Savanna Agriculture Research Institute (SARI) released materials	93	94.3
Population 5	Local cultivars	66	68.6
Mean		101.4	88.58

Table 4. Mean values of Nei's (1978) estimation of heterozygosity.

Ho	He	Ht	Dst	Gst
0.497	0.769	0.814	0.044	0.054

Ho, Observed heterozygosity; He, expected heterozygosity; Ht, total heterozygosity; Dst, interpopulation gene diversity; Gst, coefficient of genetic differentiation.

Table 5. Unbiased heterozygosity of individual populations.

Populations	Heterozygosity
Pop. 1	0.795
Pop. 2	0.643
Pop. 3	0.800
Pop. 4	0.729
Pop. 5	0.481
Mean	0.699

Population 1, *CRI* released materials; population 2, *SARI* released materials; population 3, *Prof. J.P. Tettey* released materials; population 4, *Prof. Safo Kantanka* released materials; population 5, local cultivars.

alleles per polymorphic population ranged from 66 to 163 with an average of 101.4 whereas the percentage of polymorphic loci population varied from 68.6 to 100% and averaging 88.58% (Table 3).

Nei's diversity indices estimation showed a low value of 0.497 for the observed heterozygosity (HO) whereas the expected heterozygosity (He) on the other hand recorded a high value of 0.769 (Table 4). However, a high total heterozygosity (Ht) of 0.814 was observed in all the 13

cassava cultivars. The results also established inter-population gene diversity (Dst) of 0.044 and coefficient of genetic differentiation (Gst) value of 0.054. A heterozygote deficit of 0.37 was observed. Hardy-Weinberg equilibrium also recorded 0.273. The unbiased heterozygosities of the individual populations recorded values ranging from 0.4810 to 0.800 (Table 5). The local cultivars recorded the lowest heterozygosity while *Prof. Tettey* released materials recorded the highest (Table 5). Nei (1978) genetic distance ranged from 0.1440 to 1.0057 (Table 6) whereas the genetic diversity revealed a high genetic variability range of 0.500 to 0.94.3 (Table 7).

Table 8 contains outputs of the Wright's F-statistical analysis. The results established a high correlation of genes within individuals (Fit), heterozygote deficit among individuals within each population (Fis) and a high occurrence of cross pollination (C) at 0.404, 0.372 and 0.663 respectively. The correlation of genes of different individuals in the same population (Fst) and the fixation index coefficient (F) obtained were however low at 0.052 and 0.373 respectively. The analysis also showed low occurrence (1.028) of inter-population gene flow (Nm).

DISCUSSION

The study has clearly shown high genetic variability among the five Ghanaian cassava populations assessed by 36 SSR with an average of seven (7) alleles per locus. The observed genetic variability among the cultivars contrasts sharply with an earlier work conducted by Okai et al. (2003) in which they claimed to have observed low genetic variability among 320 cassava landraces from Ghana. The high variability among the cassava populations studied attest to the use of genetically diversified progenitors for the breeding and establishment of the cassava cultivars bred for their unique nutritional and industrial benefits (Manu-Aduening et al., 2005).

Table 6. Nei (1978) unbiased distance.

Population	1	2	3	4	5
1	*****				
2	0.5748	*****			
3	0.2732	0.5541	*****		
4	0.1440	0.3408	0.4053	*****	
5	0.6477	0.7757	1.0057	0.5197	*****

Population 1, *CRI* released materials; population 2, *SARI* released materials; population 3, *Prof. J.P. Tettey* released materials; population 4, *Prof. Safo Kantanka* released materials; population 5, local cultivars.

Table 7. Gene diversity per locus and population.

Pop 1	Pop 2	Pop 3	Pop 4	Pop 5
0.835	0.676	0.943	0.850	0.500

Pop 1, *CRI* released materials; Pop 2, *SARI* released materials; Pop 3, *Prof. J.P. Tettey* released materials; Pop 4, *Prof. Safo Kantanka* released materials; Pop 5, local cultivars.

Table 8. Output of the Wright's F statistics.

Fit	Fst	Fis	F	Nm	C
0.404	0.052	0.372	0.373	1.028	0.663

Means were obtained for F-statistics parameters. The parameters analyzed were correlation of genes within individuals overall population (Fit), correlation of genes of different individuals in the same population (Fst), heterozygote deficit observed in individuals within each population (Fis), fixation index coefficient (F), estimation of gene flow (Nm) and frequency of cross pollination (C).

The high number of alleles per locus and the corresponding high number of alleles per population averaging over 100, contributed to the high polymorphic loci observed (88.58%). A collection of cassava cultivars with such high values is expected to have wide genetic distances between their constituting cultivars. Nei (1978) genetic distance of 0.1440 to 1.0057 was however expected. This wide genetic distance implies that the cassava varieties from the five populations used in the study possess very diverse genetic backgrounds (da Costa et al., 2013). Extensive diversity in their progenitor is most likely source of genetic variations in the collection considering the young age of the varieties. The large genetic distances observed among the cultivars support effective breeding of new varieties from the existing ones. These cultivars still remain potential progenitors for breeding of new cassava cultivars for various purposes.

It is clear in this study that the probability of any two randomly selected alleles from any given population

could be different that is (He) 0.769 or 76.9%. This value was observed to be higher than what was obtained in other related works on cassava (Lokko et al., 2006; Fregene et al., 2003). The value is far higher than what have been observed in outcrossing species of dicots (0.159) and all other plant species (0.205) (Hamrick and Godt, 1997). These, notwithstanding Nei's estimation of observed heterozygosity, recorded a smaller value of 49.7% resulting in a heterozygote deficit of 27.2% among the cultivars. The F statistics analysis, Fis 0.372, Fit 0.404 and F at 0.373 confirmed the observed deficit (Gehring and Lindhart, 1992; Dolan et al., 1999). However, the deviation from Hardy-Weinberg equilibrium (0.273) indicating excess of homozygosity and existence of a non-random mating among the cultivars explained the low observed heterozygosity which resulted in the 27.2 % loss of genetic diversity.

These observations could be influenced by the extensive vegetative cultivation of cassava and the rare incidence of genetic self-incompatibility within the crop. Although inflorescences are metandrous, seed dispersal in cassava via explosive mechanism does not promote long distance gene flow (Olsen and Schaal, 2001). These reasons could favor inbreeding and thus affecting cassava breeding due to narrowing of the cassava genetic base (Okai et al., 2003). A narrow genetic base restricts the progress of cultivar improvement by compounding the difficulty involved in choosing parental materials for breeding. That is, narrow genetic base is undesirable for breeding because a certain level of parental divergence is needed to create productive hybrids (Miranda et al., 2008).

Though, Nei (1978) genetic diversity revealed high genetic variability (0.500 to 0.94.3) among the populations studied in this work, a higher intra-population diversity was recorded (HS = 0.552) as compared to inter-population diversity (low Gst 0.054, FST 0.052 and Dst 0.044). This suggests that the populations assessed in our study have a low rate of inter-population gene flow, that is Nm 1.028, which could possibly lead to a low genetic overlap as a result of minimal gene flow through seed and pollen. Faraldo et al. (2000), Mühlen et al. (2000),

Asante and Offei (2003), Peroni et al. (2007) and Lokko et al. (2006) all observed similar results as they reported high intra-genetic variability in Ghanaian cassava varieties. A similar pattern was observed with sweet potato landraces from the Vale do Ribeira (Veasey et al., 2008). On the contrary, the high genetic variation within the Ghanaian released cassava populations indicates the usefulness of each of the four local populations as a valuable genetic resource for the selection of superior genotypes, as seen in the case of 'CRI released materials, Prof. J.P. Tettey released materials and Prof. Safo Kantanka released materials' showing 83.5, 85.0 and 94.3% genetic diversity respectively.

Conflict of Interests

The author(s) have not declared any conflict of interest.

ACKNOWLEDGEMENT

We wish to acknowledge the Roots and Tuber Division of the Council for Scientific and Industrial Research - Crops Research Institute (CSIR-CRI) - Fumesua - Kumasi, Ghana and the Ministry of Food and Agriculture (MOFA) Multiplication Station - Ashanti Mampong, Ghana for supplying us with the plant materials.

REFERENCES

Acquah EW, Quain MD and Twumasi P (2011). Genetic relationships between some released and elite Ghanaian cassava cultivars based on distance matrices. Afr. J. Biotechnol. 10(6):913-921.

Asante IK, Offei SK (2003). RAPD-based genetic diversity study of fifty cassava (Manihot esculenta Crantz) genotypes. Euphytica 131(1): 113-119.

Awoleye F, Van Duren M, Dolezel J Novak FJ (1994). Nuclear DNA content and in vitro induced somatic polyploidization (Manihot esculenta Crantz) cassava breeding. Euphytica 76: 195-202.

Bennett MD, Smith JB Heslop-Harrison JS (1992). Nuclear DNA amounts in Rice. Soc. London Ser. B. 216:179-1999.

da Costa TR Filho PSV, Gonçalves-Vidigal MC, Galván MZ, Lacanallo GF, da Silva LI Kvitschal MV (2013). Genetic diversity and population structure of sweet cassava using simple sequence repeat (SSR) molecular markers. Afr. J. Biotechnol. 12(10): 1040-1048.

Dolan RW, Yahr R, Menges ES, Halfhill MD (1999). Conservation implications of genetic variation in three rare species endemic to Florida rosemary scrub. Am. J. Bot. 86: 1556-1562.

Donini P, Stephenson P, Bryan GJ, Koebner RMD (1998). The potential of microsatellites for high throughput genetic diversity assessment in wheat and barley. Gen. Res. Crop Evol. 45: 415-421.

Egnin M, Mora A, Prakash CS (1998). Factors enhancing Agrobacterium tumefaciens - Mediated Gene Transfer in peanut (Arachis hypogeal L.). In vitro Cell. Dev. Bio. Plant 34: 310-318.

Fregene M, Angel F, Gomez R, Rodriguez F, Chavarriaga P, Roca WM, Tohme J, Bonierbale MW (1997). A molecular genetic map for cassava (Manihot esculentus Crantz). Theor. Appl. Genet. 95:431441

Falahati-Anbaran M, Habashi AA, Esfahany M, Mohammadi SA, Ghareyazie B (2006). Study of genetic diversity and relationships of diploid and tetraploid annual medics using microsatellite markers. J.

Sci.Technol. Agric. Nat. Res. 10:349-358.

Faraldo MIF, Silva MR, Ando A Martins PS (2000). Variabilidade genética de etnovariedades de mandioca em regiões geográficas do Brasil. Sci. Agric. 57: 499-505.

Fregene M, Suárez M, Mkumbira J, Kulembeka H, Ndedya E, Kulaya A, Mitchel S, Gullberg U, Rosling, H, Dixon A, Dean R Kresovich S (2003). Simple sequence repeat marker diversity in cassava landraces: genetic diversity and differentiation in an asexually propagated crop. Theor. Appl. Gen. 107: 1083-1093.

Gehring JL, Linhart YB (1992). Population structure and genetic differention in nature and introduced population of Deschampsia caespitosa (Poaceaee) in the Colorado Alpine. Am. J. Bot. 72(12): 1337-1343.

Goudet J (1995). FSTAT (Version 1.2): A Computer Program to Calculate F-Statistics J. Hered. 86: 485-486.

Hamrick JL, Godt MJW (1997). Allozyme diversity in cultivated crop. Crop Sci. 37:26-30.

Jarvis A, Ramirez-Villegas J, Campo BVH, Navarro-Racines C (2012). Is Cassava the Answer to African Climate Change Adaptation? Trop. Plant Biol. 5(1):9-29

Kawuki RS, Herselman L, Labuschagne MT, Nzuki I, Ralimanana I, Bidiaka M, Kanyange MC, Gashaka G, Masumba E, Mkamilo G, Gethi J, Wanjala B, Zacarias A, Madabula F, Ferguson ME (2013). Genetic diversity of cassava (Manihot esculenta Crantz) landraces and cultivars from southern, eastern and central Africa. Plant Genet. Res. 11(2): 170-181.

Lokko Y, Dixon A, Offei S, Danquah E, Fregene M (2006). Assessment of genetic diversity among African cassava Manihot esculenta Crantz accessions resistant to the cassava mosaic virus disease using SSR markers. Genet. Res. Crop Evol. 53:1441-1453.

Manu-Aduening JA, Lamboll RI, Dankyi AA, Gibson RW (2005). Cassava diversity in Ghanaian farming systems. Euphytica 144(3): 331-340.

Maughan PJ, Saghai Maroof MA, Buss GR (1995). Microsatellite and amplified sequence length polymorphisms in cultivated and wild soybean. Genome 38: 715-723.

Miller MP (1997). Tools for Population Genetic Analysis. Version 1.3. Department of Biological Sciences, Northern Arizona University, Flagstaff.

Miranda GV, Vagno de Souza L, Galvão JCC, Guimarães LJM, Vaz de Melo A, dos Santos IC (2008). Genetic variability and heterotic groups of Brazilian popcorn populations. Euphytica 162(3): 431-440.

Morgante M, Rafalsk, A, Biddle P, Tingey S, Olivier AM (1994). Genetic mapping and variability of seven soybean simple sequence repeat loci. Genome 37: 763-769.

Mühlen GS, Martins PS, Ando A (2000). Variabilidade genética de etnovariedades de mandioca, avaliada por marcadores de DNA. Sci. Agric. 57: 319-328.

Nei M. (1978). Estimation of average heterozygosity and genetic distance from a small number of individuals. Genetics 89:583-590.

Nweke FI, Haleegoah J, Dixon AGO, Ajobo U, Al-Hassan R (1999). Cassava Production and Processing Technologies. Collaborative Study of Cassava in Africa. COSCA Working Paper No. 21, IITA, Ibadan, Nigeria.

Okai E, Otoo J, Buitrago C, Fregene M, Dixon A (2003). Simple Sequence Repeat (SSR) Assessment of Genetic Diversity of Local Cassava Varieties from Ghana and Prediction of Heterosis. IITA annual report. pp. 30.

Olsen KM, Schaal BA (2001). Microsatellite variation in cassava (Manihot esculenta, Euphorbiaceae) and its wild relatives: further evidence for a southern Amazonian origin of domestication. Am. J. Bot. 88 (1):131-142

Peroni N, Kageyama PY, Begossi A (2007). Molecular differentiation, diversity, and folk classification of "sweet" and "bitter" cassava (Manihot esculenta) in Caiçara and Caboclo management systems (Brazil). Genet. Res. Crop Evol. 54:1333-1349.

Prochnik S, Marri PR, Desany B, Rabinowicz PD, Kodira C, Mohiuddin M, Rodriguez F, Fauquet C, Tohme J, Harkins T, Rokhsar DS, Rounsley S (2012).The Cassava Genome: Current Progress, Future Directions. Trop. Plant Biol. 5:88-94.

Raghu D, Saraswathi N, Raveendran M, Venkatacholam R, Shanmugasundaram P, Mohan C (2007). Morphological and simple sequence repeats (SSR) based finger printing of south Indian cassava germplasm. Int. J. Integr. Biol. 141-149.

Romanoff S, Lynam J (1992). Commentary Cassava and African Food Security: Some ethnographic examples. Ecol. Food Nutr. 27:29 - 41.

Veasey EA, Borges A, Rosa MS, Queiroz-Silva JR, Bressan EA, Peroni N (2008). Genetic diversity assessed with microsatellites in Brazilian sweet potato (Ipomoea batatas (L.) Lam) landraces. Genet. Mol. Biol. 31:725-733.

Wareing PW, Westby A, Gibbs JA, Allotey LT, Halm M (2001). Consumer preferences and fungal and mycotoxin contamination of dried cassava products from Ghana. Int. J. Food Sci. Technol. 36(1): 1-10.

8

Rapid targeting and isolation of the β-like globin gene cluster fragment from AA, AS and SS genotypes using *Bam*HI restriction enzyme

Aimola Idowu Asegame[1]*, Inuwa Hajia Mairo[1], Nok Andrew Jonathan[2] and Mamman I. Aisha[3]

1Department of Biochemistry, Ahmadu Bello University, Zaria, Nigeria.
2Centre for Biotechnology Research and Training, Ahmadu Bello University, Zaria, Nigeria.
3Hematology Department, Ahmadu Bello University Teaching Hospital, Zaria, Nigeria.

The β-globin gene cluster contains the β-globin gene on which the substitution of the 17th nucleotide gives rise to sickle cell anemia and other β-globin gene variants with varying severity. Recent therapy measures focus on understanding the structure and mechanism of expression of the β-globin gene. In this study, we attempted to isolate a gene fragment containing the complete set of the human β-globin gene cluster from AA, AS and SS blood types, using the restriction enzyme *Bam*HI which has a recognition site within the 19-kb 3' cluster region downstream the β-globin gene cluster. DNA fragment size of 69.65 kb was generated by the enzyme from the genome of AS and SS blood types while the fragment generated from the AA blood type was 70.28 kb. Fragments generated using *Bam HI* was all comparable and close to the documented size of the β-globin gene cluster. In addition to the 69.65 kb fragment generated from the DNA from SS blood type, a novel 5.73 kb fragment was also visualized which was absent in the lanes containing restricted DNA from the AA blood type.

Key words: *Bam*HI, beta-globin, genotype, isolation.

INTRODUCTION

The β-globin gene cluster lies on the short arm of chromosome 11 within a 70 kb region and contains 5 coordinately regulated genes arranged in the sequence of their expression (5'-ε-Gγ-Aγ-δ-β-3'), and is flanked 6 to 18 kb upstream the ε-globin gene by a regulatory region which contains a series of DNAse I hypersensitive sites known as the locus control region (Langdon and Kaufman, 1998). The β-globin gene cluster has a well studied restriction map, and some restriction enzymes; including; Hind II, Taq I, Pst I, Hinf I and *Bam*HI have been shown to posses recognition sequences within and immediately outside the gene locus (Aliyu et al., 2008). Several hemoglobin variants have arisen over the years as a result of mutations in the beta globin gene cluster (Smith et al., 1998). One of the most significant of these

variants is β[s] mutation which results from a substitution of adenine by thymine on the 17th nucleotide of the β-globin gene in the β-globin gene cluster.

Genotype is known to be the major risk factor in the severity of sickle cell anemia (Smith et al., 1998). Carriers of the β[s] gene are known to lead normal lives and exhibit little or no anemia (Smith et al., 1998). Homozygous for the β[s] usually exhibit chronic anemia and other symptoms known to be associated with sickle cell anemia. Sub-Saharan Africa has the highest burden of sickle cell anemia with Nigeria alone accounting for over two thirds of the sickle cell anemia population in Africa (Ashley-Koch et al., 2000). Studies on the influence of genotype on the expression of the β-globin gene cluster are scanty. Understanding mechanisms of β-globin gene expression, and similarly diagnosis of the disease requires an efficient and rapid method for the preparation of the β-globin gene cluster. In order to understand variations in the β-globin gene cluster, it is also necessary to be able to success-fully prepare the β-globin gene cluster both rapidly and

*Corresponding author. E-mail: idowuaimola@gmail.com.

precisely.

Most methods of isolation of the β-globin gene cluster rely on primer specific PCR-amplification of the β-globin gene cluster from genomic DNA (Saraswathy et al., 2010; Cao et al., 2004; Langdon and Kaufman, 1998). Conventional β-globin gene isolation requires the use primer sequences complementary to the gene (Kaufman et al., 1999; Kaufman et al., 1998; Smith et al., 1998) and in the case of Detloff et al. (1994) PCR was also used to screen for the targeting of the β-globin gene.

In this research, we attempted to isolate and compare within genotypes the β-globin gene cluster from humans using BamHI restriction endonuclease with recognition site within the 19-kb 3' cluster region downstream the β-globin gene cluster.

MATERIALS AND METHODS

Sample selection

Peripheral blood was obtained from adult volunteers who had previously signed consent forms approved by the Ahmadu Bello University Health Service Unit. 4 ml Blood samples were obtained from volunteers with AA, AS and SS genotypes and used for genomic DNA preparation. None of the volunteers of blood samples were undergoing treatment for any disease at the time of this study.

Genomic DNA preparation

DNA was prepared from 0.5 ml whole blood by an adaptation of the method of DNA isolation as described (Roe et al. 1995). Briefly, whole blood was collected in EDTA vacutainer tubes containing 50 mM EDTA, cells were washed with 1× SSC pH 7.6 and subsequently pelleted at 16, 000 rpm. The cells were then lysed with 10% SDS at 70°C for 20 min in the presence of 0.2 M sodium acetate pH 5.4. The mixture was then partitioned with phenol chloroform iso-amyl alcohol (8:7:1) mixture at 16,000 rpm for 1 min. DNA was then recovered from the aqueous phase by precipitation with cold absolute propanol at -20°C for 15 min and centrifugation at 16,000 rpm for 2 min. DNA precipitate was washed in 75% propanol and reconstituted in 10:1 TE buffer pH 7.6.

DNA quantification and purity assessment

Quantity of genomic DNA was assessed by taking absorbance at 260 nm, and the purity of genomic DNA was calculated using the ratio of DNA absorbance at 260 and 280 nm. Purity of 1.6 to1.8 was taken as optimal purity.

Restriction digestion of genomic DNA

BamHI restriction enzyme digestion was carried out at 37°C for 24 h in a total volume of 50 µl containing 2 µg of human genomic DNA and 50 units of BamHI restriction enzyme, 100 mM Tris-HCl pH 8.0, 100 mM MgCl$_2$, 100 mM NaCl, and 100 mM β-mercaptoethanol.

Isolation of generated fragments

Agarose gel electrophoresis was used to separate fragments generated. 91.2 ng of restricted DNA was electrophoresed along-side a 5 to 50 kb DNA size marker on 0.8% agarose gel. The electrophoresis was carried out at 100 to 120 mA for 30 min. Resolved fragments were visualized under a UV-transilluminator and the fragment pattern photographed with a digital Polaroid camera.

RESULTS

Yield of genomic DNA from 0.5 ml whole blood

The mean yield of genomic DNA obtained from 0.5 ml of whole blood differed between genotypes as shown in Figure 1. The highest mean quantity obtained was 7.06 µg obtained from whole blood of the SS genotype and the lowest was 5.80 µg obtained from the AA genotype.

Purity of genomic DNA from 0.5 ml whole blood

All the samples produced genomic DNA of acceptable purity as shown in Figure 2

Digestion with *Bam HI* generated the β-like globin gene cluster

In order to isolate the entire β-globin gene cluster we subjected the *2µg of human genomic DNA to 50 units of Bam HI. A characteristic 69.65 kb DNA fragment was visualized from the genome of the SS and AS genotype, while the fragment visualized by the genome of the AA genotype used in this study was 70.28 kb the fragment pattern is shown in Figure 3. In addition, a novel 5.75 kb was visualized in from the SS genome.

DISCUSSION

The β-globin gene is a well studied gene and had been well characterized with well defined restriction sites (Langdon and Kaufman, 1998). Restriction enzymes with recognition sequences outside the β-globin gene clusters include; Hind II which has a recognition site 5' upstream the ε-globin gene of the β-globin gene cluster, and the BamHI restriction enzyme which has a recognition site within the 19-kb 3' cluster region downstream the β-globin gene of the β-globin gene cluster. In addition to the recognition site 5' of the beta globin gene, Hind II is also known to have recognition sites within the β-globin gene cluster (Langdon and Kaufman, 1998). The major reason for the choice of BamHI is to produce a gene fragment which would contain the entire β-globin gene cluster and the locus control region of the gene.

Similar fragment sizes were generated for both the AA, AS and SS genotypes digested with BamHI all close to the documented size of the β-globin gene cluster (Langdon and Kaufman, 1998). The fragments generated from the sickle cell patient and AS was 69.65 kb, while

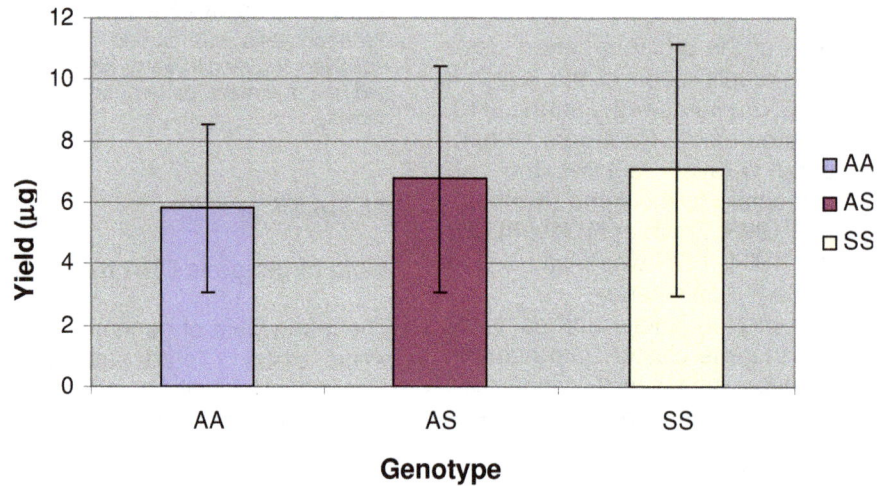

Figure 1. Mean yield of genomic DNA from the three genotypes. DNA was quantified by reading absorbance of the samples at 260 nm. The mean of three readings of three independent isolation events was plotted against the corresponding genotype as shown above. A significantly higher yield of genomic DNA was obtained from the SS genotype $P < 0.05$.

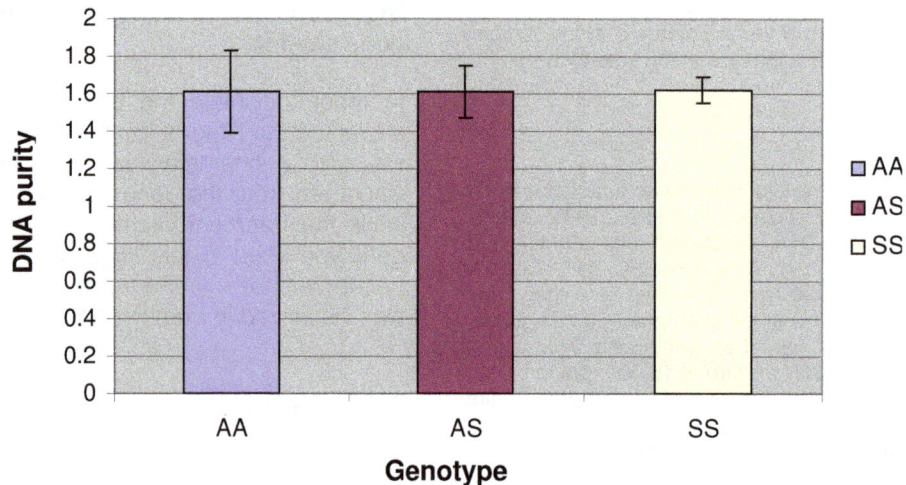

Figure 2. Mean purity of isolated genomic DNA for the three genotypes. The purity was calculated using the of absorbance at 260 and 280 nm of the DNA samples. The mean was computed from the calculated purity of each isolation event

the fragment generated from the AA patient correspondded to a fragment size of 70.28 kb. These observed differrences could be attributable to the polymorphi differences although the presence of the same sized fragment from the AS and SS genotype suggest a significant characteristic peculiar to AS and SS genotypes which is similar to the findings in the study by Lin et al. (2004) and Fullerton et al. (2000). Similarly, 70.28 kb fragment generated was similar in all the AA patients used for this study this could be clinically significant because it offers a rapid and precise way to molecularly determine the presence of the β^S gene. The presence of the 5.73 kb observed in the lanes with genomes of the AS and SS genotypes digested with *Bam*HI, suggests the presence of a control sequence upstream the β^S gene. The 5.73 kb fragment could also be of diagnostic significance as the fragment was not visible in the lanes containing AA digested with *Bam*HI restriction enzyme and could also suggest a polymorphic region present in the AS and SS patient used for the study but absent in the AA patient as previously demon strated (Currat et al., 2001; Fullerton et al., 2000; Chang et al., 1983).

The difference in the fragment sizes of the β-like globin cluster could reveal conservation in certain sequences among the genotypes although further studies would be required to confirm this.

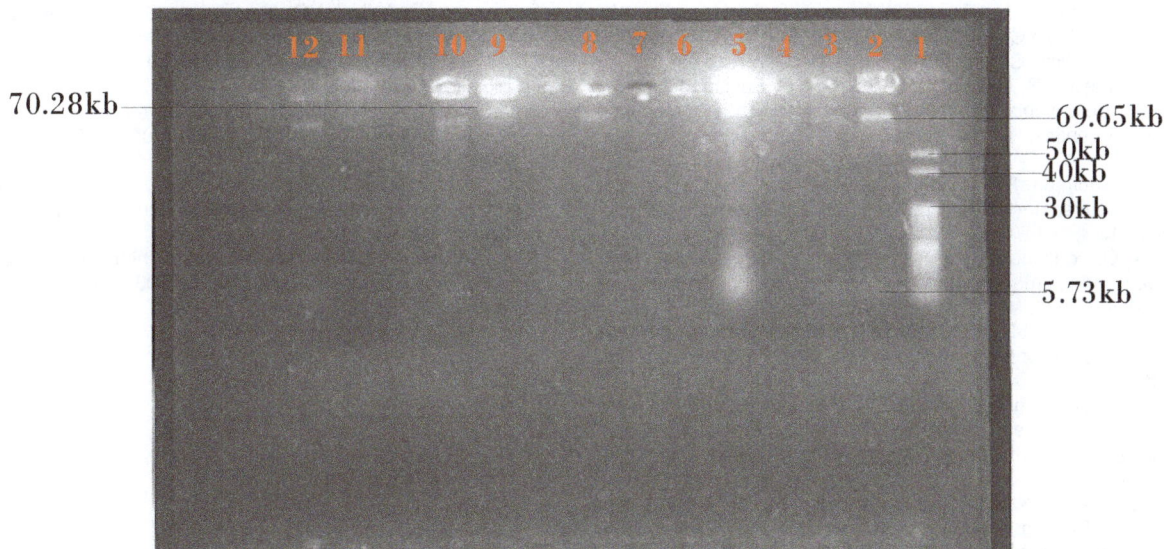

Figure 3. Electrophoregram of genomic DNA samples digested with *Bam*HI: 2 µg of human genomic DNA was digested with 50 units of Eco RI restriction endonuclease and 91.2 ng electrophoresed on 0.8% agarose gel for 30 min alongside a 10 to 50 kb Hyper DNA ladder. Photograph was taken with a digital Polaroid camera and documented. Lane 1 represents the 5 to 50 kb ladder; Lanes 2, 3, 4 represents the SS genotype digested with *Bam*HI; Lanes 10 and 12 represents the AS genotype digested with *Bam*HI; Lanes 5, 9 and 11 represents the AA genotype digested with *Bam*HI; Lanes 6 and 7 contain undigested genomic DNA (control)

Comparative studies of several human populations have concentrated on the diversity associated with β^A chromosomes. Several mutations associated with haemoglobinopathies have also been documented in the β-globin gene cluster (Weatherall, 2010; Strouboulis et al., 1992). The β^S (sickle cell) mutation is very common in countries where *Plasmodium falciparum*-mediated malaria is endemic (Currat et al., 2001; Orkin et al., 1993; Antonarakis et al., 1984; Pagnier et al., 1984). This study demonstrated a rapid and precise method to isolate the β-globin gene circumventing the rigors of PCR and expense of primers which are common problems which plague the application of molecular methods to clinical diagnosis in third world countries although more studies is required to completely sequence the gene fragments isolated.

To our knowledge, this is the first report of the isolation of the beta-like globin gene cluster among the different genotypes in Nigeria which is known to have the highest occurrence of the β^S gene.

The quantity of DNA obtained from 0.5 ml of blood using this protocol varied clearly between genotypes, the highest quantity of DNA obtained using this method was from the SS genotype. In comparison of the mean quantities of DNA obtained from the different genotypes, the amount of DNA obtained from AA blood samples were the lowest using this method. This could be due to the fact that the AA genotype exhibit the lowest turnover rate for RBCs as compared to AS and SS (Orkin et al., 1983), and the higher content of immature RBC content in AS and SS content as compared to AA patients could be responsible for the higher DNA content as it has been reported that immature RBCs contain nucleus and as such would also contain DNA.

REFERENCES

Aliyu ZY, Gordeuk V, Sachdev V, Babadoko A, Mamman AI, Akpanpe P, Attah E, Suleiman Y, Aliyu N, Yusuf J, Mendelsohn L, Kato GJ, Gladwin MT (2008). Prevalence and risk factors for pulmonary artery systolic hypertension among sickle cell disease patients in Nigeria. Am. J. Hematol., 83(6): 485-90.

Antonarakis SE, Boehm CD, Serjeant GR, Theisen CE, Dover GJ, Kazazian HH (1984), Origin of the beta-S-globin gene in blacks: the contribution of recurrent mutation or gene conversion or both. Proceed. Nat. Acad. Sci., 81: 853-856.

Ashley-Koch A, Yang Q, Olney RS (2000). Sickle haemoglobin (Hb S) Allele and Sickle Cell disease. Human Genome Epidemiol. (HuGE) Rev., 151: 839-845.

Cao H, Stamatoyannopoulos G, Jung M (2004), Induction of human γglobin gene expression by histone deacetylase inhibitors, Blood, 103(2): 701-709.

Chang JC, Alberti A, Kan YW (1983). A β-thahssemia lesion abolishes the same Mst II site as the sickle mutation. Nucl. Acids Res. 11(22): 7789-7794.

Currat M, Trabuchet G, Rees D, Perrin P, Harding RM, Clegg JB, Langaney A, Excoffier L (2001). Molecular Analysis of the β-Globin Gene Cluster in the Niokholo Mandenka Population Reveals a Recent Origin of the β^S Senegal Mutation. Am. J. Hum. Genet., 70(1): 207-223.

Detlof PJ, Lewis J, John SWM, Shehee RW, Langenbach R, Maeda N, Smithies O (1994). Deletion and Replacement of the Mouse Adult β-Globin Genes by a "Plug and Socket" Repeated Targeting Strategy. Mole. Cell. Biol., 14(10): 6936-6943.

Fullerton SM, Bond J, Schneider JA, Hamilton B, Harding RM, Boyce AJ, Clegg JB (2000) Polymorphism and divergence in the β-globin replication origin initiation region. Mole. Biol. Evol., 17: 179-188.

Kaufman RM, Pham CTN, Ley TJ (1999). Transgenic Analysis of a 100-kb Human beta globin Cluster-Containing DNA Fragment Propagated as a Bacterial Artificial Chromosome. Blood, 94: 3178-3184.

Kaufman L, Carnese FR, Goicoechea A, Dejean C, Salzano FM, Hutz MH (1998). Beta-globin gene cluster haplotypes in the Mapuche Indians of Argentina. Genet. Mole. Biol., 21.

Langdon DS, Kaufman RE (1998). Gamma-Globin Gene Promoter Elements Required for Interaction With Globin Enhancers, Blood, 91(1): 309-318.

Lin Z, Suzow JG, Fontaine JM, Naylor EW (2004). A high throughput β-globin genotyping method by multiplexed melting temperature analysis. Mole. Genet. Metabol., 81(3): 237-243.

Orkin SH, Antonarakis SE, Kazazian HH (1983). Polymorphism and Molecular Pathology of the Human β-globin gene. Program Hematol., 13: 49-73.

Pagnier J, Mears JG, Dunda-Belkhodja OK, Schaefer-Rego E, Beldjord C, Nagel RL, Dominque L (1984). Evidence for the Multicentric Origin of the Sickle Cell Haemoglobin Gene in Africa, Proceed. Nat. Acad. Sci., 81: 1771-1773.

Roe BA, Crabtree JS, Khan AS (1995), Protocols For Recombinant DNA Isolation, Cloning, and Sequencing. A Laboratory Manual John Wiley and Sons Publishers. Pp. 25-68.

Saraswathy R, Abilash VG, Manivannan G, George A, Babu KT (2010). Four novel mutations detected in the exon 1 of MBL2 gene associated with rheumatic heart disease in South Indian patients. Intl. J. Genet. Mole. Biol., 2(8): 165-170.

Smith RA, Joy P, John H, Clegg B, Kidd JR, Thein SL (1998). Recombination Breakpoints in the Human β-globin Gene Cluster Blood, 92(11): 4415-4421.

Strouboulis J, Dillon N, Grosveld F (1992). Developmental regulation of a complete 70kb human beta-globin locus in transgenic mice. Genes Dev., 6: 1857-1864.

Roe D (2010). The inherited diseases of hemoglobin are an emerging global health burden. Blood, 115(22): 4331-4336.

9

Isolation and sequence analysis of a putative MerR-type-transcriptional regulator and a multidrug efflux protein of *Bacillus circulans* ATCC 21588: As potential targets of therapeutics

Khaled Mohamed Anwar Aboshanab[1] and **Mostafa Mahmoud Elshafey**[2]

[1]Department of Microbiology and Immunology, Faculty of Pharmacy, Ain Shams University, Organization of African Unity St., POB: 11566, Abbassia, Cairo, Egypt.
[2]Department of Biochemistry, Faculty of Pharmacy, Al-Azhar University (Boys), Nasr City, Cairo, Egypt.

Mercury-type transcriptional regulators (MerR-transcriptional regulator) and major facilitator superfamily (MFS) transporters usually form an important sensor-response transport system in many microorganisms. This system has been shown to be involved in the regulation and transport (efflux) of a wide and diverse array of secondary metabolites including antimicrobial agents, dyes, chemicals, metals and evem harmful oxygen radicals. Inhibition or inactivation of this transport system is considered a promising approach for controlling microbial resistance, and thus may become a promising target of therapeutics particularly for the clinically relevant pathogens. However, the genetic and proteomics of this system have not been fully studied. In this work, a DNA segment (1.926 kb) from *Bacillus circulans* ATCC 21588 harboring the two genes, *bciR and bciT* arranged in an operon was amplified using PCR, analyzed and submitted into the GenBank database (accession code, KR049081). A two open reading frames (ORFs), namely BciR and BciT were found to encode a putative MerR-transcriptional regulator (BciR; 153 aa) and a putative MFS transporter (BciT; 392 aa), respectively. Analysis of the conserved domains and modeled tertiary structures revealed that, BciR possesses an N-terminal H-T-H motive (HTH type) region with possible transcriptional related activity and a conserved metal binding site at the C-terminal end. BciT was likely an MFS protein with nine transmembrane helices. This is the first report about detection of a *bciR/bciT* operon that putatively encode a sensor-response transport system in *Bacillus circulans* ATCC 12588.

Key words: MerR-type transcription regulator, multidrug efflux protein, major facilitator superfamily MFS, *Bacillus circulans* ATCC 21588.

INTRODUCTION

Mercury regulatory (MerR) family transcription regulators have been shown to mediate responses to stress such as exposure to drugs, heavy metals, or harmful oxygen radicals in various microorganisms (Helmann et al., 1990 Baranova et al., 1999). Their regulations were elicited by reconfiguring the promoter elements of many transporter proteins leading to suppression of the transcription process of the respective proteins (Helmann et al., 1989; Ahmed et al., 1995). A typical MerR regulator is comprised of two distinct domains that harbor the regulatory (effector-binding) site and the active (DNA-binding) site. Their N-terminal domains are homologous and contain a DNA-binding helix-turn-helix (HTH) motif, while the C-terminal domains are often dissimilar and bind specific co-activator molecules such as metal ions, drugs, and other organic substrates. In previous studies, it was confirmed that a MerR transcription regulator (BmrR) protein activates expression of a downstream multidrug efflux transporter *(bmr)* upon binding the transporter substrates (Ahmed et al., 1994; Zheleznova et al., 1999).

Bacterial transporters can be grouped based on energy sources. These groups are primary active transporters and the secondary transporters (Zhang et al., 2015). The primary transporters use energy generated by ATP hydrolysis while secondary transporters mainly rely on the electrochemical gradient across the cell membrane (Maloney, 1992; Floyd et al., 2010). The major facilitator super family (MFS) is a family of secondary transporters of usually transmembrane alpha-helices. The MFS transport diverse substrates, such as the ions, drugs, sugars, nucleosides, amino acids, small peptides, and other small molecules (Yan, 2013). Multi drug efflux functions of some MFS tranporters of many microorganisms have been studied (Wisedchaisri et al., 2014; Hinchliffe et al., 2014; Xu et al., 2014; Shilton, 2015). The MFS transporter was found in several pathogens such as *Escherichia coli, Klebsiella pneumoniae, Staphylococcus aureus, Bacillus cereus* and *Mycobacterium sp.* as integral membrane proteins involved in nonspecific antibiotic resistance to various antibacterial and anti-fungal agents (Changela et al., 2003; Floyd et al., 2010; Simm et al., 2012; Srinivasan et al., 2014; Zhang et al., 2015; Ogasawara et al., 2015). MFS transporters therefore, appear to contribute to intrinsic resistance to antibiotics in bacteria. Since, antibiotic resistance by the clinically relevant pathogens to the conventional antimicrobial agents are mediated by some MFS transporters, MFS transports become potential targets for the development of new antibacterial drugs (Saidijam et al., 2006). In a previous study, a MerR-type transcriptional regulator (Mta) was found to activate both bacillibactin secretion and an MFS transporter (YmfE) gene expression confirming involvement of both proteins in the bacillibactin biosynthetic pathway in *Bacillus subtilus* (Miethke et al., 2008). Therefore, molecular and 2008).

Therefore, molecular and proteomic studies of efflux pumps, their substrates as well as their regulatory mechanisms may lead to the discovery of new therapeutics and pump inhibitors. A side from the use of inhibitors, photodynamic inactivation using the synergistic action of efflux pump inhibitors is an alternative (Wasaznik et al., 2009). *Bacillus circulans*, the Gram positive spore forming rod was shown to be a potential pathogen to both plants and humans. In plants, it causes rapid and destructive soft rot of the tissues of Date Palm (Leary et al 1986) and a case report identified this organism in the setting of fatal sepsis in an immunocompromised patient (Alebouyeh et al., 2011).

Resistance of *Bacillus* spores to ultraviolet light, disinfectants and some other sterilizing agents as well as resistant of clinical isolates to many prescribed antibiotics suggest new therapeutic agents are needed (Alebouyeh et al., 2011). This study focuses on the identification, phylogenetic and sequence analysis of a putative MerR-type-transcriptional regulator and a multidrug efflux protein of *B. circulans* ATCC 21588 that is expected to be involved in intrinsic resistance to commonly used antibiotics, chemicals, disinfectants and metals. This study is a first step in the quest for inhibitors of this regulatory/transport system.

MATERIALS AND METHODS

Bacterial strains, culture media

B. circulans ATCC 21588 was cultured on tryptic soy broth (TSB) or on solid or liquid LB culture medium at 37°C (Kieser et al., 2000).

Extraction and manipulation of genomic DNA

Chromosomal DNA of *B. circulans* was prepared according to the method of Pospiech and Neumann (1995) with the following modifications. Strain inoculation was done in 25 ml TSB in 250 ml-volume flask and grown at 37°C on a shaker (250 rpm) for 24 h. The cells were then harvested by centrifugation at 13,000 rpm for 10 min and washed twice with 10.3% sucrose, resuspended in 20 ml of sodium chloride-EDTA-Tris (SET) buffer with 1.5 mg/ml lysozyme and incubated for 1 h at 37°C. About, 1/10 volume of 10% SDS and proteinase K (final concentration of 0.5 mg/ml) were added and incubated at 55°C for 1 - 2 h with frequent gentle inversion. About 1/3 volume of 5 M NaCl was added and an equal volume of phenol/chloroform was added and incubated at room temperature for 20 min with gentle inversion. The mixture was then centrifuged at 4,000 rpm for 10 min, and the aqueous phase was further extracted with an equal volume of chloroform/isoamyl alcohol (24:1), incubated at room temperature for 20 min with gentle inversion, and centrifuged at 10,000 rpm for 10 min. The DNA was precipitated by the addition of an equal volume of

Table 1. Oligonucleotides used in this study.

Primer	Target	Primer sequences [1]	Annealing temperature(°C), Annealing time (t)
Designation	**Gene**		
PBciR-F	*bci*R	5' ATGACACGGTTAAAAATTGATGATGTC 3'	53°C, 45 s
PBciR-R	(0.5 kb)	5' CTATTGTCCGGAAGACGGG 3'	
PBciT-F	*bci*T	5' CAGCGGTGACCGGCCCGCTC 3'	55°C, 1 min
PBciT-R	(1.1 kb)	5' CGAGGTCATCGCGTCCCCCTGC 3'	
PBciR3'	bciR(5')/T(3')	5' TCGTCGTCCGGCAAGACC 3'	55°C, 1 min
PBciT5'	(0.8 kb)	5' GCTGGATGGCCGACATATGAAGC 3'	

isopropanol, centrifuged at 10,000 rpm for 5 min. DNA was then washed using 70% ice cold ethanol, dried and finally dissolved in 1000 μl TE buffer with RNase 100 μg/ml. Agarose gel electrophoresis was carried out essentially as described by Sambrook and Russell (2001) using 0.8% agarose gels containing 0.1 μg/ml ethidium bromide. DNA fragment size was determined by comparison to a conventional 1 Kb DNA ladder (Sigma-Aldrich co, Egypt).

Polymerase chain reaction (PCR) and Recovery of DNA fragments from agarose gels

Amplification of different probes by PCR was performed using 200 - 400 ng of the genomic DNA as a template and the selected primers for each probe (Table 1). PCR was performed in a Nyx-Technik Inc. Personal Cycler (ATC401, USA). Each assay (50 μl) consisted of 200 ng chromosomal DNA, 100 pmole of each appropriate primer, 0.2 mM dNTPs (Invitrogen, Karlsruhe, Germany), 3 mM $MgCl_2$, 10% DMSO to improve the denaturation of the template DNA and 2 U *Taq* DNA polymerase (Sigma, USA). PCR general conditions were: 98°C for 5 min; then 30 cycles [95°C for 1 min; annealing temperatures and time according to Table 1, 72°C for 1 min (normally 1 min for 1 kb)]; and 72°C for 5 min (ramping rate 1°C/s).

DNA sequencing, assemble and detection of possible open reading frames (ORFs)

The PCR products were purified using GeneJET™ purification kit at Sigma Scientific Services Company, Egypt. Afterwards, samples were sent for sequencing at GATC co, Germany using ABI 3730xl DNA sequencer. The PCR products were sequenced from both forward and reverse directions. The obtained sequence files were assembled into a final contig using Staden Package program version 3 (http://staden.sourceforge.net/) (Staden, 1996). The resulting contig was analyzed for ORFs using FramePlot2.3.2 (http://www0.nih.go.jp/~jun/cgi-bin/frameplot.pl) (Ishikawa and Hotta, 1999), annotated and submitted into the GenBank database. Restriction analysis of the final contig was carried out using RestrictionMapper version 3 (http://www.restrictionmapper.org).

Nucleotide accession codes

The nucleotide sequence reported in this study was submitted in the GenBank database under the accession code: KR049081

Computer-assisted analysis of DNA sequences

Multiple alignment and phylogeny analysis of the obtained ORFs

were carried out using ClustalW2 (http://www.ebi.ac.uk/Tools/msa/clustalw2 (Thompson et al., 1994). Structure of proteins and conserved domain analysis were conducted using Basic Local Alignment Search Tool (NCBI): http://www.ncbi.nlm.nih.gov/Structure/index.shtml (Marchler-Bauer et al., 2015).

Analysis and prediction of the tertiary structure of encoded proteins

The putative tertiary structure of the obtained ORFs were predicted and analyzed using the Swiss-Model software (http://swissmodel.expasy.org; Arnold et al., 2006, Kiefer et al., 2009; Guex et al., 2009; Biasini et al, 2014). The QMEAN4 score of the predicted protein model was also calculated (Benkert et al., 2011). This was done to visualize the predicted conformation of the protein and the possible metal-binding residues which might have an effect on the enzyme activity.

RESULTS

Sequence analysis of the DNA segment (final contig) was submitted into the GenBank database (Accession code, KR049081). As depicted in Figure 1, two complete open reading frames (ORFs) were detected and annotated BciR (153 aa) and BciT (392 aa) on the submitted DNA segment (1.926 kb) of *B. circulans* ATCC 12588. The *bci*R (462 bp) and *bci*T (1179 bp) genes were found to encode a predicted MerR family transcriptional regulator of 153 amino acids (aa) and a major facilitator transporter (1179 bp, 392 aa), respectively.

Both BciR and BciT were encoded by the parent DNA strand and BciT was located downstream of BciR. A possible strong ribosomal binding site (RBS) for each ORF was detected and annotated as 5'-AGGAG-3' located at position -7 from the predicted start codon (ATG) of BciR and 5'-GAAGGGG-3' located at position -12 from the predicted start codon (ATG) of BciT. Restriction analysis profile of the respective DNA segment using some selected restriction endonucleases is also illustrated (Figure 1).

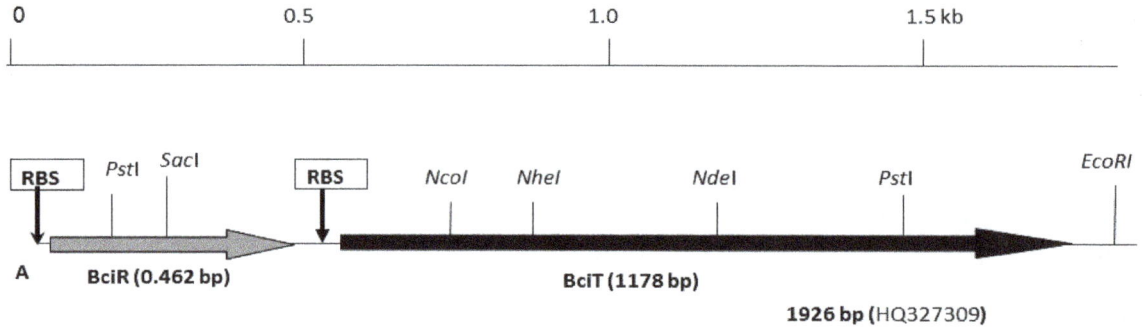

Figure 1. Restriction analysis profile of the DNA segment of *Bacillus circulans* ATCC 21588 submitted to GenBank database (accession code: KR049081) using some selected restriction endonucleases. BciR = predicted MerR family transcriptional regulator (462 bp, 153 aa); BciT = predicted major facilitator transporter (1179 bp, 392 aa). Arrows indicate direction of the oper reading frames (ORFs). RBS= possible ribosomal binding site.

Figure 2. Multiple sequence alignment of the conserved domains of the transcriptional regulator protein of *Bacillus circulans* ATCC 21588 (BciR; query) and its homologs using http://www.ncbi.nlm.nih.gov/Structure/cdd/cddsrv.cgi (Marchler-Bauer et al., 2015). The numbers indicate the position within the corresponding proteins: 2JML_A = a transcription regulator from the MerR superfamily of *Myxococcus Xanthus*, accession code (AC)= 2JML_A; gi 218284090 = hypothetical protein of *Eubacterium biforme* DSM 3989, AC= ZP_03489918; gi 262204171 = MerR family transcriptional regulator of *Gordonia bronchialis* DSM 43247, AC= YP_003275379; gi 418251195 = MerR family transcriptional regulator of *Mycobacterium abscessus* 47J26, AC= ZP_12877392. Amino acids within the indicated rectangules are required for DNA binding (DNA binding-residues).

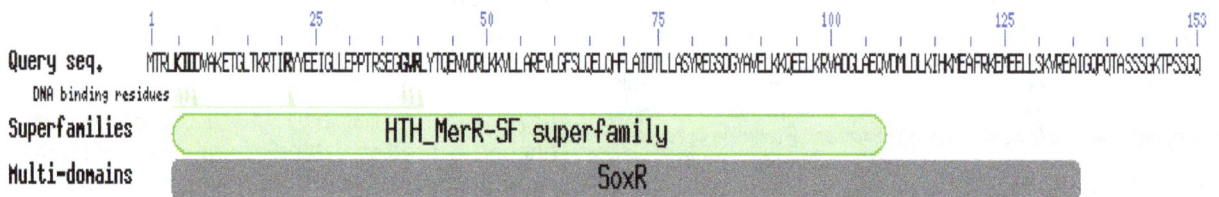

Figure 3. Putative conserved domain of BciR transcriptional regulator protein (http://blast.ncbi.nlm.nih.gov/Blast.cgi) HTH-MerR-SF = helix-turn-helix DNA binding domain of transcriptional metalo-regulatory protein superfamily; accession cl02600.SoxR= Predicted transcriptional regulators (COG0789).

Multiple alignments and domain analysis of BciR (a predicted member of the MerR family transcriptional regulator) and homologous proteins

As shown in Figure 2, BciR showed more than 85, 83, 82, 80, 78.3% similarities in the amino acid sequences of homologous proteins encoding diverse MerR transcription regulators of *Myxococcus Xanthus*, accession code (AC= 2JML_A), *Eubacterium biforme* DSM 3989 (AC = ZP_03489918), Gordonia *bronchialis*

```
gi 74626962  86 FFIVIFITAFTAAT.[1].NDAGSI.[3].LNEK.[2].ISYDAMNTGAGVLFL.[2].GWGTFFL    TPA.[2].L 145
query         15 TVFATFLAFMGIGV    VDPILP.[1].IAEK.[2].ANPAQIELLFTAYIL.[2].AIMMIPS   GLA.[2].R  71
gi 81550827   46 LFLTIFIAMLGLSV    LFPIIG.[1].LGRQ.[2].LTPAQIGWFSTSYSL.[2].FIFSPIW   GNR.[2].H 102
gi 81760282   10 VLVAVAAAFGSWSL    LLPVVP.[3].LNNG   GSSAVAGATTGIFMA.[2].VITQIFT.[1].AAL       R  65
gi 81622273   29 GLLGVLLASLTAGL.[2].HVTEIA.[3].VRGA.[2].IGHDEGTWLTVLYEA.[4].AMAFAPW    CSV       T  89


gi 74626962  146 .[1].GRKITYFICIFLGLLG.[1].VWFA.[5].SD     SIWSQLFVGISE.[1].CAEAQVQLSLSELYFAH 204
query         72 .[1].GDKKMMVAGLAVVTVF.[1].FLCG.[2].GG.[3].LALFRAGWGFGN   SMFFATAMTLLIALTPS 129
gi 81550827  103 .[1].GRRPTLLMGLVGFSVS    FGLF.[9].GG.[9].LVGTRIIGGIFS   SATLPTAQAMMADISSE 172
gi 81760282   66 .[2].GYTPVMAFAAFMLGVP.[1].IGYI.[1].SV.[4].VLVVSALRGIGF   GALTVAESALVAELVPV 124
gi 81622273   90 .[1].SLRRFTLFAIGGFALL.[1].LLCP.[5].ES     LLVLRTLQGLMA    GCLPPMLMTVALRFLPP 147


gi 74626962  205    N.[2].SVLTSYIVATSVGTYL.[1].PLIAAFIVQNI.[1].FRWVGWIAAIISGAL.[3].IV.[5].TYF 264
query        130 .[1].N   TAVGMYEAAIGLGMAG   GPLVGGLLGGI.[1].WRLPFIATGCFVLIA    FL.[3].FMI 182
gi 81550827  173 .[1].D.[2].ASMGLIGAAFGLGVVF.[1].PAIGGLLSGIS.[1].VAPVYFSAGLGLFTA    AL   AYF 225
gi 81760282  125 .[1].F.[2].KASGMLGVFIGLSQML.[1].LPAGLALGDQF.[1].YNVVYVLGAVIALVA    AV.[2].LRI 179
gi 81622273  148 .[1].I.[2].YGLGAYALTATFGPNL.[1].LPLAAFWFEYV.[1].WQWVFWQVIPLCLVA    MA.[3].HGI 203


gi 74626962  265 .[11].GIAPFDSSYS.[48].KRTAMIT.[14].NQL.[ 2].LLKVFLYP.[7].CWGI.[4].LTFYLTV.[9].YS 400
query        183    QEPEKKSVRK.[ 5].ELLHLAT   HKP.[ 3].VAGSSMFY.[2].GFFV   VLAYSPL.[1].LH 234
gi 81550827  226    MVPETRHPGQ   TRTATVD.[ 5].SRG.[ 4].FLAASALT.[2].ASVG   MEQTISF.[5].MK 282
gi 81760282  180    PQVKAAAKQQ.[ 5].QERSVST.[ 5].VPS   LAVTSLSM   TFGA   VSSFLPA.[4].LD 234
gi 81622273  204    PQDPVRLERF.[30].FESPLIC.[14].VNE.[17].LTFALITL.[2].VLVV.[4].MGVPTEF.[4].RG 315


gi 74626962  401 YG    NIAV.[2].MNIPCLIGAII.[3].FAGT   LSDY.[9].K.[7].RLWFLLPPA.[3].PVGLILFAV 468
query        235 MS.[ 1].IQLG.[2].FFGWGLMLAYG.[2].KLAH.[1].LEER.[2].P.[2].IIPWSLGAF   CLILLLLFL 288
gi 81550827  283 LD.[10].MLAI   FGILAALVQGG   AIRP   LSKK.[2].P   TPLILVGLV.[3].AGMFLLPQM 341
gi 81760282  235 PG    LGAA.[2].GIILSITGGSS.[5].LSGV   IADR   R.[2].PGTTMIPAQ.[3].FLGVVLITV 290
gi 81622273  316 YR.[ 1].LQSV.[2].VLLVALPQLVA.[2].LVAA   LCNI.[2].V.[2].RWVLACGLC.[3].GACLSFSQL 371


gi 74626962  469 G.[8].PTYI.[4].IGFGYGCAGDVSMSYLMDSYPNA.[1].IETMTVVAVINNCIGCVFTF 529
query        289 V.[8].LIIV   IGLFCGLNNALFTSHVMEVSPFE   RSITSGAYNFVRWLGAAIAP 344
gi 81550827  342 A.[7].ALAL   IGVGSAILSPTLSAALSLSVGKD.[1].QGAVAGLNSSALALGRMVGP 397
gi 81760282  291 T.[8].LLII.[4].FGGAFGMVQNEALLSMFFRLPRT.[1].VSEASAIWNIAFDSGTGIGS 351
gi 81622273  372 T.[6].DFYL.[4].LVVGQPMAVIPLLMLSTSVIAPI.[1].GPFASAWFNTVRGFSGVVAT 430
```

Figure 4. Multiple amino acid sequences of the conserved domains of the major facilitator transporter of *Bacillus circulans* ATCC 21588 (BciT; query) and its homologous using http://www.ncbi.nlm.nih.gov/Structure/cdd/cddsrv.cgi (Marchler-Bauer et al., 2015). The numbers indicate the position within the corresponding proteins: gi 74626962 = Transport protein of *Candida albicans*, accession code (AC)= O94019; gi 81550827 = Drug transport protein of *Deinococcus radiodurans*, AC= Q9RSF5; gi 81760282 = Permeases of the major facilitator superfamily of *Corynebacterium glutamicum*, AC= Q8NNT7; gi 81622273 = Probable major facilitator superfamily (MFS) transporter of *Pseudomonas aeruginosa*, AC= Q9I008.

DSM 43247 (AC= YP_003275379) and *Mycobacterium abscessus* 47J26 (AC= ZP_12877392), respectively. The N-terminal region of BciR and the respective homologous proteins were highly conserved (more than 95%) at the indicated catalytic sites. As depicted in Figures 2 and 3, the N-terminal domains of the BciR protein and its homologous proteins showed conservation of the amino acid moieties and the helix-turn-helix (H-T-H) motive required for DNA binding (DNA binding residues).

Multiple alignment and domain analysis of BciT (predicted Major Facilitator Superfamily, MFS) and homologous proteins

As shown in Figure 4, BciT showed more than 83% sequence similarities to homologous proteins encoding diverse major facilitator transporters (multidrug efflux proteins) such as the multidrug transport protein of *Candida albicans* (AC, O94019), drug transport protein of *Deinococcus radiodurans* (AC, Q9RSF5), drug permeases of the major facilitator superfamily of *Corynebacterium glutamicum* (AC, Q8NNT7) and a probable major facilitator superfamily transporter of *Pseudomonas aeruginosa* (AC, Q9I008). As delineated in Figure 5, domain analysis of the BciT transporter protein revealed a putative conserved domain (specific hits) with the major facilitator superfamily, cd06174 (pfam07690).

Phylogram analysis of BciR and BciT

A cladogram of BciR in relation to other MerR transcription proteins is shown in Figure 6. BciR of *B. circulans* clustered closely with two homlogous proteins of two *Paenbacillus* species (AC, WP-009673983.1 & AC, WP-042231219.1) with pairwise score ranging from

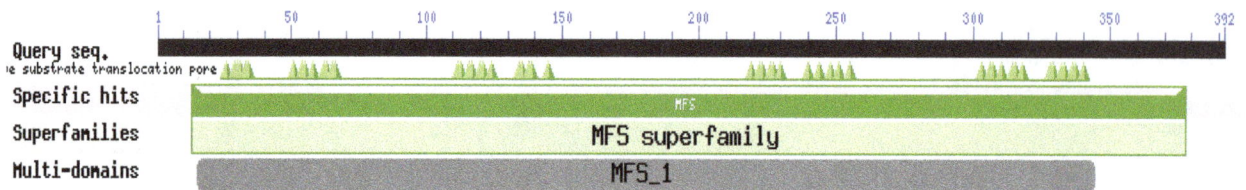

Figure 5. Putative conserved domain of BciT transporter protein (http://blast.ncbi.nlm.nih.gov/Blast.cgi). MFS= major facilitator superfamily; cd06174: MFS-1 = Major Facilitator Superfamily (pfam07690).

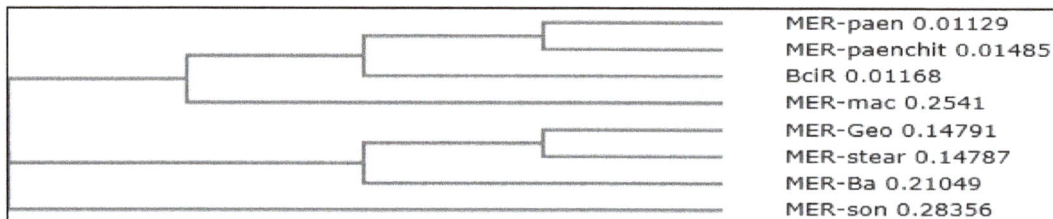

Figure 6. Cladogram of the phylogram analysis of BciR and its homologous proteins from the GenBank database. Mer-paen = MerR family transcriptional regulator of *Paenibacillus*, AC: WP_009673983.1; MER-Paenchit= MerR family transcriptional regulator of *Paenibacillus chitinolyticus*, AC: WP_042231219.1; MER-mac= MerR family transcriptional regulator of *Fictibacillus macauensis*, AC: WP_007202128; MER-Geo= MerR family transcriptional regulator of *Geobacillus* sp. JF8, Ac: WP_020958590.1; MER-Stear= MerR family transcriptional regulator of *Geobacillus stearothermophilus*, AC: WP_043903498.1; MER-Ba = MerR family transcriptional regulator of *Bacillus sp.* SJS, AC: WP_035412317.1;. MER-son= MerR family transcriptional regulator of *Bacillus sonorensis*, AC: WP_006637406.1. AC= Protein accession code within the GenBank database.

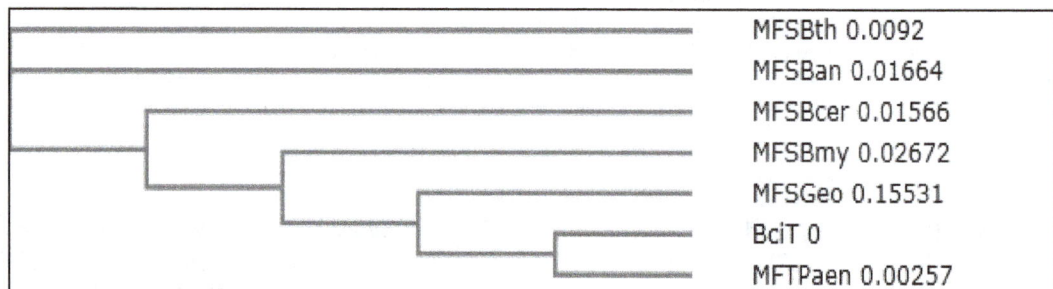

Figure 7. Cladogram of the phylogram analysis of BciT and its homologous proteins frm the GenBank database. MFSBth = major facilitator transporter of *Bacillus thuringiensis*, AC: WP_038413898.1; MFSBan= major facilitator transporter of *Bacillus anthracis*, AC: WP_003159546.1; MFSBcer = major facilitator transporter of *Bacillus cereus*, AC: WP_000444231.1; MFSBmy = major facilitator transporter of *Bacillus mycoides*, AC: WP_042981104.1; MFSGeo = major facilitator transporter of *Geobacillus stearothermophilus*, AC: WP_043903497.1; MFTPaen = major facilitator transporter of *Paenibacillus*, AC: WP_009673982.1. AC = Protein accession code within the GenBank database.

0.01129-0.01485 (MerR transcription proteins from other *Bacillus* species formed distinct clusters however, they were relatively related (pairwise scores ranged from 0.14791- 0.28356).

As depicted in Figure 7, a cladogram showing BciT in relation to other MFS transporter proteins showed, BciT

of *Bacillus circulans* was clustered almost closely with a homologous protein of *Paenbacillus* species (AC, WP_009673982.1) of pairwise score that ranged from 0.0-0.00257. MFS transporter proteins from other *Bacillus* species were also closely related with pairwise scores that ranged from 0.0092- 0.01664.

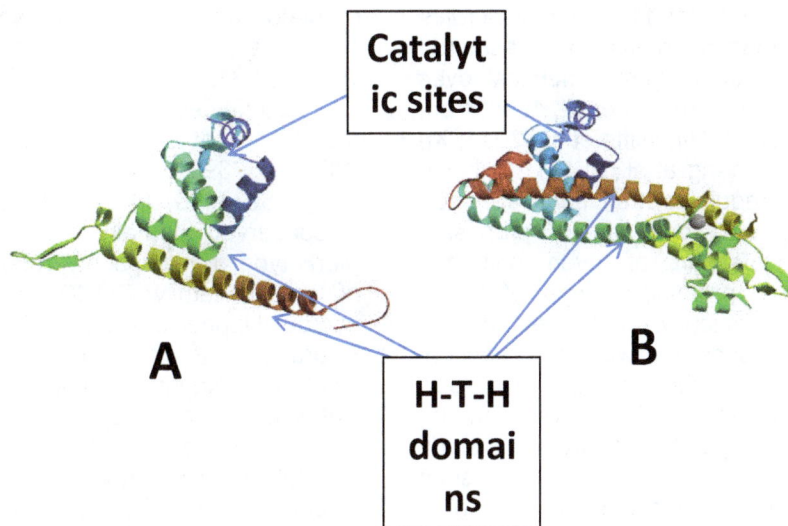

Figure 8. The predicted tertiary structure using SWISS-MODEL homology modeling report of: **A,** putative MerR family transcriptional regulator of *Bacillus circulans* ATCC 12588 (BciR), accession code GenBank KR049081 and **B**, MerR the transcriptional regulator family from *Bacillus cereus* crystal structure using X-RAY DIFFRACTION 2.67 Å. arrows show the suggested metal legends for binding (catalytic sites) and N-terminal Helix-Turn-Helix domains (H-T-H domains).

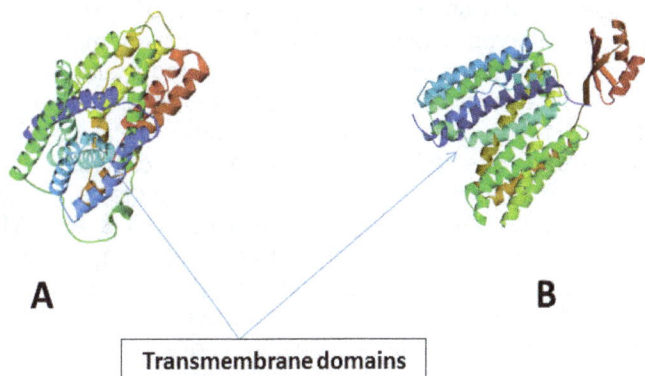

Figure 9. The predicted tertiary structure using SWISS-MODEL homology modeling report of: A; putative major facilitator superfamily transporter (MFS) of *Bacillus circulans* ATCC 12588 (BciT), accession code GenBank KR049081 and B; major facilitator superfamily transporter (YajR) Crystal structure using X-RAY DIFFRACTION 2.67 Å .YajR transporter suggests a transport mechanism based on the conserved motif A (Jiang, D. et al., 2013). Arrows indicate nine conserved transmembrane domains.

Prediction of the tertiary structures of BciR and BciT proteins using SWISS-MODEL homology modeling report

As depicted in Figure 8, the three dimensional structure of BciR was predicated using a standard template model

of a HTH-type transcriptional activator TIPA of *Bacillus cereus* (crystal structure using X-RAY DIFFRACTION 2.67Å). A model of BciR was built and revealed high degree of similarities to the template model as well as a conserved N-terminal helix-turn-helix domain (H-T-H domains) required for DNA binding in addition to a metal binding sites (catalytic sites) conserved at the C-terminal of the BciR protein.

The model showed also a QMEAN4 score of -1.08. As shown in Figure 9, the three dimensional structure of BciT was predicated using a standard template model of major facilitator superfamily transporter (YajR) crystal structure using x-ray diffraction 2.67 Å. A model of BciT was built and revealed conserved nine transmembraneous domains.

DISCUSSION

With the increase of antibiotic-resistant pathogens, inadequate discovery of new antibiotics, as well as the extreme cost required for isolation of new antimicrobial agents, new approaches may be needed. Recently, with the extensive knowledge and profound use of gene manipulations worldwide, it became apparent that new approaches can be discovered. Transporter proteins particularly those involved in the transport/efflux of primary or secondary metabolites or those that regulate their expression have been confirmed to play crucial roles in cell growth, spore formation, intrinsic and acquired resistance to various antimicrobial agents, chemical, dyes and even toxic oxygen radicals (Ahmed et al., 1995; Yan, 2013; Wisedchaisri et al., 2014; Hinchliffe et al., 2014; Xu et al., 2014; Shilton, 2015; Zhang et al., 2015). Inhibition or inactivation of this regulatory/transport transport system would be considered one of the promising approaches to control infection or even microbial resistance, and become a promising target of thera-peutics particularly for the clinically relevant pathogens. Therefore, this study was concerned with the detection, sequence analysis of a model regulatory/transport system in a model of Gram positive, spore forming bacterium, *B. circulans*. Findings obtained from this study can be applied to those closely related pathogens such as *Bacillus anthracis*, *Bacillus cereus* and *Mycobacterium tuberculosis*. In this work various primers were designed based on the conserved amino acid sequences of some selected metalo-regulatory transcriptional regulators (MerR-Type) and major facilitator superfamily transporters (MFS-type) located in the GenBank database. The designed primers were used for amplification of the target sequences using chromosomal DNA of *B. circulans* ATCC 21588 as a PCR template. The PCR products obtained were analyzed using agarose gel electrophoresis, purified and sequenced in both forward and reverse directions. The obtained sequence files were assembled into a final consensus sequence (contig) of 1.926 kb using StadenPackage. The resulted consensus sequence was analyzed using the frameplot programme to detect the ORFs and was submitted into the GenBank database under accession code, KR049081.

Two complete open reading frames (ORFs) on the respective DNA segment (1.926 kb) of *Bacillus circulans* ATCC 12588 were detected and annotated BciR and BciT. The *bciR* (462 bp) and *bciT* (1179 bp) were found to encode a predicted MerR family transcriptional regulator (153 aa) and a major facilitator transporter (1179 bp, 392 aa), respectively. Both BciR and BciT were encoded by the parent DNA strand where BciT was located downstream BciR suggesting an operon of sensor/response. Analysis of the DNA segment harboring BciR and BciT ORFs showed a possible strong ribosomal binding site (RBS) for each ORF. The RBS (5'-AGGAG-3') of BciR was located at position-7 from the predicted start codon (ATG) of respective ORF while RBS of BciT (5'-GAAGGGG-3') located at position -12 from the predicted start codon (ATG) indicated that they are not coupled translated.

The first ORF (BciR; 153 aa) was found to encode a putative Metalo- transcriptional regulator (MerR-type) with aa identities to the multispecies Metalo- transcriptional

regulator of: *Paenibacillus* sp. (97%, WP_009673983; 153 aa); *Paenibacillus chitinolyticus* (97%, WP_042231219; 153 aa); *Geobacillus* sp. JF8 (56%, WP_020958590; 142 aa); HTH-type transcriptional regulator of *Geobacillus* sp. GHH01 (64%, WP_015373901; 142 aa); and *Bacillus sonorensis* (50%, WP_006637406; 140 aa). Detection of the BciR showed a conserved domain with proteins encoding diverse MerR-type transcriptional regulators (MerR superfamily) of protein family COG0789 from wide varieties of microbial species such as N-Terminal domain of Cara Repressor *of Myxococcus xanthus* DK 1622 (AC, 2JML_A; Navarro-Aviles et al., 2007; 81 aa), *Eubacterium biforme* DSM 3989 (Ac, ZP_03489918; 211aa), *Gordonia bronchialis* DSM 43247 (AC, YP_003275379; Ivanova et al., 2010; 264 aa), and *Mycobacterium abscessus* 47J26 (AC, ZP_12877392; 246 aa). Multiple amino acid sequence alignment revealed presence of the amino acids moieties required for DNA binding (DNA binding residues) and located the N-terminal part of BciR as well as in the respective BciR homologous proteins (Marchler-Bauer et al., 2015). Moreover, the three dimensional structure of BciR was predicted via SWISS-MODEL Homology Modeling Report using a standard template model of MerR transcriptional regulator family of *Bacillus cereus* crystal structure. A model of BciR showed a QMEAN4 score of -1.08 as well as a conserved N-terminal Helix-Turn-Helix domain (H-T-H domains) required for DNA binding and metal binding sites (catalytic sites) conserved at the C-terminal (Ahmed et al., 1994; Zheleznova et al., 1999). Phylogenetic analaysis revealed that, BciR of *B. circulans* was clustered closely with two homlogous proteins of two *Paenbacillus* species (AC, WP-009673983.1 & AC, WP-042231219.1) of pairwise score that ranged from 0.01129-0.01485 (MerR transcription proteins from other *Bacillus* species formed distinct clusters however, they were relatively related (pairwise scores ranged from 0.14791- 0.28356).

The second ORF (BciT; 392 aa) was found to encode a putative major facilitator superfamily (MFS) with aa identities to the mutispecies major facilitator transporter of: *Paenibacillus* sp. (100%, WP_009673982); *Paenibacillus chitinolyticus* (99%, WP_042231220), *Geobacillus stearothermophilus* (66%, WP_043903497); *Bacillus cereus* (64%, WP_000444231); and *Bacillus thuringiensis* (64%, WP_000444219); *Bacillus mycoides* (64%, WP_042981104). Detection of the BciT showed that, it shared conserved domains with proteins encoding diverse major facilitator transporters (multidrug efflux proteins) of the protein family 07690 (pfam07690) from a wide varieties of microbial species such as *Candida albicans* (AC, O94019; Tait et al., 1997), *Deinococcus radiodurans* (AC, Q9RSF5; White et al., 1999),

Corynebacterium glutamicum (AC, Q8NNT7) and *Pseudomonas aeruginosa* (AC, Q9I008; Stover et al., 2000). The tertiary structure of BciT was predicted via SWISS-MODEL Homology Modeling Report using a standard template model of major facilitator superfamily transporter (YajR) crystal structure. A model of BciT revealed presence of nine transmembrane alpha helices and the majority of MFS proteins contained from 6-12 transmembrane alpha helices (TMs) connected by hydrophilic loop (Yan, 2013). BciT was predicted to be a MFS transporter protein in the form of monomer oligostate (Jiang et al., 2013). The BciT clustered almost closely with a homologous protein of *Paenbacillus* species (AC, WP_009673982.1) of pairwise score that ranged from 0.0-0.00257. MFS transporter proteins from other *Bacillus* species were also closely related with pairwise scores that ranged from 0.0092- 0.01664. The results obtained indicates that, MFS of *B. circulans* is closely related to those of *Paenbacillus* sp. than to those MFS of other *Bacillus sp.* Several studies showed presence of this regulatory/transport system where a MerR-type transcriptional regulator regulated a downstream multidrug-efflux transporter in *Bacillus subtilis* (Ahmed et al., 1994, 1995; Baranova et al., 1999). Other studies have confirmed the role of MerR-transcriptional regulator proteins in response to several metal ions conferring bacterial resistance to such toxic ions (Helmann et al., 1989, 1990) and to 2-nitroimidazole, the antifungal and antibacterial agent (Ogasawara et al., 2015). Zheleznova et al. (1999) revealed that up on drug (or toxin) binding to the transcription regulator, BmrR (MerR-type) of *Bacillus subtilis*, it activated expression of the multiple transporter (Bmr) that demonstrated an unusual ability to recognize multiple structurally dissimilar toxins. Therefore, BciR encodes a putative MerR-type transcriptional regulator that putatively regulate expression of the downstream located BciT. Both BciR and BciT that putatively encode a MFS-secondary transporter were shown to be arranged in an operon. This operon encodes proteins putatively involved in the assistance of transport across cytoplasmic or internal membranes of a variety of substrates. Therefore, inhibition or quenching the activity of this sensor/-response regulator system becomes a promising target of developing new therapeutics

Conflict of interests

The authors did not declare any conflict of interest.

REFERENCES

Ahmed M, Borsch CM, Taylor SS, Vázquez-Laslop N, Neyfakh AA (1994). A protein that activates expression of a multidrug efflux transporter upon binding the transporter substrates. J. Biol. Chem. 269(45): 28506-28513.

Ahmed M, Lyass L, Markham PN, Taylor SS, Vázquez-Laslop N, Neyfakh AA (1995). Two highly similar multidrug transporters of Bacillus subtilis whose expression is differentially regulated. J. Bacteriol. 177(14):3904-3910.

Alebouyeh M, Gooran Orimi P, Azimi-rad M, Tajbakhsh M, Tajeddin E, Jahani SS, Nazemalhosseini EM, Zali MR (2011). Fatal sepsis by Bacillus circulans in an immunocompromised patient Iran J. Microbiol. 3(3):156–158.

Arnold K, Bordoli L, Kopp J, Schwede T (2006). The SWISS-MODEL Workspace: A web-based environment for protein structure homology modelling. Bioinformatics 22:195-201. http://dx.doi.org/10.1093/bioinformatics/bti770

Baranova NN, Danchin A, Neyfakh AA (1999). Mta, a global MerR-type regulator of the Bacillus subtilis multidrug-efflux transporters. Mol. Microbiol. 5:1549-1559. http://dx.doi.org/10.1046/j.1365-2958.1999.01301.x

Benkert P, Biasini M, Schwede T (2011). Toward the estimation of the absolute quality of individual protein structure models. Bioinformatics 27(3):343-350. http://dx.doi.org/10.1093/bioinformatics/btq662

Biasini M, Bienert S, Waterhouse A, Arnold K, Studer G, Schmidt T, Kiefer F, Cassarino TG, Bertoni M, Bordoli L, Schwede T (2014). SWISS-MODEL: modelling protein tertiary and quaternary structure using evolutionary information. Nucleic Acids Research; 42 (W1): W252-W258; http://dx.doi.org/10.1093/nar/gku340

Changela A, Chen K, Xue Y, Holschen J, Outten CE, O'Halloran TV, Mondragón A (2003) Molecular basis of metal-ion selectivity and zeptomolar sensitivity by CueR. Science 301(5638):1383-1387. http://dx.doi.org/10.1126/science.1085950

Floyd JL, Smith KP, Kumar SH, Floyd JT, Varela MF (2010). LmrS is a multidrug efflux pump of the major facilitator superfamily from Staphylococcus aureus. Antimicrob. Agents Chemother. 54(12):5406-12. http://dx.doi.org/10.1128/AAC.00580-10

Guex N, Peitsch MC, Schwede T (2009). Automated comparative protein structure modeling with SWISS-MODEL and Swiss-PdbViewer: A historical perspective. Electrophoresis 30(1):162-173. http://dx.doi.org/10.1002/elps.200900140

Helmann JD, Ballard BT, Walsh CT (1990). The MerR metalloregulatory protein binds mercuric ion as a tricoordinate, metal-bridged dimer. Science 247(4945):946-948. http://dx.doi.org/10.1126/science.2305262

Helmann JD, Wang Y, Mahler I, Walsh CT (1989). Homologous metalloregulatory proteins from both gram-positive and gram-negative bacteria control transcription of mercury resistance operons. J. Bacteriol. 171(1):222-9.

Hinchliffe P, Greene NP, Paterson NG, Crow A, Hughes C, Koronakis V (2014). Structure of the periplasmic adaptor protein from a major facilitator superfamily (MFS) multidrug efflux pump. FEBS Lett. 588(17):3147-3153.

Hopwood DA, Wright HM (1978). Bacterial protoplast fusion: recombination in fused protoplasts of Streptomyces coelicolor. Mol. Gen. Genet. 162: 307-317. http://dx.doi.org/10.1007/BF00268856

Ishikawa J, Hotta K (1999). Frameplot: a new implementation of the Frame analysis for predicting protein-coding regions in bacterial DNA with a high G+C content. FEMS. Microbiol. Lett. 174: 251-253.. http://dx.doi.org/10.1111/j.1574-6968.1999.tb13576.x

Ivanova N, Sikorski J, Jando M, Lapidus A, Nolan M, Lucas S, Del Rio TG, Tice H, Copeland A, Cheng JF, Chen F, Bruce D, Goodwin L, Pitluck S, Mavromatis K, Ovchinnikova G, Pati A, Chen A, Palaniappan K, Land M, Hauser L, Chang YJ, Jeffries CD, Chain P, Saunders E, Han C, Detter JC, Brettin T, Rohde M, Goker M, Bristow J, Eisen JA, Markowitz V, Hugenholtz P, Klenk HP, Kyrpides NC (2010). Complete genome sequence of Gordonia bronchialis type strain (3410). Stand Genomic Sci. 2 (1): 19-28 http://dx.doi.org/10.4056/sigs.611106

Jiang D, Zhao Y, Wang X, Fan J, Heng J, Liu X, Feng W, Kang X, Huang B, Liu J, Zhang XC (2013). Structure of the Yajr transporter suggests a transport mechanism based on the conserved motif A. Proc. Natl. Acad. Sci. 110(36):14664-14669. http://dx.doi.org/10.1073/pnas

Kiefer F, Arnold K, Künzli M, Bordoli L, Schwede T (2009). The SWISS-MODEL Repository and associated resources. Nucleic Acids Res. 37:D387-D392. http://dx.doi.org/10.1093/nar/gkn750

Kieser T, Bibb MJ, Buttner MJ, Chater KF, Hopwood DA (2000). Practical Streptomyces genetics. The John Innes Foundation, John Innes centre, Norwich, UK.

Leary JV, Nelson N, Tisserat B, Allingham EA (1986). Isolation of Pathogenic Bacillus circulans from Callus Cultures and Healthy Offshoots of Date Palm (Phoenix dactylifera L). Appl. Environ. Microbiol. 52(5):1173-1176

Maloney PC (1992). The molecular and cell biology of anion transport by bacteria. Bioessays 14:757-762. http://dx.doi.org/10.1002/bies.950141106

Marchler-Bauer A, Derbyshire MK, Gonzales NR, Lu S,Chitsaz F, Geer LY, Geer RC, He J, Gwadz M, Hurwitz DI, Lanczycki CJ, Lu F, Marchler GH, Song JS, Thanki N, Wang Z, Yamashita RA, Zhang D, Zheng C, Bryant SH (2015). CDD: NCBI's conserved domain database. Nucleic Acids Res. 43. http://dx.doi.org/10.1093/nar/gku1221

Miethke M, Schmidt S, Marahiel MA (2008). The major facilitator super family type transporter YmfE and the multidrug efflux activator Mta mediate bacillibactin secretion in Bacillus subtilis. J. Bacteriol. 190(15):5143-5152. http://dx.doi.org/10.1128/JB.00464-08

Navarro-Aviles G, Jimenez MA, Perez-Marin MC, Gonzalez C, Rico M, Murillo FJ, Elias-Arnanz M, Padmanabhan S (2007). Structural basis for operator and antirepressor recognition by Myxococcus xanthus CarA repressor. Mol. Microbiol. 63 (4): 980-994. http://dx.doi.org/10.1111/j.1365-2958.2006.05567.x

Ogasawara H, Ohe S, Ishihama A (2015). Role of transcription factor NimR (YeaM) in sensitivity control of Escherichia coli to 2-nitroimidazole. FEMS Microbiol. Lett. 362(1):1-8. http://dx.doi.org/10.1093/femsle/fnu013

Pospiech A, Neumann B (1995). A versatile quick-prep of genomic DNA from Gram-positive bacteria. Trends Genet. 11: 217-218. http://dx.doi.org/10.1016/S0168-9525(00)89052-6

Saidijam M, Benedetti G, Ren Q, Xu Z, Hoyle CJ, Palmer SL, Ward A, Bettaney KE, Szakonyi G, Meuller J, Morrison S, Pos MK, Butaye P, Walravens K, Langton K, Herbert RB, Skurray RA, Paulsen IT, O'reilly J, Rutherford NG, Brown MH, Bill RM, Henderson PJ (2006). Microbial drug efflux proteins of the major facilitator superfamily. Curr. Drug Targets 7(7):793-811. http://dx.doi.org/10.2174/138945006777709575

Sambrook J, Russell DW (2001) Molecular cloning: a laboratory manual, 3rd Eds. Cold Spring harbor laboratory Press, Cold Spring Harbor, New York.

Shilton BH (2015). Active transporters as enzymes: an energetic framework applied to major facilitator superfamily and ABC importer systems. Biochem. J. 467(2):193-199. http://dx.doi.org/10.1042/BJ20140675

Simm R, Vörös A, Ekman JV, Sødring M, Nes I, Kroeger JK, Saidijam M, Bettaney KE, Henderson PJ, Salkinoja-Salonen M, Kolstø AB (2012). C4707 is a major facilitator superfamily-multidrug resistance transport protein from Bacillus cereus implicated in fluoroquinolone tolerance. PLoS One. 7(5):e36720.. http://dx.doi.org/10.1371/journal.pone.0036720

Srinivasan VB, Singh BB, Priyadarshi N,, Chauhan NK, Rajamohan G (2014). Role of novel multidrug efflux pump involved in drug – resistance in Klebsiella pneumoniae. PLoS One. 9(5): http://dx.doi.org/10.1371/journal.pone.0096288

Staden R (1996). The Staden sequence analysis package. Mol. Biotechnol. 5:233-241. http://dx.doi.org/10.1007/BF02900361

Stover CK, Pham XQ, Erwin AL, Mizoguchi SD, Warrener P, Hickey MJ, Brinkman FS, Hufnagle WO, Kowalik DJ, Lagrou M, Garber RL, Goltry L, Tolentino E, Westbrock-Wadman S, Yuan Y, Brody LL, Coulter SN, Folger KR, Kas A, Larbig K, Lim R, Smith K, Spencer D, Wong GK, Wu Z, Paulsen IT, Reizer J, Saier MH, Hancock RE, Lory S, Olson MV (2000). Complete genome sequence of Pseudomonas aeruginosa PA01, an opportunistic pathogen. Nature 406 (6799):959-964. http://dx.doi.org/10.1038/35023079

Tait E, Simon MC, King S, Brown AJ, Gow NA, Shaw DJ (1997). A Candida albicans genome project: cosmid contigs, physical mapping, and gene isolation. Fungal Genet. Biol. 21(3):308-314. http://dx.doi.org/10.1006/fgbi.1997.0983

Thompson JD, Higgins DG, Gibson TJ (1994). CLUSTAL W: improving the sensitivity of progressive multiple sequence alignment through sequence weighting, position-specific gap penalties and weight matrix choice. Nucleic Acids Res. 22:4673-4680. http://dx.doi.org/10.1093/nar/22.22.4673

Wasaznik A, Grinholc M, Bielawski KP (2009). Active efflux as the multidrug resistance mechanism. Postepy Hig Med Dosw. 63:123-133.

White O, Eisen JA, Heidelberg JF, Hickey EK, Peterson JD,Dodson RJ, Haft DH, Gwinn ML, Nelson WC, Richardson DL, Moffat KS, Qin H, Jiang L, Pamphile W, Crosby M, Shen M, Vamathevan JJ, Lam P, McDonald L, Utterback T, Zalewski C, Makarova KS, Aravind L, Daly MJ, Minton KW, Fleischmann RD, Ketchum KA, Nelson KE, Salzberg S, Smith HO, Venter JC, Fraser CM (1999).Genome sequence of the radioresistant bacterium Deinococcus radiodurans R1. Science 286 (5444):1571-1577. http://dx.doi.org/10.1126/science.286.5444.1571

Wisedchaisri G, Park MS, Iadanza MG, Zheng H, Gonen T (2014). Proton-coupled sugar transport in the prototypical major facilitator superfamily protein xyle. nat. commun. 4;5:4521. http://dx.doi.org/10.1038/ncomms5521

Xu X, Chen J, Xu H, Li D (2014). Role of a majorfacilitator superfamil transporter in adaptation capacity of Penicillium funiculosum under extreme acidic stress. Fungal Genet. Biol. 69:75-83. http://dx.doi.org/10.1016/j.fgb.2014.06.002

Yan N (2013). Structural advances for the major facilitator superfamily (MFS) transporters. Trends Biochem. Sci. 38:151–159. http://dx.doi.org/10.1016/j.tibs.2013.01.003

Zhang Z, Wang R, Xie J (2015). Mycobacterium smegmatis MSMEG_3705 Encodes a Selective Major Facilitator Superfamily Efflux Pump with Multiple Roles. Curr. Microbiol. http://dx.doi.org/10.1007/s00284-015-0783-0

Zheleznova EE, Markham PN, Neyfakh AA, Brennan RG (1999). Structural basis of multidrug recognition by BmrR, a transcription activator of a multidrug transporter. Cell 96(3):353-62. http://dx.doi.org/10.1016/S0092-8674(00)80548-6

10

Corticosteroid binding globulin and glucocorticoid receptor genotypes influence body composition in a male population

Elodie Richard[1], José-Manuel Fernandez-Real [2], Abel Lopez-Bermejo[2], Wifredo Ricart[2], Henri Déchaud[3], Michel Pugeat[3] and Marie-Pierre Moisan[1]*.

[1]Laboratoire PsyNuGen, INRA, UMR 1286, CNRS 5226, Université de Bordeaux, 146 rue Léo Saignat, F-33077 Bordeaux, France.
[2]Unitat de Diabetologia, Endocrinologia i Nutricio, University Hospital of Girona, "Dr. Josep Trueta" and CIBER Fisiopatología de la Obesidad y Nutrición CB06/03/010, 17007 Girona, Spain.
[3]Fédération d'Endocrinologie, INSERM U863, IFR62, Groupe Hospitalier Est F-69437 Lyon, France.

Glucocorticoid Receptor (GR) polymorphisms have been repeatedly associated with obesity and metabolic parameters in man. We have previously shown the genetic influence of a polymorphism in the gene encoding CBG on some obesity parameters in a small female population. In this study we have explored possible genetic associations between obesity and metabolic measures and CBG or GR polymorphisms in a new male population. Two hundred and ninety-five men with body mass index (BMI) ranging from 19 to 55 kg/m^2 were studied. Serum CBG levels, CBG and GR gene polymorphisms in relation with anthropometric and biochemical parameters were analysed. GR BclI polymorphism was found to influence weight, BMI, waist circumference and glucose levels. CBG polymorphism showed a significant effect for BMI and waist circumference. The frequency of CBG allele 90 was markedly increased among men with morbid obesity compared to the rest of the population (30 versus 18%, p=0.02). The influence of GR BclI polymorphism is replicated in an additional population and CBG polymorphism has a small but significant influence on obesity in men. Further studies are needed to understand the mechanism by which these polymorphisms impact on cortisol and obesity.

Key words: Obesity, genetics, cortisol, transcortin, glucocorticoids.

INTRODUCTION

The metabolic syndrome which includes insulin resistance, glucose intolerance, dyslipidemia, hypertension and type II diabetes arises from disproportionate accumulation of visceral fat mass. Chronic excessive cortisol secretion such as in Cushing's syndrome leads to abdominal obesity and metabolic syndrome (Bjorntorp and Rosmond, 2000; Rosmond et al., 1998). As cortisol is known to regulate adipose tissue differentiation, function and distribution in the presence of high insulin levels, a rise in cortisol concentration or bioavailability may result into fat mass deposits in hyperinsulinemic patients (Bjorntorp and Rosmond, 2000). However, no clear relationship between visceral obesity and the hypothalamo-pituitary adrenal (HPA) axis activity has been reported so far. Indeed, most patients with visceral obesity and metabolic syndrome have normal or even low circulating cortisol levels. These findings have suggested that an altered peripheral metabolism of cortisol, increased clearance, differences in cortisol feedback sensitivity or variation in glucocorticoid target tissue sensitivity may contribute to the development of obesity (Walker, 2001).

As glucocorticoids exert most of their effects by binding to the glucocorticoid receptor (GR), a high number of studies have examined polymorphisms in the gene encoding the GR in relation to obesity and metabolic parameters to explain individual vulnerability. Indeed, several polymorphisms within the GR gene that influence glucocorticoid sensitivity have been described and associated with body composition and metabolic parameters (Van Rossum and Lamberts, 2004). However, many other genes are susceptible to play a role in cortisol

*Corresponding author. E-mail: mpmoisan@bordeaux.inra.fr.

driven obesity. Corticosteroid Binding Globulin (CBG) is of particular interest because of its physiological role and the recent reports highlighting its putative role in obesity. CBG is a glycoprotein synthesized mainly in liver and secreted in blood where it forms a high affinity complex with glucocorticoids (Breuner and Orchinik, 2002). Serum CBG levels have been shown to be negatively correlated with body mass index (BMI), waist-to-hip ratio, blood pressure and HOMA (homeostasis model assessment) in a human healthy Spanish population (Fernandez-Real et al., 2002) with metabolic syndrome markers such as plasma triacylglycerol concentrations and fasting blood glucose in a population of French obese women (Duclos et al., 2005). Additionally, patients with a null mutation in the CBG gene tend to be obese (Torpy et al., 2001) and it is hypothesized that absence of CBG induces locally, in adipose tissue, increased cortisol levels. Finally, by genetic mapping analysis on a pig intercross, we recently reported that CBG gene is a strong candidate for a quantitative trait locus associated with cortisol level and fat deposits (Ousova et al., 2004). CBG genetic polymerphisms have been tested in human obesity only in one small women population. In that report, we had examined the influence of CBG genotype in a female obese population extensively phenotyped for the HPA axis activity. We had shown that a strong correlation between obesity parameters (waist-to-hip ratio, waist circumference) and cortisol level was observed only for patients harbouring CBG allele 90 suggesting that this polymorphism influences cortisol driven fat distribution (Barat et al., 2005). To dissect further the putative role of CBG in obesity, we studied the influence of CBG polymorphism on a large male population phenotyped for obesity, visceral fat mass and metabolic parameters. We analysed also the role of the BclI polymorphism within the intron 2 of the GR gene, as the importance of this polymorphism in obesity and insulin resistance has been replicated in several population studies (Van Rossum and Lamberts, 2004).

MATERIAL AND METHODOLOGY

Inclusion and exclusion criteria

This is a retrospective study on a total of 295 men consisting of two populations. First, 253 consecutive, unselected (except for inclusion criteria, see below) Caucasian subjects, participants in an ongoing epidemiological study that began on 2003 of risk factors for cardiovascular disease in Northern Spain, were included in the study. Subjects were randomly localized from a census and they were invited to participate. The participation rate was 71%. None of the subjects was taking any medication or had any evidence of metabolic disease other than obesity. All subjects reported that their body weight had been stable for at least three months before the study. All subjects underwent a 75 g oral glucose tolerance test to define glycemic status. Inclusion criteria were:

1) Absence of any systemic disease.
2) Absence of clinical symptoms and signs of infection in the previous month by structured questionnaire to the patient.
3) Hepatitis C virus antibody sero-negative.

Cushing's syndrome was routinely excluded by using normal 24 h urinary free cortisol as exclusion criteria. Second, we also studied an additional series of 42 consecutive men with morbid obesity from the outpatients' clinic. These obese patients are all Caucasians and come from the same geographical area as the first population. Informed consent was obtained from all subjects. Local Ethics Committee approved the study.

Measurements

Fat mass and percent fat mass were calculated using bioelectric impedance (Holtain BC Analyzer, UK). Blood pressure was measured in the supine position on the right arm after a 10 min rest; a standard sphygmomanometer of appropriate cuff size was used and the first and fifth phases were recorded. Values used in the analysis are the average of three readings taken at 5 min intervals. Patients were requested to withhold alcohol and caffeine during at least 12 h prior to the different tests.

Assays

Blood samples were drawn from each subject after an overnight fasting period. Serum glucose concentrations were measured in duplicate by the glucose oxidase method with the use of a Beckman Glucose Analyser II (Beckman Instruments, Brea, CA). The coefficient of variation was 1.9%. Glycated hemoglobin (HbA1c) was measured by high performance liquid chromatography by means of a fully automated glycated hemoglobin analyser system (Hitachi L-9100, USA). Normal range among 774 subjects with normal glucose tolerance was $4.71 \pm 0.46\%$. Total serum cholesterol was measured through the reaction of cholesterol esterase /cholesterol oxidase/peroxidase. Total serum triacylglycerol was measured through the reaction of glycerol-phosphate-oxidase and peroxidase. The plasma CBG concentration was measured by radio-immunoassay (Radim, KP31, Angleur, Liege, Belgium). Intra- and inter-assays coefficients of variation were 3.6 and 7.5%, respectively.

Genetic analysis

Genomic DNA was isolated from peripheral blood samples using the commercially available extraction kit QIAamp DNA blood (Qiagen, France). CBG and GR polymorphisms were analysed by PCR as described in reference (Barat et al., 2005).

Statistical analysis

Data are expressed as the mean $\pm$ SEM. Statistica v.6 software was used. ANOVA or non parametric Kruskal-Wallis tests were performed to compare variables between groups. Correlations between variables were performed using Spearman's correlation test. Fisher's exact test was used to compare allele frequencies among groups. The level of statistical significance used was a p value less than 0.05. The statistical power for the difference in BMI between carriers and non carriers of the polymorphism was 99% for both polymorphisms (considering a mean difference between groups of 3 points of BMI and an overall SD of 0.7).

RESULTS

The frequency of CBG genotypes was in Hardy-Weinberg equilibrium in the whole population and did not differ from previous report [21], 86/86: 63%, 86/90: 34% and 90/90: 2.7%. First, patients of the whole cohort were grouped according to their CBG or GR genotype and compared for

Table 1. Influence of genotype on patient's obesity.

	Weight (kg)	BMI (kg/m2)	Waist (cm)	Glucose (mM)	TG (mM)	CBG (nM)
CBG 86/86 (n)	87.7±1.6 (186)	29.7± 0.5 (186)	97.7±1.3 (181)	5.7±2.2 (185)	1.3±0.1 (178)	744±11 (169)
CBG 90/-(n)	95.1± 2.8 (109)	32.1±0.9 (107)	102.9± 2.0 (106)	5.7±0.1 (108)	1.3±0.1 (103)	714 ±12 (101)
p	ns	0.039	0.049	ns	ns	0.09
GR C/C(n)	87.3±2.2 (122)	29.6±0.7 (122)	96.6±1.6 (118)	5.4±0.1(121)	1.2±0.1 (118)	731±11 (120)
GR G/- (n)	94.6±2.2 (144)	31.9±0.7 (144)	103.0±1.7 (140)	5.9±0.1(143)	1.4±0.1(134)	744±12 (105)
p	0.014	0.017	0.009	0.004	0.064	ns
86/C	83.7±2.9 (77)	28.5±0.7 (77)	94.3±1.8 (76)	5.3±0.1(77)	1.2±0.1 (75)	744.4±15 (75)
86/G	92.2±3.0 (73)	31.1±0.9 (73)	100.8±2.3 (69)	5.9±0.2 (72)	1.3±0.1 (68)	738.8±16 (66)
90/C	93.4±3.8 (45)	31.4±1.3 (45)	100.9±3.1 (42)	5.5±0.1 (44)	1.2±0.1 (43)	729.4±15 (42)
90/G	97.1±3.0 (71)	32.8±1.1 (71)	105.2±2.5 (71)	5.9±0.2(71)	1.5±1.1 (66)	714.4±18 (64)
p	0.046	0.025	0.01	0.016	ns	ns
p(86C vs 90G)	0.049	0.017	0.006	0.015	NA	NA

Data are presented as means ± SEM. **BMI**; body mass index, **TG**; triacylglycerols, **n**; number of subjects, **p**; p value of non parametric Kruskal-Wallis analysis of variance test, p (86C vs90G) and p value for bilateral comparison between groups 86C and 90G. ns= non significant; NA= not applicable.

for every phenotype measured. Subjects with 86/90 and 90/90 CBG genotypes, named "90/-"were combined in statistical analysis because of the small number of 90/90 patients. Non parametric Kruskal-Wallis test was used to compare the groups as the variances of groups were often not homogenous. The main results are summarized in Table 1. CBG genotype was found to influence signifycantly BMI and waist circumference. A trend toward lower serum CBG levels was found in 90/- patients compared to 86/86 ones (CBG (nM): 714 ± 12, n=101 vs. 744 ± 11, n=169 respectively, p=0.07 by ANOVA, p=0.09 by Kruskal-Wallis test). This trend became significant when subjects with morbid obesity were excluded from the analysis (701.0 ± 11.4, n=83 vs. 741.8 ± 11.2, n=159 respectively, p=0.012). No difference was found for blood pressure, plasma glucose, cholesterol, HDL and fasting triacylglycerols between CBG genotypes. The frequency of GR genotypes was also found in Hardy-Weinberg equilibrium in the whole population C/C: 45.9%, C/G: 44.3%, G/G: 9.8%, that is, 68.9% allele C and 31.1% allele G which is comparable to previous reports [24]. Subjects with G/C and G/G GR genotypes, named "G/-"were combined in statistical analysis because of the small number of G/G patients (Table 1). Weight, BMI, waist circumference, and plasma glucose were found significantly increased in G/- compared to C/C patients and they also show a trend towards higher fasting triacyl-glycerols. No difference was found for blood pressure between GR genotypes. Finally, subjects were stratified in four groups according to their genotype at both CBG and GR loci. Overall the same results (p value) were found as for GR analysis but the groups means are more contrasted between 86/C and 90/G suggesting an additive effect of genotypes at CBG and GR loci.

In a second step, patients were divided in different groups according to Garrow's classification: lean (BMI below 25), overweight (BMI between 25 and 30), obese (BMI between 30 and 40) and morbidly obese patients (BMI greater than 40). Anthropometrical and biochemical data are summarized in Table 2. Fifty-six percent of morbidly obese subjects had three or more components of the metabolic syndrome according to ATP III criteria. We compared allelic distribution between these patients groups. Although no difference was found between lean, overweight and obese subjects, the frequency of CBG allele 90 rose from 18 to 30% in morbidly obese patients (Table 3). Among these morbidly obese subjects, serum CBG concentration was negatively associated with BMI (n=28, r =-0.4133, p = 0.029) and waist circumference (n= 27, r =-0.4163, p = 0.031). These correlations were not found when the total population is considered. A trend towards increased frequency of GR G allele was observed in obese subjects (p = 0.07) that became significant in morbidly obese subjects (Table 3).

Finally, when serum CBG concentration was analysed as dependent variable and BMI, fasting glucose and CBG gene polymorphism as independent variables, stepwise multiple linear regression analysis showed that in non-obese subjects, only BMI (not glucose or CBG gene polymorphism) contribute to almost 6% of the serum CBG variance (R^2 = 0.059, p < 0.0001). In obese subjects, both BMI (p = 0.001) and CBG gene polymerphism (p=0.023) independently contribute to 5.4 and 2.5%, res-pectively, of serum CBG variance (r^2 in additive models of 0.054 and 0.079).

DISCUSSION

In this study, we have found that CBG genotype is associated with BMI and waist circumference in a male population. Furthermore, the frequency of CBG allele 90 is markedly increased in the morbid obese population.

Our data suggest that this CBG polymorphism is associated with decreased CBG levels. Indeed, in the

Table 2. Anthropometrical and biochemical variables of the study subjects according to BMI.

	Normal	Overweight	Obese	Morbidly obese
n	65	118	67	45
Age (year)	47.1 ± 1.6	52.8 ± 1.02	51.2 ± 1.8	42.3 ± 1.8
Obesity parameters				
Weight (kg)	68.5 ± 0.8	65.3 ± 0.6	99.3 ± 1.8	137.3 ±3.0
BMI (kg/m²)	23.0 ± 0.2	27.3 ± 0.1	33.9 ± 0.4	45.5 ± 0.7
Visceral fat mass parameters				
Waist circumference (cm)	82.1 ± 0.8	91.7 ± 0.5	108.2 ± 1.3	134.7 ± 1.6
Free Fat mass (kg)	68.6 ± 1.2	72.0 ± 1.0	75.2 ± 1.3	85.8 ± 3.1
Fat mass (kg)	8.0 ± 1.2	7.99 ± 1.0	38.4 ± 1.5	47.6 ± 3.2
Metabolic parameters				
SBP (mmHg)	119.2 ± 1.6	126.2 ± 1.5	135.0 ± 2.1	137.4 ± 2.8
DBP (mmHg)	74.2 ± 1.0	77.9 ± 1.0	84.7 ± 1.4	85.0 ± 1.8
Plasma glucose (mM)	5.3 ± 0.1	5.4 ± 0.07	5.6 ± 0.08	6.9 ± 0.4
Cholesterol (mM)	5.13 ± 0.1	5.43 ± 0.09	5.46 ± 0.12	5.28 ± 0.20
HDL (mM)	1.41± 0.05	1.36 ± 0.03	1.27 ± 0.04	1.11 ± 0.04
LDL (mM)	3.23 ± 0.09	3.48 ± 0.09	3.43 ±0.11	3.19 ± 0.17
Triacylglycerols (mM)	1.06± 0.09	1.12 ± 0.05	1.50 ± 0.09	1.83 ± 0.17
HbA$_{1c}$ (%)	4.81 ± 0.1	4.88 ± 0.1	5.17 ± 0.1	5.91 ± 0.2
Serum CBG (nM)	760±20 n = 64	710± 10 n = 115	720 ± 20 n = 63	780 ± 4 n = 28

Data are presented as means ± SEM. **BMI**; body mass index, **SBP**; systolic blood pressure, **DBP**; diastolic blood pressure, **HDL**; high-density lipoprotein, **LDL**; low-density lipoprotein and **HbA$_{1c}$**; glycated hemoglobin.

Table 3. Allelic distribution in different groups of patients classified by BMI.

CBG gene	Allele 86		Allele 90		Odds ratio		
	n	%	n	%		95%CI	p Fisher
BMI < 25	107	82.3	23	17.7			
BMI 25 – 30	193	81.8	43	18.2	1.036	0.59-1.81	0.51
BMI 30 – 40	110	82.1	24	17.9	1.015	0.54-1.91	0.54
BMI > 40	63	70	27	30	1.994	1.054-3.77	0.02*
GR gene	Allele C		Allele G		Odds ratio		
	n	%	n	%		95%CI	p Fisher
BMI < 25	83	71.6	33	28.5			
BMI 25 – 30	152	73.8	54	26.2	0.89	0.54-1.49	0.38
BMI 30 – 40	78	61.9	48	38.1	1.55	0.90-2.65	0.07
BMI > 40	49	58.3	35	41.7	1.80	0.99-3.25	0.03*

Allelic distribution analysis was performed with a Fisher's exact test with one group of men patients compared to normal patients (BMI<25).

morbidly obese male population, serum CBG was negatively correlated to BMI and waist circumference. These results are consistent with those obtained from other independent studies in which serum CBG concentration was also negatively correlated with several markers of obesity including BMI in individuals of both sexes from a healthy population (Fernandez-Real et al., 2002; Lapidus et al., 1986). These data also confirm our previous finding that subjects with CBG allele 90 are more associated with cortisol driven fat distribution than allele 86 patients

(Barat et al., 2005). CBG protein concentration also tended to be lower among CBG 90/- subjects when the whole population was examined, a finding that became significant when morbidly obese patient are excluded from the analysis (p = 0.016). Thus, CBG levels appear to decrease with BMI until the patients become morbidly obese. It is well known that normal homeostatic mechanisms are lost in morbid obesity; this may explain why CBG levels are higher in this group.

Although some reports suggest that CBG may help to

capture low concentrations of cortisol and facilitate its active transfer into the cells, several studies on human patients with either variant or total deficiency of CBG strengthen the hypothesis of the cortisol sequestering effect of CBG (Torpy et al., 2001; Joyner et al., 2003; Emptoz-Bonneton et al., 2000; Roitman et al., 1984). In particular, ligand-binding experiments demonstrated that in the CBG-deficient patient peaks in cortisol secretion during the circadian cycle are accompanied by large increases in free cortisol (not yet bound to albumin) that are even higher after stress-induced naloxone stimulation (Lewis et al., 2005). To compare cortisol levels between patients, strict conditions should be applied in order to avoid variations due to the circadian cycle of cortisol secretion as well as stress levels. Unfortunately, cortisol levels could not be measured in this retrospective study but only CBG that does not vary so rapidly.

Our study further replicates the association between GR BclI polymorphism and obesity markers in middle-aged population. The molecular mechanism of the BclI polymorphism is not known but it has been associated with increased glucocorticoid sensitivity that leads to different consequences during life. Thus, BclI G-allele carriers show higher abdominal fat mass when young, whereas late in life they show lower lean mass due to muscle atrophy (Van Rossum and Lamberts, 2004). When we grouped the patients according to their genotype at both CBG and GR loci, we observed that the subjects harbouring CBG allele 90 and GR allele G are the most obese whereas patients with CBG allele 86 and GR allele C show the lowest values of obesity parameters. Interestingly, waist circumference came out as the most significant phenotype in this genetic analysis. This suggests that CBG and GR polymorphisms would influence central rather than peripheral obesity which fits with the hypothesis of the role of hypercortisolemia in central obesity. Although the contrast is apparently mostly driven by GR genotype, both loci would be interesting to consider in future studies.

In conclusion, our data suggest that polymorphisms located in intron 1 of the CBG gene and in intron 2 of the GR gene are associated with morbid obesity in a Spanish middle aged male population. Selection bias of the morbid obese group is unlikely as subjects come from the same geographical area as the rest of the population and also because the frequency of GR allele G is already increased in the obese group. However, our sample size is still too small to prove a direct effect of CBG or GR genotypes in such complex disease traits. Additional work is required to unravel the mechanism by which CBG and GR polymorphisms influence free cortisol concentrations in a larger sample size. Our data suggest that lowered plasma CBG levels may be one mechanism.

REFERENCE

Barat P, Duclos M, Gatta B, Roger P, Mormede P, Moisan MP (2005). Corticosteroid binding globulin gene polymorphism influences cortisol driven fat distribution in obese women. Obes. Res. 13: 1485-1490.

Bjorntorp P, Rosmond R (2000). Obesity and cortisol. Nutrition 16: 924-936.

Breuner CW, Orchinik M (2002). Plasma binding proteins as mediators of corticosteroid action in vertebrates. J. Endocrinol. 175: 99-112.

Duclos M, Marquez PP, Barat P, Gatta B, Roger P 2005 Increased cortisol bioavailability, abdominal obesity, and the metabolic syndrome in obese women. Obes. Res. 13: 1157-1166.

Emptoz-Bonneton A, Cousin P, Seguchi K, Avvakumov GV, Bully C, Hammond GL, Pugeat M (2000) Novel human corticosteroid-binding globulin variant with low cortisol- binding affinity. J. Clin. Endocrinol. Metab 85: 361-367.

Fernandez-Real JM, Pugeat M, Grasa M, Broch M, Vendrell J, Brun J, Ricart W (2002). Serum corticosteroid-binding globulin concentration and insulin resistance syndrome: a population study. J. Clin. Endocrinol. Metab. 87: 4686-4690.

Joyner JM, Hutley LJ, Bachmann AW, Torpy DJ, Prins JB (2003). Greater replication and differentiation of preadipocytes in inherited corticosteroid-binding globulin deficiency. Am. J. Physiol. Endocrinol. Metab. 284 E1049-E1054.

Lapidus L, Lindstedt G, Lundberg PA, Bengtsson C, Gredmark T (1986). Concentrations of sex-hormone binding globulin and corticosteroid binding globulin in serum in relation to cardiovascular risk factors and to 12-year incidence of cardiovascular disease and overall mortality in postmenopausal women. Clin. Chem. 32:146-152.

Lewis JG, Bagley CJ, Elder PA, Bachmann AW, Torpy DJ (2005). Plasma free cortisol fraction reflects levels of functioning corticosteroid-binding globulin. Clinica Chimica Acta 359: 189-194.

Ousova O, Guyonnet-Duperat V, Iannuccelli N, Bidanel JP, Milan D, Genet C, Llamas B, Yerle M, Gellin J, Chardon P, Emptoz-Bonneton A, Pugeat M, Mormede P, Moisan MP (2004). Corticosteroid binding globulin: a new target for cortisol-driven obesity. Mol. Endocrinol. 18: 1687-1696.

Roitman A, Bruchis S, Bauman B, Kaufman H, Laron Z (1984). Total deficiency of corticosteroid-binding globulin. Clin. Endocrinol. (Oxf) 21: 541-548.

Rosmond R, Dallman MF, Bjorntorp P (1998). Stress-related cortisol secretion in men: relationships with abdominal obesity and endocrine, metabolic and hemodynamic abnormalities [see comments]. J. Clin. Endocrinol. Metab. 83:1853-1859.

Torpy DJ, Bachmann AW, Grice JE, Fitzgerald SP, Phillips PJ, Whitworth JA, Jackson RV (2001). Familial corticosteroid-binding globulin deficiency due to a novel null mutation: association with fatigue and relative hypotension. J. Clin. Endocrinol. Metab. 86: 3692-3700.

Van Rossum EF, Lamberts SW (2004). Polymorphisms in the glucocorticoid receptor gene and their associations with metabolic parameters and body composition. Recent Prog. Horm. Res. 59: 333-357.

Walker BR (2001). Activation of the hypothalamic-pituitary-adrenal axis in obesity: cause or consequence? Growth Horm. IGF. Res. 11 Suppl A S91-S95.

11

Application of subspecies-specific marker system identified from *Oryza sativa* to *Oryza glaberrima* accessions and *Oryza sativa* × *Oryza glaberrima* F$_1$ interspecific progenies

Isaac Kofi Bimpong[1,2], Joong Hyoun Chin[2]*, Joie Ramos[2] and Hee-Jong Koh[3]

[1]Africa Rice Centre, BP 96. St Louis, Senegal.
[2]International Rice Research Institute (IRRI), DAPO Box 7777, Metro Manila, Philippines.
[3]Department of Plant Science, College of Agriculture and Life Sciences, Seoul National University, Seoul, 151-921, Korea.

Interspecific hybrids (F$_1$'s) between Asian rice (*Oryza sativa* 2n=24 AA) and African rice (*Oryza glaberrima* 2n=24 AA) are almost completely sterile. This hybrid sterility barrier is mainly caused by an arrest of pollen development at the microspore stage. Intersubspecific F$_1$ hybrid sterility is mainly caused by cryptic chromosomal aberrations and allelic interaction between *indica* and *japonica*. To identify *O. glaberrima* specific loci, 67 subspecies-specific (SS) sequenced-tagged site (STS) marker were used to evaluate 30 *O. glaberrima* accessions, which could be classified into sub eleven groups. SPI (subspecies-prototype index) of *O. glaberrima* accessions ranged from 51.67 to 60.00, suggesting intermediate subspecific type based on whole-genome. Some informative markers for classifying O. *glaberrima* accessions, called reference markers, S01054, S01160, S02085, S02140, S03041, and S08107, showed indica allele, which might have contributed to genomic diversification of *O. glaberrima*. Ten (14.9%) SS markers generated *glaberrima*-specific allele, implying loci adjacent with these markers could be a key for interspecific hybrid sterility. Only 40 (59.7%) SS markers might be useful in *O. glaberrima* analysis, as other markers did not amplify heterozygous allele in F$_1$ of *O. sativa* × *O. glaberrima*.

Key words: *Oryza glaberrima*, *Oryza sativa*, sequenced-tagged site, subspecies-specific, interspecific progenies.

INTRODUCTION

The genus *Oryza* is known to consist of two cultivated species, Asian rice (*O. sativa* 2n=24=AA) and African rice (*O. glaberrima* 2n=24=AA) and 22 wild species (2n=24, 48) representing 10 genomic types namely, AA, BB, CC, BBCC, CCDD, EE, FF, GG, HHJJ and HHKK (Vaughan, 1994; Aggarwal et al., 1997). Unlike *O. sativa*, *O. glaberrima* has no known subspecies; it might have arisen from an African wild species, *O. barthii* independently of the origin of *O. sativa* from the Asian form of *O.*

perennis (Morishima et al., 1963; Semon et al., 2005). The two species are adapted to diverse environ-ments and has its own ecologically adapted and useful traits (Glaszmann, 1987; Sarla et al., 2005).

A number of different markers such as isozyme (Glaszmann, 1987), protein (Bi et al., 1997), RFLP (Qian et al., 1995) RAPD (Chin et al., 2003; Subudhi et al., 1999), simple sequence repeat (SSR) (McCouch et al., 2002; Chen et al., 2002; Ni et al., 2002), AFLP (Cho et al., 1999), STS (Chin et al., 2007; Edwards et al., 2004), SNPs (McNally et al., 2009; Feltus et al., 2004), and chloroplast DNA (Sun et al., 2002) have been utilized to estimate the extent of genetic diversity in *O. sativa*. In *O.*

*Corresponding author. E-mail: kofibimpong@yahoo.com.

glaberrima, estimates of genetic diversity based on molecular markers are comparatively few, markers such as isozyme, RFLP SSR and SNPs have been used to estimates genetic diversity in *O. glaberrima* and its close genetic relationship to *O. barthii* (Lorieux et al., 2000; Semon et al., 2005). Even though diversity in *O. glaberrima* are significantly lower than those in *O. sativa,* it had been shown to harbors genes that have allowed the species to survive and prosper in West Africa with minimal human intervention (Barry et al., 2006; Wang et al., 2001)

Even though *O. glaberrima* and *O. sativa* share the same genome, with minor sub-genomic differences which do not hinder normal chromosome pairing and gamete formation in the hybrids (Nayar, 1973); yet, F₁ hybrids between them shows complete sterility irrespective of the combinations of parental varieties (Pham and Bougero, 1993). Various causes such as meiotic irregularities (Heuer et al., 2003), low proportion of viable pollen, low pollen germination, cytoplasm and its interaction effects from male side and early elimination of female gametes and zygotes from female side (Porteres, 1956; Kitampura, 1962) have been ascribed for sterility. Other causes of sterility are due to segregation distortion (Causse et al., 1994; Lorieux et al., 2000), presence of sterility loci in *O. glaberrima*, (Koide et al., 2008; Sano, 1986; Doi et al., 1999; Li et al., 2008), hybrid breakdown (Li et al., 1997; Kubo and Yoshimura, 2005) and suppressed recombination (Ikehashi, 1982; Neiman and Linksvayer, 2006); hindering easy transfer of useful genes between the two species. Some QTLs and epistatic interaction controlling hybrid-sterility have also been identified (Li et al., 2008).

It is of interest to understand the genetic structure of *O. glaberrima* as information on its diversity and structure is expected to assist plant breeders in the selection of parents for hybridization and also to identify materials that harbor allele of value for plant improvement. Molecular markers have increasingly become useful tools for evaluating genetic diversity and determining cultivar identity. Compared to morphological markers, molecular markers can reveal differences among accessions at the DNA level and thus provide a more direct, reliable, and efficient tool for germplasm conservation and management.

Subspecies-specific (SS) or species-specific genomic regions could be inherited in a conserved manner to each of subspecies and species from which the SS regions were originated in the progenies of inter-subspecific or inter-specific crosses (Tanksley et al., 1992; Wang et al., 2001). Thus SS regions may provide a clue in understanding the mechanisms for reproductive barriers including inter-subspecific hybrid sterility and for the differentiation of rice subspecies.

The purpose of this study was to evaluate the extent of genetic differentiation between diverse collections of *O. glaberrima* accessions using 67 subspecies-specific (SS) markers. We were interested in developing molecular markers

in identifying *O. sativa* and/or *O. glaberrima* loci, and it usefulness in interspecific crosses.

MATERIALS AND METHODS

Plant material

Thirty accessions of *O. glaberrima* were obtained from the germplasm collection at International Rice Research Institute (IRRI) in the Philippines and Africa Rice Centre, Benin in West Africa, 12 accessions of *O. sativa*, representing both *indica* and *japonica* subspecies and 16 F₁ progenies from cross between *O. sativa × O. glaberrima* were used. The F₁s were produced by making crosses in the screenhouse of IRRI between 4 elite indica cultivars (IR64, PSBRc 18 -irrigated, IR 69502-6-SRN-3-UBN-1-B – rainfed and IR55423-01-upland cultivar) and several accessions of *O. glaberrima* referred to as RAM which were received from Mali in West Africa and have been field tested as drought tolerance. *O. sativa* was used as female parent in crosses with *O. glaberrima*. The F₁ plants were intermediate in morphological characteristics but were highly sterile. Variety names source/origin are given in Table 1.

Primer designing

A set of 67 STS markers used in this study were design by Chin et al. (2007) using an online-service software Primer3 (http://frodo.wi.mit.edu/cgibin/primer3/primer3_www.cgi) to detect the insertion/deletion (InDel) polymorphism between the genome sequence of Nipponbare (*japonica*) and 9311 (*indica*). The amplicon size for each primer set was determined so that the amplicon contained at least 5% In Del difference of its whole size, 100 to 400 bp. These markers covers the whole chromosomes at an 2 to 3 cm interval based on the sequence information available at RGP for *japonica* and NCBI for *indica* and are distributed throughout the 12 chromosomes.

PCR amplification

The protocols for PCR amplification and detection for STS markers were similar as described in Temnykh et al. (2000) with some modifications. Each 25 µl reaction mixture contained 50 ng DNA, 5 pmol of each primer, 2 µl PCR buffer [100 mM Tris (pH 8.3), 500 mM KCl, 15 mM MgCl₂, 2 µg gelatin], 250 µM of each dNTPs and 0.5 unit *Taq* polymerase. The thermocycler profile was: 5 min at 94°C, 35 cycles of 1 min at 94°C, 1 min at 48°C or 55°C, 2 min at 72°C, and 5 min at 72°C for final extension using the MJ research PCR system. PCR amplicons were resolved by electrophoresis in 3% agarose gels and marker bands were revealed using the silver-staining protocols as described by Panaud et al. (1996).

Scoring of the SS –STS markers

The SS STS markers were scored as 'a' (*japonica* allele) or 'b' (*indica* allele) for each marker locus. The total number of 'a' from *japonica* varieties and 'b' from *indica* were counted. Since some markers showed variation in generating SS allele among varieties within and inter-subspecies, the concept of subspecies-specificity (SS) was employed as follows:

Subspecies-specificity (SS) score of each marker = (Total number of expected allele in each subspecies) / (Total number of varieties/accessions tested) × 100%.

For example, if a marker has a SS score of 100%, it means that

Table 1. Plant materials in this study.

Species/ subspecies	Entry no.	Name	Source	Entry No.	Cross combination*	Source
	1	IR55423-01	IRRI	43	IR64 × RAM54	IRRI
	2	IR60080-46-A		44	IR64 × RAM86	
	3	IR64		45	IR64 × RAM90	
indica	4	IR68703-AC-24-1		46	IR64 × RAM120	
	5	IR69502-6-SRN		47	IR64 × RAM134	
	6	PSBRC18		48	IR64 × RAM131	
	7	PSBRC82		49	PSBRC18 × RAM111	
	8	Hwacheongbyeo	Korea	50	IR55423-01 × RAM3	
	9	Ilpumbyeo		51	IR55423-01 × RAM24	
Japonica	10	Jinmibyeo		52	IR55423-01 × RAM163	
	11	Junambyeo		53	IR69502 × RAM118	
	12	TR22183	China	54	IR69502 × RAM121	
	13	RAM3	Mali	55	IR69502 × RAM163	
	14	RAM24		56	IR60080-46-A × IG10	
	15	RAM54		57	IR68703-AC-24-1 × CG14	
	16	RAM86		58	IR60080-46-A × CG14	
	17	RAM90				
	18	RAM111				
	19	RAM118				
	20	RAM120				
	21	RAM121				
	22	RAM131				
	23	RAM134				
	24	RAM152				
	25	RAM163				
	26	IG10	Ivory coast			
	27	CG14				
O. glaberrima	28	CG17				
	29	CG20				
	30	Acc.103477				
	31	TOG5674				
	32	TOG5681				
	33	TOG5860	Africa			
	34	TOG6472				
	35	TOG6508				
	36	TOG6589				
	37	TOG6597	Ivory coast			
	38	TOG6629				
	39	TOG6631	Africa			
	40	TOG7235				
	41	TOG7291				
	42	TOG7442				

Entry no. 43- 58 are F₁ progenies between *O. sativa* × *O. glaberrima*

the SS marker generated SS allele for all the accessions without exception. A marker with SS scores equal to or higher than 93.3 (up to 2 exceptions out of the total number of accessions) was regarded as SS marker. In addition, the subspecies-prototype (SP) degree for each accession was calculated in order to describe the relative genomic inclination of each accession toward either subspecies as follows:

Subspecies prototype (SP) degree of each accession = (Total number of *japonica* SS allele in each accession - Total number of *Indica* SS allele in each accession) / Total number of SS markers tested If a variety has a SP degree close to 1 or -1, the variety is estimated to have the genomic constitution close to the prototype of *japonica* or *indica*, respectively.

RESULTS

Genotyping by subspecies-specific (SS) markers

The information of 67 SS STS markers used in this study is summarized in supplementary Table 1. For a marker to be confirmed as SS, a threshold of 93.3% was set. There was only one SS markers detected on chromosome 6, while 11 SS markers were identified on chromosome 3. The average number of SS markers was 5.6 per chromosome. The BAC/PAC clones from which STS markers were originated and the marker location within BAC/PAC clones were denoted in a sequence order of base pairs. The SS markers which showed perfect SS scores were S01022, S02026, S02140, S03020, S03041, S04128, S06001, S07011, S09026B, S10003A, S11004A, S11006A, and S12011B.

Classification of *O. glaberrima* accessions by SS markers

Figure 1 show the gel profile of an SS- STS marker applied to amplify 12 *O. sativa* varieties, 30 *O. glaberrima* accessions, and 16 F_1 progenies between *O. sativa* × *O. glaberrima*. As expected, most of the SS markers detected *O. sativa* allele with only 7% (10 markers) detecting *O. glaberrima*-specific allele (Table 2). Estimated size of allele present in only *O. glaberrima* ranged from 160 bp in SS marker S10003A to 610 bp in marker S11004A. Two SS markers S02085 and S02140 did not detect any indica- allele among the *O. glaberrima* accessions (Table 2).

The average value of *Indica*-prototype index which is similar to *indica* varieties of *O .sativa* (IPI) was about 50% for each *O. glaberrima* accessions and between 80 to 90% among *indica* varieties from IRRI (Figure 2). No IPI was observed in the *japonica* varieties of Korean origin while the IRRI type had about 15%. The 2 japonicas and the 5 indica varieties were similar in allelic composition to the IRRI varieties, even though they might have different plant types (Figures 2, supplementary Tables 2 and 3); The subspecies-prototype index (SPI) of *O. glaberrima* accessions ranged between 51.67 to 60.00, while the japonica species had very low SPI (0 to 13.24) (supplementary Table 2).

A total of 10 subgroups were identified based on 6 informative markers, called reference markers, S01054, S01160, S02085, S02140, S03041, and S08107 (Table

3). Each of the 10 subgroups revealed different markers showing the kind of mutation occurring at that specific locus (either as an *indica/japonica* allele mutating to *japonica*, *indica* or *O. glaberrima* allele). For example in group IV, the *O. glaberrima* allele mutated from *japonica* allele as revealed by marker S01160. Also *O. glaberrima* accession TOG5674 could be distinguished from CG14 by the presence of additional *indica* allele revealed by marker S02085 and specific allele by S02140. Twenty-nine of the 30 *O. glaberrima* accessions (except TOG 6629) were observed to be segregating for different allele (Table 3).

Forty-two percent (42%) SS markers detected heterozygous allele between *japonica/glaberrima* in the F_1 progenies; whilst 34% markers also detected heterozygous allele between *indica/glaberrima* in the F_1 progenies (Table 4 and supplementary Table 3). Some makers (13%) did not detect heterozygous allele in the F_1 between *O. sativa* and *O. glaberrima* species; whilst others such as S09093A could not distinguish the heterozygous allele between *indica and glaberrima*.

Comparative view of genome of *O.glaberrima* based on *O.sativa* spp.*japonica* genome

A total of 38 loci among the *O. glaberrima* accessions had only indica allele and 26 loci had only Japonica allele, whilst only 1 loci showed both *indica* and japonica allele (Figure 3). Some non-sativa allele were detected on chromosome 1, 2, 9, 10, 11, and 12. Heterozygous allele of h (G+I) were identified on 3 loci on chromosomes 1, 2 and 3. Markers on inter-sub specific hybrid sterility QTLs, S05014B and RM413 on chromosome 5 and S08066 on chromosome 8, showed *indica* allele in 29 *O. glaberrima* accessions.

DISCUSSIONS

Allele frequency of 30 *O. glaberrima* accessions

A small proportion of the SS-STS markers tested (10 in all=14.9% 10 in 67) did not amplify *O. sativa* allele but rather *O. glaberrima* specific allele, and consist of 6 to 7 glaberrima specific allele. Polymorphism between *O. glaberrima* at the DNA level has been reported to be low; few polymorphisms (37 to 4%) could be detected by Enriquez et al. (2001) using SSR markers. Further, Bimpong et al. (2004) observed 38% polymorphism between *O. sativa* × *O. glaberrima* parents using SSR markers. Those SS markers that detected *O. glaberrima* allele might be related to the evolution of *O. glaberrima*. Very few STS markers have been used to detect polymorphism among *O. glaberrima* species. SSR have been used widely in genetic diversity studies (Semon et al. 2005; Garris et al., 2004; Senior, 1998); however, not much work has been done on the use of STS to detect

Supplementary table 1. List of subspecies-specific primers

Chromosome	Marker	Physical location in Nipponbare pseudomolecule(bp)		Expected size of amplicon (bp)		Primer sequence		BAC clone 1)
		start	stop	Japonica	Indica	Forward	Reverse	
1	S01022	4384676	4384975	300	312	catgatgatgcttccctct	ttgacagtggctccacaaag	AP002484
	S01054	10309451	10309687	237	266	gcgaagcctgcttttgat	cggagattttccctaaaacaa	AP002070
	S01140	35147646	35147820	175	187	gctaggcagactctagctcatca	tggaacaagtagaagcagaagtca	AP003411
	S01157B	39802962	39803196	235	223	ccctcaatcatcgcaactgt	cagatgcagaaaagcgcata	AP006531
	S01160	40802478	40802660	183	179	ttgcgatttatttgccagtg	ccaggcatccaatgctatt	AP004672
2	S02026	5345560	5345726	167	180	tggtccatcatattgccaac	tcctctcagatccgatttca	AP004184
	S02052	12020182	12020373	192	201	gcagtcggttcaattggt	gattttccagccattctca	AP005743
	S02054	14117145	14117316	172	157	tttgaagcacgagggatctt	ataaagaccgatgcaaacg	AP004856
	S02057B	17440500	17440729	230	241	agcctcttcctcctcctcac	tgcaaacaccataacaaccaa	AP004999
	S02081B	20964659	20964883	225	201	agcggcatatttgcatagc	tgtttgcaggacgcagtag	AP004876
	S02085	21636396	21636559	164	153	gcgagagtgtacccctttga	tgtgtaccttgcacccctgaa	AP006068
	S02140	32850429	32850633	205	220	tgggaggaggatattgtgga	tgacaggttgatgtgatggaa	AP005538
3	S03010B	2098371	2098585	215	203	gtgcggatttggttcgttt	gagggagaggccagattctt	AC118132
	S03020	4302802	4302984	183	168	tttctaggtacatttaagcaagca	catgaatttgaagctgcgagta	AC126223
	S03027	5713283	5713531	249	232	tgaacattttggtcgtgtcgg	ttgacgaagtcaccatagacg	AC105928
	S03041	8900833	8901024	192	201	gctgacattgtccgaggttt	ccgacgtccaacctaagc	AC139168
	S03046	10137125	10137381	257	248	tcacagttacaggcggaatc	gcaccatgtatagaccattcca	AC137634
	S03048	10754658	10754836	179	159	gggatgggagaaaggaataa	gccagctaggatgttgaagg	AC137267
	S03096	24299548	24299716	169	183	cacttgcaagctaagcacca	ccttcctgcttgacgagaaa	AC120505
	S03099	24995662	24995883	222	233	ctccaggatgctcactcag	ataatccaaggcacagcac	AL731878
	S03120	27366848	27367089	242	254	tgtgcgctcgtgattatttt	aagggagcagataatgcag	AC092779
	S03136	30109023	30109222	200	219	gcattaaggcacacaaagca	tgtttgtaatccgcatggaa	AC118133
	S03145	32174684	32174922	239	251	tcacctacagagcagcag	gccgtcgttgaagagtagc	AC091494
4	S04058	20162588	20162832	245	226	gatccatgcagttgattgtga	tcgtcttatctaaaaagaaaatttga	AL662947
	S04060	20474915	20475135	221	203	tatggtttatccgccaacc	gctacaactaaaaacaagaaaacgtga	AL606598
	S04077B	24949310	24949483	174	201	atgtggatgtgggtcctat	agggttcatcgttgaagca	AL606604
	S04077A	24958459	24958724	266	247	tccagggaactacgcgact	cagcatttcagttggaagca	AL606604
	S04087A	27761378	27761633	256	248	atgttggcaatccgctaag	aaagatggttgagcggaaga	AL606682
	S04097B	29346635	29346813	179	189	tccacagtcgtccgtgaaa	ctccttgtgctgctcagaattg	AL662957
	S04128	34569925	34570087	163	181	tcacgggaaagctttggtat	aacttatgcagccaccatcc	AL606456

Supplementary table 1

	Marker	Position 1	Position 2	Size 1	Size 2	Sequence 1	Sequence 2	Accession
	S04129B	35102075	35102256	182	203	aatcgattcattgcacaaa	cttcatgtctgccattga	AL606637
5	S05064	16966554	16966786	245	257	aaagcaagtcacaacaaaataaa	tgcctcgatttcataagca	AC104272
	S05080A	20663155	20663383	229	254	tggccaacttgggaattta	aagagtcgtgcaaatgaaaaga	AC109595
6	S06001	546207	546437	231	244	agtcaatatcaggcaagcag	aaatgacacagttgacctttgaa	AP000616
7	S07011	2543283	2543511	229	205	ctggatccaaggcatcattc	cttcgctctcaccatcaaca	AP004263
	S07048	8487589	8487745	157	172	catggcaccttgagagttga	acacatggagctggcttctc	AP005824
	S07050C	13437934	13438142	209	225	ctccacttatggcagcgaat	caagtgaagtgggagcaggt	AP003745
	S07050A	14531440	14531638	199	211	tacacgaacgaacgacaagg	cgctgatttgggtaggtctc	AP005200
	S07101	26848149	26848350	202	216	gcatgccaggatatggtctc	tcggtacacacctcctgtga	AP003832
	S07103	27558599	27558809	211	223	agcatggatccttcatccaa	actccgattttgcacttcg	AP005182
8	S08066	18904657	18904874	218	238	ttgttccgttgtgtgtgcaact	gatgcagcgacgtgaaatc	AP003947
	S08090	23079842	23080054	213	231	gcgtgtgaagaggagagaaag	cagtgagaatctgcagtcg	AP004693
	S08106	25773305	25773524	220	194	ttacggattgtcacggtttt	ggaatttgtcactggttcca	AP005509
	S08107	25956924	25957152	229	240	ttggtaatgccatgctaga	cacgattcggtcatttcaga	AP003888
9	S09000A	244321	244528	208	221	ccaattcacggtttaacaagg	gccatgaagcttcgttagga	AP006058
	S09026B	9142928	9143141	214	189	gggaggcagaggggaactact	ttatcaggccaggtcctttg	AP005780
	S09040B	12641376	12641601	226	214	taatatcgcatggcaagacg	actttgcagaggcgacaaac	AP005637
	S09058	15942709	15942930	222	233	cgtgagagtccagtccaca	attgatcgattgggggattt	AP005551
	S09062B	16864607	16864856	250	236	acgcataccgaatgtgacag	gttggcactccgattaaaa	AP005559
	S09065	17914403	17914639	237	246	tgtgttgacgtttgaccat	gggccagggtacattgaata	AP005574
	S09073	19077948	19078180	233	250	accacctgaaccacacacat	tcactggttctgtgtccaa	AC099403
	S09075B	19519638	19519866	229	211	gactaacgaacggggcctat	ggcagcccacactatttagg	AC108753
	S09075A	19575874	19576047	174	154	cctcactcacctggagaagg	cgtccacactaacggacaca	AC108753
	S09093A	22803693	22803950	258	232	caccgctctcactgtcattc	tccctcagccataaaaccag	AP006162
10	S10001	992379	992586	208	229	atcgtgtgtcgggattatgag	gcatcatggctttgttgttg	AC078891
	S10003A	1715981	1716214	234	246	ataagacggacggtcaaacg	atcttgtgtgggctttgtgg	AC025098
	S10013A	5180767	5180949	183	170	agtcgggtcatttcttagcc	ctacgtcctccgtttcacaa	AC083944
	S10019	10299169	10299319	151	163	atgcatctacatggcatttg	gatgctgagatgcgattgaa	AC123594
	S10026C	13594825	13595071	247	227	tacgtgtccttgtgcctgaa	tttcaccccactgtaaagg	AC021893
	S10071	20926684	20926850	167	158	tatggtcaaccctggaaac	cgtgctagtttgttcactgga	AC051633
	S10072	21129266	21129444	179	203	tgagtgttgcgttgtcttcc	tggtaaggcctgaagatgg	AC020666
11	S11004A	1081615	1081787	173	157	tctctggccttcactcatgg	ttgtgtttctacttggactctttt	AC136970
	S11006A	1270331	1270591	261	248	atgcgccgtccaacttatac	tggtcaaagaattgaacaac	AC123525
	S11028	5463772	5463946	175	186	attccctggggtagctaga	atgggtgaattgcagagaat	AC123523

Supplementary table 1

12	S12011B	1884649	1884804	156	177	tggggggagttctgaaatctg	ttaagttcggtgcccataa	AL935154
	S12030	3843516	3843732	217	235	tccacatgtaaaccgctgaa	tgagtgatataacaacacacaacca	AL732376
	S12109B	27415607	27415770	164	173	ggactcggtaaccgcatta	ggaacgcagcgaaagaat	AL732378

1) BAC clone information is available at International Rice Genome Sequencing Project (IRGSP).

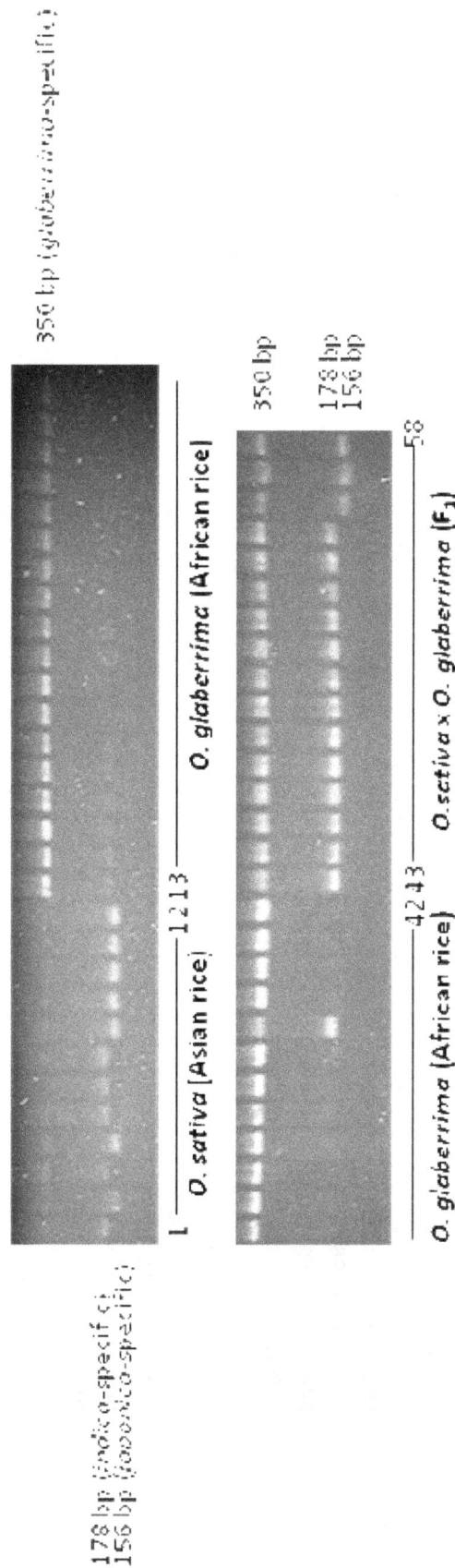

Figure 1. Genotyping of O. glaberrima accessions by subspecies-specific (SS) STS markers.
(a) 1-12: O. sativa (indica allele: 178 bp and japonica allele: 156 bp); 13-42: O.glaberrima (350 bp non-O. sativa allele). (b) 43-58: Showed both the alleles of sativa and glaberrima, but the band size of F1's were different, implying the amplified regions of O. sativa and O. glaberrima were neighbored.

Whole genome analysis has revealed the structural similarities between O. sativa and O. glaberrima species (Park et al., 2003), but due to the high frequency of polymorphism in subspecies-specific (SS) loci, it is assumed that there is a relationship between sativa × glaberrima F1 hybrid sterility and SS loci. O. glaberrima had indica allele at two loci associated with inter-subspecific F₁ hybrid sterility on chromosome 5 and chromosome 8 (Figure 3).

Application of SS-STS markers to study the relations within O. glaberrima species

The SS STSs markers used in the study were able to revealed different allele both from japonica and indica sources among the O. glaberrima accessions. The efficiency of STS markers to determine relations within the AA genome species is well documented. Robeniol et al. (1996) using only 14 STS markers accurately determined the genome composition of O. meridionalis as an AA genome species and most distantly related species to O. sativa, and O. longistaminata the second most distantly related.

The SS markers which detected glaberrima-specific allele suggest that loci adjacent to these markers could be a key for interspecific hybrid sterility. It may be interesting to compare these SS markers with other allele of other wild rice species. The detection of heterozygous allele

Table 2. Subspecies-specific STS markers generating *glaberrima* (G)-specific alleles.

Name of varieties/lines	S01160	S02085	S02140	S03145	S09093A	S10003A	S10013A	S11004A	S12011B
IR55423-01	J[1)]	—	—	J	J	J	—	J	J
IR60080-46-A	—	J	—	J	J	J	—	—	—
IR64	—	J	—	—	—	—	—	—	J
IR68703-AC-24-1	—	J	—	J	J	J	—	—	—
IR69502-6-SRN	—	—	—	—	—	—	—	—	—
PSBRC18	J	J	—	J	J	J	—	J	—
PSBRC82	—	J	—	J	J	J	—	—	J
Hwacheongbyeo	J	J	—	J	J	J	J	J	—
Ilpumbyeo	J	J	—	J	J	J	J	J	J
Jinmibyeo	J	J	—	J	J	J	J	J	J
Junambyeo	J	J	—	—	J	J	J	J	J
TR22183	J	G	—	G+J	G	G	G	G	G
RAM3	G	G	—	G+I	G	G	G	G	G
RAM24	G	G	—	G+I	G	G	G	G	G
RAM54	G	G	—	G+I	G	G	G	G	G
RAM86	G	G	—	G+I	G	G	G	G	G
RAM90	G	G	—	G+I	G	G	G	G	G
RAM111	G	G	—	G+I	G	G	G	G	G
RAM118	G	G	—	G+I	G	G	G	G	G
RAM120	G	G	—	G+I	G	G	G	G	G
RAM121	G	G	—	G+I	G	G	G	G	G
RAM131	G	G	—	G+I	G	G	G	G	G
RAM134	G	G	—	G+I	G	G	G	G	G
RAM152	G	G	—	G+I	G	G	G	G	G
RAM163	G	G	—	G+I	G	G	G	G	G
IG10	G	G	—	G+I	G	G	G	G	G
CG14	G	G	—	G+I	G	G	G	G	G
CG17	G	G	—	G+I	G	G	G	G	G
CG20	J	G	—	G+I	G	G	G	G	G
Acc.103477	G	G	—	G+I	G	G	G	G	G
TOG5674	—	G+I	G	G+I	G	G	G	G	G
TOG5681	—	G+I	—	G+I	G	G	G	G	G
TOG5860	—	G	—	G+I	G	G	G	G	G
TOG6472	—	G	—	G+I	G	G	G	G	G
TOG6508	G+I	G	—	G+I	G	G	G	G	G
TOG6589	G+I	G	—	G+I	G	G	G	G	G

Table 2 Cont.

	176	NULL	270 and 290	240	300	160	610	350
TOG6597	G+I	NULL	-	G	G	G	G	G
TOG6629	G+I	NULL	-	-	G+I	G+I	J	G+I
TOG6631	-	G	G+I	G	G	G	G	G
TOG7235	-	G	G+I	G	G	G	G	G
TOG7291	-	G	G+I	G	G	G	G	G
TOG7442	-	G	G+I	G	G	G	G	G
IR64 × RAM54	G+I	-	G+I	-	G+I	G+I	G+I	G+I
IR64 × RAM86	G+I	-	G+I	-	G+I	G+I	G+I	G+I
IR64 × RAM90	G+I	-	G+I	-	G+I	G+I	G+I	G+I
IR64 × RAM120	G+I	-	G+I	-	G+I	G+I	G+I	G+I
IR64 × RAM134	G+I	-	G+I	-	G+I	G+I	G+I	G+I
IR64 × RAM131	G+I	-	-	-	G+I	G+I	G+I	G+I
PSBRC18 × RAM111	G+J	-	G+I	-	G+I	G+I	G+I	G+I
IR55423-01 × RAM3	G+J	-	G+I	-	G+I	G+I	G+J	G+I
IR55423-01 × RAM24	G+J	-	G+I	-	G+I	G+I	G+J	G+I
IR55423-01 × RAM163	G+J	-	G+I	-	G+I	G+I	G+J	G+I
IR69502 × RAM118	G+I	-	G+I	-	G+I	G+I	G+I	G+I
IR69502 × RAM121	G+I	-	G+I	-	G+I	G+I	G+I	G+I
IR69502 × RAM163	G+I	-	G+I	-	G+I	G+I	G+I	G+I
IR60080-46-A × IG10	G+I	H	G+J	G+J	G+J	G+J	G+J	G+J
IR68703-AC-24-1 × CG14	I	H	G+J	G+J	G+I	G+I	G+I	G+J
IR60080-46-A × CG14	I	H	G+J	G+J	G+J	G+J	G+J	G+J
Alleles in glaberrima[2] (bp)	176	NULL	270 and 290	240	300	160	610	350

J: japonica-specific allele, I: indica-specific allele, G: alleles-present in O.glaberrima accessions, G+I: alleles in glaberrima and indica allele, G+J: alleles in glaberrima and japonica allele. Estimated size of alleles only found in O. sativa not in O. glaberrima in basepair (bp). 'NULL' represents no amplification of indica-japonica alleles in O. glaberrima accessions.

between japonica/glaberrima and between indica/glaberrima in the F1s, suggest caution when applying some SS markers to other rice species and implying their distinguished association to O. glaberrima genome. Only 40 (59.7%) of the SS markers might be useful in the O. glaberrima analysis, as other markers did not detect any amplification of heterozygous allele in F1 progenies between O. sativa × O. glaberrima.

This might be due to minute genomic differences between O. sativa and O. glaberrima (Ohmido and Fukui, 1995; Park et al., 2003). Also, some markers did not generate heterozygous allele in the F1's, suggesting that those loci are unique in O. sativa or O. glaberrima (S01160, S02052, S03099, S07048, S07050C, S08090, S08107, S09000A and S09058). Minor difference in the sequence of some markers might have affected

the recombination in PCR amplicon region of some markers such as S02054, S02081, S04060, and S08106), and are allelic-specific (S03020 and S08106). Some small cryptic changes and mutations in PCR amplification region of some markers might have caused some markers not to align well during PCR amplification, that is, S03048. O. glaberrima had indica allele at two loci associated with intersubspecific F1 hybrid sterility

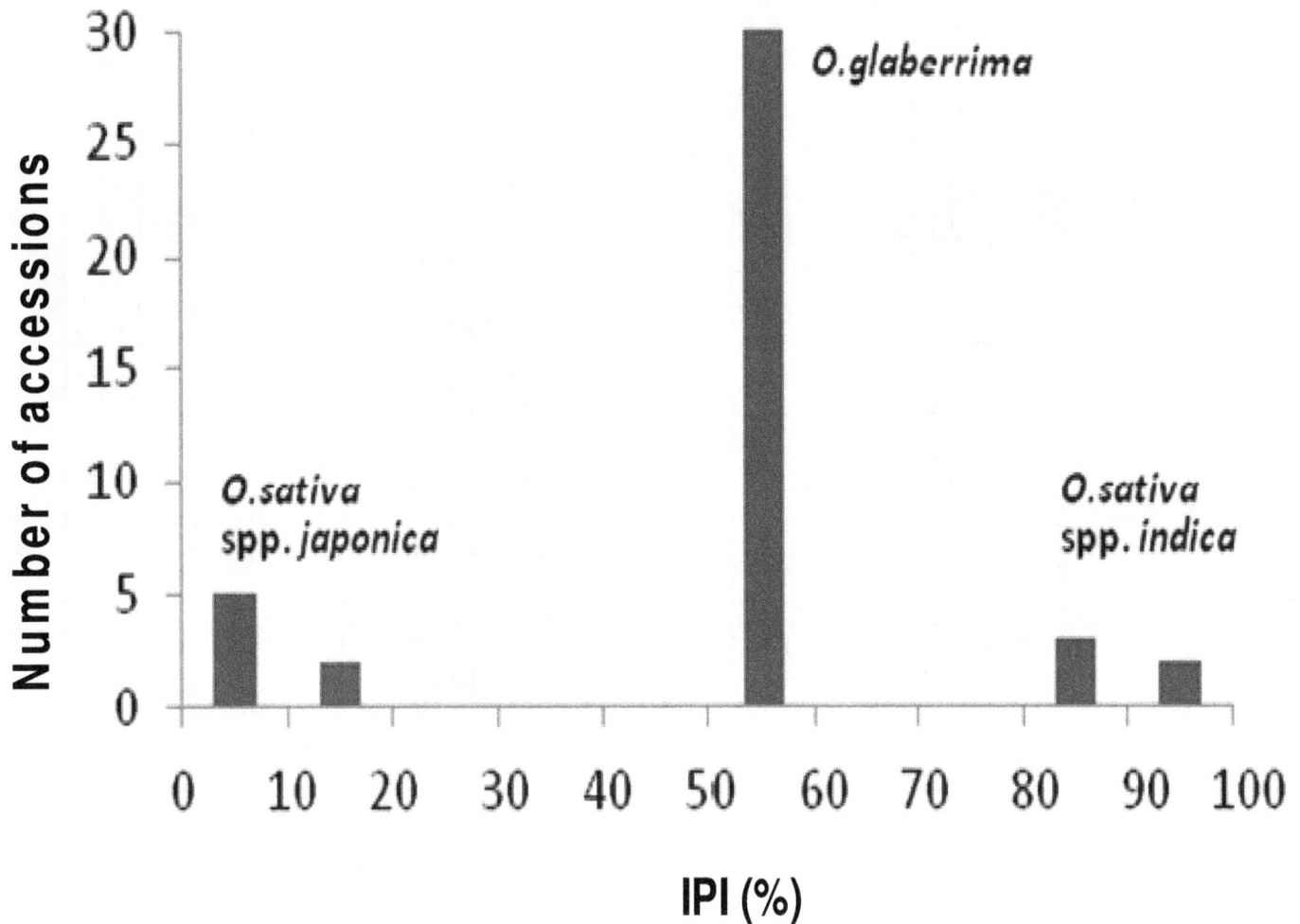

Figure 2. IPI (*Indica*-prototype index: Similarity to *indica* varieties of *O. sativa*) of 30 *O. glaberrima* accessions.

on chromosome 5 and chromosome 8. Interspecific hybrids (F₁'s) between *O. sativa* and *O. glaberrima* are almost completely sterile. This hybrid sterility barrier is mainly caused by an arrest of pollen development at the microspore stage (Heuer et al., 2003; Peltier, 1953). Intersubspecific F_1 hybrid sterility is mainly caused by cryptic chromosomal aberrations and allelic interaction between *indica* and *japonica* (Chin et al., 2007). The SS (Subspecies-specific) STS marker were able to classify the *O. glaberrima accessions* into 10 sub-groups. Subspecies-prototype index (SPI) of *O. glaberrima* accessions ranged from 51.67 to 60.00, suggesting intermediate subspecific type based on whole-genome.

Comparative view of genome of *O. glaberrima* based on *O. sativa* spp. *japonica* genome

A total of 23 and 22 loci showed only indica and japonicas allele respectively whilst 4 loci showed both indica and japonica allele. Some non-sativa allele which were detected on chromosomes 1, 2, 3, 9, 10, 11, and 12

might be *O. glaberrima* specific allele (Figure 3) , The heterozygous allele of indica and *O. glaberrima* (G+I) identified on 3 loci on chromosomes 1, 2 and 3, suggests that non-sativa regions might be located on aligned BAC clones of *O. glaberrima*. This information can be useful in further studies involving the F_1 hybrid sterility between *O. sativa* and *O. glaberrima*.

Conclusion

The informative markers identified in this study might be very useful in studying the diversification of *O. glaberrima*, Loci adjacent to the SS markers which detected *glaberrima*-specific allele could be a key for interspecific hybrid sterility between *O. sativa* × *O. glaberrima*. The detection of heterozygous allele between *japonica* and *glaberrima* and between *indica* and *glaberrima* by some SS markers suggest that caution must be taken when applying some SS markers to other rice species and implying their distinguished association to *O. glaberrima* genome.

Supplementary Table 2. *O.sativa O.glaberrima* genotyping by 68 SS markers and SPI (Subspecies-prototype index).

Description	Entry No.	S01022	S01054	S01140	S01157B	S01160	S02026	S02052	S02054	S02057B	S02081B	S02085	S02140	S03010B	S03020	S03027	S03041	S03046
IR55423-01	1	I	I	I	I	J	I	I	I	I	I	J	I	I	I	I	I	I
IR60080-46-A	2	I	I	I	I	I	I	I	I	I	I	I	I	I	I	I	I	I
IR64	3	I	I	I	I	I	I	I	I	I	I	I	I	I	I	I	I	I
IR68703-AC-24-1	4	J	J	J	J	I	I	I	J	J	J	I	I	J	J	J	I	J
IR69502-6-SRN	5	I	I	I	I	I	I	I	I	I	I	I	I	I	I	I	I	I
PSBRC18	6	I	I	I	I	I	I	I	I	I	I	I	I	I	I	I	I	I
PSBRC82	7	I	I	I	I	I	I	I	I	P	I	I	I	I	I	I	I	I
Hwacheongbyeo	8	J	J	J	J	J	J	J	J	J	J	J	J	J	J	J	J	J
Ilpumbyeo	9	J	J	-	J	J	J	J	J	J	J	J	J	J	J	J	J	J
Jinmibyeo	10	J	J	J	J	J	J	J	J	J	J	J	J	J	J	J	J	J
Junambyeo	11	J	J	J	J	J	J	J	J	J	J	J	J	J	J	J	J	J
TR22183	12	J	J	J	J	J	J	J	J	J	J	J	J	J	J	J	J	J
RAM3	13	J	J	J	J	G	I	I	J	J	J	G	I	J	J	I	I	J
RAM24	14	J	J	J	J	G	I	I	J	J	J	G	I	J	J	I	I	J
RAM54	15	J	J	J	J	G	I	I	J	J	J	G	I	J	J	I	I	J
RAM86	16	J	J	J	J	G	I	I	J	J	J	G	I	J	J	I	I	J
RAM90	17	J	J	J	J	G	I	I	J	J	J	G	I	J	J	I	I	J
RAM111	18	J	J	J	J	G	I	I	J	J	J	G	I	J	J	I	I	J
RAM118	19	J	J	J	J	G	I	I	J	J	J	G	I	J	J	I	I	J
RAM120	20	J	J	J	J	G	I	I	J	J	J	G	I	J	J	I	I	J
RAM121	21	J	J	J	J	G	I	I	J	J	J	G	I	J	J	I	I	J
RAM131	22	J	J	J	J	G	I	I	J	J	J	G	I	J	J	I	I	J
RAM134	23	J	J	J	J	G	I	I	J	J	J	G	I	J	J	I	I	J
RAM152	24	J	J	J	J	G	I	I	J	J	J	G	I	J	J	I	I	J
RAM163	25	J	J	J	J	G	I	I	J	J	J	G	I	J	J	I	I	J
IG10	26	J	J	J	J	G	I	I	J	J	J	G	I	J	J	I	I	J
CG14	27	J	J	J	J	G	I	I	J	J	J	G	I	J	J	I	I	J
CG17	28	J	J	J	-	G	I	I	J	J	J	G	I	J	J	I	I	J
CG20	29	J	J	J	J	J	I	I	J	J	J	G	I	J	J	I	I	J
Acc.103477	30	J	J	J	J	G	I	I	J	J	J	G	I	J	J	I	I	J
TOG5674	31	J	J	J	J	I	I	I	J	J	J	G+I	G	I	I	I	I	I
TOG5681	32	J	H	J	H	I	I	I	J	J	J	G+H	I	I	I	I	I	H
TOG5860	33	H	J	J	H	I	I	I	J	J	-	G	I	I	I	I	I	H
TOG6472	34	J	J	J	J	G+I	I	I	J	J	J	G	I	I	I	I	I	H
TOG6508	35	J	J	J	J	G+I	I	I	J	J	J	G	I	I	I	I	I	H
TOG6589	36	J	J	J	J	G+I	I	I	J	J	J	G	I	J	J	I	I	H
TOG6597	37	J	H	J	H	G+I	I	I	J	J	J	G	I	J	J	I	I	H
TOG6629	38	H	H	J	H	G+I	I	I	H	J	H	G+I	I	I	H	I	-	H
TOG6631	39	J	H	J	H	G+J	I	I	J	J	J	G+I	I	J	J	I	I	H
TOG7235	40	J	J	-	-	G+J	I	I	J	J	J	G	I	J	J	I	I	H
TOG7291	41	J	J	J	H	I	I	I	J	J	J	G	G	I	J	I	I	H
TOG7442	42	J	J	J	J	I	I	I	J	J	J	G	I	J	J	I	I	H
IR64 x RAM54	43	H	H	I	H	G+I	I	I	H	H	H	G+I	I	I	H	I	I	H
IR64 x RAM86	44	H	H	I	H	G+I	I	I	H	H	H	G+I	I	I	H	I	I	H
IR64 x RAM90	45	H	H	I	H	G+I	I	I	H	H	H	G+I	I	I	H	I	I	H
IR64 x RAM120	46	H	H	I	H	G+I	I	I	H	H	H	G+I	I	I	H	I	I	H
IR64 x RAM134	47	H	H	I	H	G+I	I	I	H	H	H	G+I	I	I	H	I	I	H
IR64 x RAM131	48	H	H	I	H	G+I	I	I	H	H	H	G+I	I	I	H	I	I	H
PSBRC18 x RAM111	49	H	H	I	H	G+I	I	I	H	H	H	G+I	I	I	H	I	I	H
IR55423-01 x RAM3	50	H	H	I	H	G+J	I	I	H	H	H	G+I	I	I	H	I	-	H
IR55423-01 x RAM24	51	H	H	I	H	G+J	I	I	H	H	H	G+I	I	I	H	I	I	H
IR55423-01 x RAM163	52	H	J	J	H	G+J	I	I	H	H	H	G+I	I	I	H	I	I	H
IR69502 x RAM118	53	H	H	I	H	G+I	I	I	H	H	H	G+I	I	I	H	I	I	H
IR69502 x RAM121	54	H	H	I	H	G+I	I	I	H	H	H	G+I	I	I	H	I	I	H
IR69502 x RAM163	55	H	H	I	H	G+I	I	H	H	H	H	G+I	I	I	H	I	I	H
IR60080-46-A x IG10	56	J	J	H	H	G+I	-	H	H	H	H	G+J	H	H	J	I	H	H
IR68703-AC-24-1 x CG14	57	J	J	H	H	I	-	I	J	H	J	G+J	H	H	J	H	H	H
IR60080-46-A x CG14	58	J	J	H	H	I	I	I	H	H	H	G+J	H	H	J	H	H	J
SPECIFIC ALLELE (NEITHER I NOR J)		J				G=176			I=H	P=260		NULL	NULL	I=H				

Supplementary Table 2 Cont.

Entry No.	Description	S03048	S03096	S03099	S03115	S03120	S03136	S03145	S04058	S04060	S04077A	S04077B	S04087A	S04087B	S04097B	S04128	S04129B	S05064	S05080A	S06001	S07011	S07048	S07059C	S07059A	S07101	S07103	S08066
1	IR55423-01	I	I	I	I	I	I	I	I	I	I	I	I	I	I	I	I	I	I	I	I	I	I	I	I	I	I
2	IR60080-46-A	I	I	I	I	I	I	I	I	I	I	I	I	I	I	I	I	I	I	I	I	I	I	I	I	I	I
3	IR64	I	I	I	I	I	I	I	I	I	I	I	I	I	I	I	I	I	I	I	I	I	I	I	I	I	I
4	IR68703-AC-24-1	I	I	I	I	I	I	I	I	I	I	I	I	I	I	I	I	I	I	I	I	I	I	I	I	I	I
5	IR69502-6-SRN	I	I	I	I	I	I	I	I	I	I	I	I	I	I	I	I	I	I	I	I	I	I	I	I	I	I
6	PSBRC18	I	I	I	I	I	I	I	I	I	I	I	I	I	I	I	I	I	I	I	I	I	I	I	I	I	I
7	PSBRC82	I	I	I	J	I	I	I	I	I	I	I	I	I	I	I	I	I	I	I	I	I	I	I	I	I	I
8	Hwacheongbyeo	J	J	J	J	J	J	J	J	J	J	J	J	J	J	J	J	J	J	J	J	J	J	J	J	J	J
9	Ilpumbyeo	J	J	J	J	J	J	J	J	J	J	J	J	J	J	J	J	J	J	J	J	J	J	J	J	J	J
10	Jinmibyeo	J	J	J	J	J	J	J	J	J	J	J	J	J	J	J	J	J	J	J	J	J	J	J	J	J	J
11	Junambyeo	J	J	J	J	J	J	J	J	J	J	J	J	J	J	J	J	J	J	J	J	J	J	J	J	J	J
12	TR22183	G+J	I	I	I	I	I	G+J	I	I	I	I	I	I	I	I	I	I	I	I	I	I	I	I	I	I	I
13	RAM3	J	I	I	J	J	J	G+J	I	I	I	I	I	I	I	I	I	I	I	J	J	J	I	J	J	J	I
14	RAM24	J	I	I	J	J	J	G+J	I	I	I	I	I	I	I	I	I	I	I	J	J	J	I	J	J	J	I
15	RAM54	J	I	I	J	J	J	G+J	I	I	I	I	I	I	I	I	I	I	I	J	J	J	I	J	J	J	I
16	RAM86	J	I	I	J	J	J	G+J	I	I	I	I	I	I	I	I	I	I	I	J	J	J	I	J	J	J	I
17	RAM90	J	I	I	J	J	J	G+J	I	I	I	I	I	I	I	I	I	I	I	J	J	J	I	J	J	J	I
18	RAM111	J	I	I	J	J	J	G+J	I	I	I	I	I	I	I	I	I	I	I	J	J	J	I	J	J	J	I
19	RAM118	J	I	I	J	J	J	G+J	I	I	I	I	I	I	I	I	I	I	I	J	J	J	I	J	J	J	I
20	RAM120	J	I	I	J	J	J	G+J	I	I	I	I	I	I	I	I	I	I	I	J	J	J	I	J	J	J	I
21	RAM121	J	I	I	J	J	J	G+J	I	I	I	I	I	I	I	I	I	I	I	J	J	J	I	J	J	J	I
22	RAM131	J	I	I	J	J	J	SMEAR	I	I	I	I	I	I	I	I	I	I	I	J	J	J	I	J	J	J	I
23	RAM134	J	I	I	J	J	J	G+J	I	I	I	I	I	I	I	I	I	I	I	J	J	J	I	J	J	J	I
24	RAM152	J	I	I	J	J	J	G+J	I	I	I	I	I	I	I	I	I	I	I	J	J	J	I	J	J	J	I
25	RAM163	J	I	I	J	J	J	G+J	I	I	I	I	I	I	I	I	I	I	I	J	J	J	I	J	J	J	I
26	IG10	J	I	I	J	J	J	G+J	I	I	I	I	I	I	I	I	I	I	I	J	J	J	I	J	J	J	I
27	CG14	J	I	I	J	J	J	G+J	I	I	I	I	I	I	I	I	I	I	I	J	J	J	I	J	J	J	I
28	CG17	J	I	I	J	J	J	G+J	I	I	I	I	I	I	I	I	I	I	I	J	J	J	I	J	J	J	I
29	CG20	J	I	I	J	J	J	G+J	I	I	I	I	I	I	I	I	I	I	I	J	J	J	I	J	J	J	I
30	Acc.103477	J	I	I	J	J	J	G+J	I	I	I	I	I	I	I	I	I	I	I	J	J	J	I	J	J	J	I
31	TOG5674	J	I	I	J	J	J	G+J	I	I	I	I	I	I	I	I	I	I	I	J	J	J	I	J	J	J	I
32	TOG5681	J	I	I	J	J	J	G+J	I	I	I	I	I	I	I	I	I	I	I	J	J	J	I	J	J	J	I
33	TOG5860	J	I	I	J	J	J	G+J	I	I	I	I	I	I	I	I	I	I	I	J	J	J	I	J	J	J	I
34	TOG6472	J	I	I	J	J	J	G+J	I	I	I	I	I	I	I	I	I	I	I	J	J	J	I	J	J	J	I
35	TOG6508	J	I	I	J	J	J	G+J	I	I	I	I	I	I	I	I	I	I	I	J	J	J	I	J	J	J	I
36	TOG6589	J	I	I	J	J	J	G+J	I	I	I	I	I	I	I	I	I	I	I	J	J	J	I	J	J	J	I
37	TOG6597	J	I	I	J	J	J	I	I	I	I	I	I	I	I	I	I	I	I	J	J	J	I	J	J	J	I
38	TOG6629	H	I	I	H	H	H	G+J	I	H	I	H	I	H	H	H	H	H	I	H	H	H	H	H	H	H	I
39	TOG6631	J	I	I	I	J	J	G+J	I	H	I	I	I	I	I	I	I	I	I	I	I	I	I	I	I	I	I
40	TOG7235	J	I	I	I	J	J	G+J	I	I	I	I	I	I	I	I	I	I	I	I	I	I	I	I	I	I	I
41	TOG7291	J	I	I	I	J	J	G+J	I	I	I	I	I	I	I	I	I	I	I	I	I	I	I	I	I	I	I
42	TOG7442	J	I	I	J	J	J	G+J	I	I	I	I	I	I	I	I	I	I	I	J	J	J	I	J	J	J	I
43	IR64 x RAM54	H	I	I	H	H	H	G+J	I	H	I	H	H	H	H	H	H	H	I	H	H	H	H	H	H	H	I
44	IR64 x RAM86	H	I	H	I	I	I	G+J	I	H	I	H	H	H	I	I	I	H	I	H	H	I	I	H	H	H	I
45	IR64 x RAM90	H	I	H	H	J	J	G+J	I	H	I	H	H	H	I	I	H	H	I	H	H	I	I	H	H	H	I
46	IR64 x RAM120	H	I	H	H	H	H	G+J	I	H	I	H	H	H	I	I	H	H	I	H	H	I	I	H	H	H	I
47	IR64 x RAM134	H	I	I	H	H	H	G+J	I	H	I	H	H	H	I	I	H	H	I	H	H	I	I	H	H	H	I
48	IR64 x RAM131	H	I	I	J	J	J	I	I	H	I	H	H	I	I	I	I	H	I	H	I	I	I	I	H	H	I
49	PSBRC18 x RAM111	H	I	I	H	H	H	G+J	I	H	I	H	H	H	H	H	H	H	I	H	H	H	H	H	H	H	I
50	IR55423-01 x RAM3	J	I	I	H	H	J	G+J	I	H	I	H	H	H	H	H	H	H	I	H	H	I	I	H	H	H	I
51	IR55423-01 x RAM24	J	I	I	H	H	J	G+J	I	H	I	H	H	H	H	H	H	H	I	H	H	I	I	H	H	H	I
52	IR55423-01 x RAM163	J	I	I	I	J	J	G+J	I	H	I	I	I	I	I	I	I	H	I	I	H	I	I	H	H	H	I
53	IR69502 x RAM118	H	I	I	I	J	J	G+J	G	H	I	I	I	I	I	I	I	H	I	H	H	I	I	H	H	H	I
54	IR69502 x RAM121	H	I	I	I	J	J	G+J	G	H	I	I	I	I	I	I	I	H	I	H	H	I	I	H	H	H	I
55	IR69502 x RAM163	H	I	I	J	J	J	G+J	I	H	I	H	H	H	I	I	I	H	I	I	H	I	I	H	H	J	I
56	IR60080-46-A x IG10	H	H	H	J	J	H	SMEAR	SMEAR	H	H	H	H	H	H	H	H	H	H	H	I	I	I	I	I	I	H
57	IR68703-AC-24-1 x CG14	G+J	I	H	J	J	H	SMEAR	SMEAR	H	H	H	H	H	H	H	H	H	H	H	H	H	H	H	H	H	H
58	IR60080-46-A x CG14	G+J	I	H	J	J	H	SMEAR	SMEAR	H	H	H	H	H	H	H	H	H	H	J	J	J	H	J	J	J	H
	SPECIFIC ALLELE (NEITHER I NOR J)	HETERO G						G=270,290	HETERO G																		

Supplementary Table 2 Cont.

Entry No.	Description	I	J	–	G	H	G+I	G+J	SPJ/(I+J+H)
1	IRS5423-01	63	3	2	0	0	0	0	95.45
2	IR60080-46-A	8	60	0	0	0	0	0	11.76
3	IR64	68	0	0	0	0	0	0	100.00
4	IR68703-AC-24-1	9	59	0	0	0	0	0	13.24
5	IR69502-6-SRN	67	1	0	0	0	0	0	98.53
6	PSBRC18	66	2	0	0	0	0	0	97.06
7	PSBRC82	67	0	1	0	0	0	0	100.00
8	Hwacheongbyeo	1	67	0	0	0	0	0	1.47
9	Ilpumbyeo	0	67	1	0	0	0	0	0.00
10	Jinmibyeo	0	67	1	0	0	0	0	0.00
11	Junambyeo	2	66	0	0	0	0	0	2.94
12	TR22183	4	62	1	0	0	0	1	6.06
13	RAM3	35	25	0	7	0	0	1	58.33
14	RAM24	35	25	0	7	0	0	1	58.33
15	RAM54	35	25	0	7	0	0	1	58.33
16	RAM86	36	24	0	7	0	0	1	60.00
17	RAM90	36	24	0	7	0	0	1	60.00
18	RAM111	36	24	0	7	0	0	1	60.00
19	RAM118	36	24	0	7	0	0	1	60.00
20	RAM120	36	24	0	7	0	0	1	60.00
21	RAM121	36	24	0	7	0	0	1	60.00
22	RAM131	36	24	0	7	0	0	1	60.00
23	RAM134	36	24	0	7	0	0	1	60.00
24	RAM152	36	24	0	7	0	0	1	60.00
25	RAM163	36	24	0	7	0	0	1	60.00
26	IG10	36	24	2	7	0	0	1	58.62
27	CGI4	35	25	0	7	0	0	1	58.33
28	CG17	36	24	0	6	0	0	1	60.00
29	CG20	36	25	0	6	0	0	1	59.02
30	Acc.103477	35	25	1	6	0	0	1	57.63
31	TOG5674	34	25	1	6	0	2	0	57.63
32	TOG5681	35	25	1	5	0	2	0	58.33
33	TOG5860	35	23	2	6	1	2	0	59.32
34	TOG6472	35	25	1	6	0	1	0	58.33
35	TOG6508	34	25	2	6	0	1	0	57.63
36	TOG6589	34	24	2	6	0	2	0	58.62
37	TOG6597	33	22	5	6	1	2	0	58.93
38	TOG6629	31	2	3	1	27	5	0	51.67
39	TOG6631	32	0	3	1	28	5	2	59.02
40	TOG7235	36	24	0	6	1	1	0	60.00
41	TOG7291	33	0	0	6	28	5	2	58.33
42	TOG7442	32	1	1	7	27	7	0	59.02
43	IR64 x RAM54	33	0	0	6	28	7	0	
44	IR64 x RAM86	34	0	0	7	27	7	0	
45	IR64 x RAM90	33	0	1	6	27	7	0	
46	IR64 x RAM120	34	0	0	7	27	7	0	
47	IR64 x RAM134	34	0	0	7	27	7	0	
48	IR64 x RAM131	33	0	2	6	27	7	0	
49	PSBRC18 x RAM111	31	0	2	7	28	6	1	
50	IRS5423-01 x RAM3	32	1	0	6	28	5	2	
51	IRS5423-01 x RAM24	34	0	0	6	28	5	2	
52	IRS5423-01 x RAM163	33	0	0	6	28	5	2	
53	IR69502 x RAM118	32	1	1	7	27	7	0	
54	IR69502 x RAM121	32	1	0	7	27	7	0	
55	IR69502 x RAM163	7	22	1	0	27	2	6	
56	IR60080-46-A x IG10	7	23	0	0	29	2	5	
57	IR68703-AC-24-1 x CGI0	5	23	1	0	29	2	5	
58	IR60080-46-A x CG14	7	21	0	0	29	0	7	

SPECIFIC ALLELE (NEITHER I NOR J) · G=240 · G=300 · G=160 · G=300 · G=610 · G=350

I=indica allele
J=japonica allele
G=glaberrima allele
NULL=non-amplified allele of glaberrima
P=new allele of TR22183 (O.sativa)

Supplementary Table 3. Allele constitution of parents and F$_1$ progenies to elucidate useful polymorphic markers in *O.sativa* x *O.glaberrima* breeding program

Subspecies [1]	Entry No.	Description	S01022	S01054	S01140	S01157B	S01160	S02026	S02052	S02054	S02057B	S02081B	S02085	S02140	S03010B	S03020	S03027	S03041	S03046	S03048	S03096	S03099	S03120	S03136	S03145	S04058	S04060	S04077A	S04077B	S04087A	S04097B	S04128
indica x glaberrima	3	IR64	J	J	J	J	J	J	J	J	J	J	J	J	J	J	J	J	J	J	J	J	J	J	J	J	J	J	J	J	J	J
	15	RAM54	H	H	J	H	G	J	J	J	J	J	G	J	J	J	J	J	J	J	J	J	J	J	G+J	J	J	J	J	J	J	J
	43	IR64 x RAM54	H	H	J	H	G+J	J	J	H	H	H	G+J	J	J	H	J	J	H	H	J	J	H	J	G+J	J	H	J	J	J	J	J
	3	IR64	J	J	J	J	J	J	J	J	J	J	J	J	J	J	J	J	J	J	J	J	J	J	J	J	J	J	J	J	J	J
	16	RAM86	J	J	J	J	G	J	J	J	J	J	G	J	J	J	J	J	J	J	J	J	J	J	G+J	J	J	J	J	J	J	J
	44	IR64 x RAM86	H	H	J	H	G+J	J	J	H	H	H	G+J	J	J	H	J	J	H	H	J	J	H	J	G+J	J	H	J	J	J	J	J
	3	IR64	J	J	J	J	J	J	J	J	J	J	J	J	J	J	J	J	J	J	J	J	J	J	-	J	J	J	J	J	J	J
	17	RAM90	J	J	J	J	G	J	J	J	J	J	G	J	J	J	J	J	J	J	J	J	J	J	G+J	J	J	J	J	J	J	J
	45	IR64 x RAM90	H	H	J	H	G+J	J	J	H	H	H	G+J	J	J	H	J	J	H	H	J	J	H	J	G+J	J	H	J	J	J	J	J
	3	IR64	J	J	J	J	J	J	J	J	J	J	J	J	J	J	J	J	J	J	J	J	J	J	J	J	J	J	J	J	J	J
	20	RAM120	J	J	J	J	G	J	J	J	J	J	G	J	J	J	J	J	J	J	J	J	J	J	G+J	J	J	J	J	J	J	J
	46	IR64 x RAM120	H	H	J	H	G+J	J	J	H	H	H	G+J	J	J	H	J	J	H	H	J	J	H	J	G+J	J	H	J	J	J	J	J
	3	IR64	J	J	J	J	J	J	J	J	J	J	J	J	J	J	J	J	J	J	J	J	J	J	J	J	J	J	J	J	J	J
	23	RAM134	J	J	J	J	G	J	J	J	J	J	G	J	J	J	J	J	J	J	J	J	J	J	G+J	J	J	J	J	J	J	J
	47	IR64 x RAM134	H	H	J	H	G+J	J	J	H	H	H	G+J	J	J	H	J	J	H	H	J	J	H	J	G+J	J	H	J	J	J	J	J
	3	IR64	J	J	J	J	J	J	J	J	J	J	J	J	J	J	J	J	J	J	J	J	J	J	J	J	J	J	J	J	J	J
	22	RAM131	J	J	J	J	G	J	J	J	J	J	G	J	J	J	J	J	J	J	J	J	J	J	-	J	J	J	J	J	J	J
	48	IR64 x RAM131	H	H	J	H	G+J	J	J	H	H	H	G+J	J	J	H	-	J	H	H	J	J	H	J	G+J	J	H	J	J	J	J	J
	6	PSBRC18	J	J	J	J	J	J	J	J	J	J	J	J	J	J	J	J	J	J	J	J	J	J	-	J	J	J	J	J	J	J
	18	RAM111	J	J	J	J	G	J	J	J	J	J	G	J	J	J	J	J	J	J	J	J	J	J	G+J	J	J	J	J	J	J	J
	49	PSBRC18 x RAM111	H	H	J	H	G+J	J	J	H	H	H	G+J	J	J	H	J	J	H	H	J	J	H	J	G+J	J	H	J	J	J	J	J
	13	IR55423-01	J	J	J	J	J	J	J	J	J	J	J	J	J	J	J	J	J	J	J	J	J	J	-	J	J	J	J	J	J	J
	13	RAM3	J	J	J	J	G	J	J	J	J	J	G	J	J	J	J	J	J	J	J	J	J	J	G+J	J	J	J	J	J	J	J
	50	IR55423-01 x RAM3	H	H	J	H	G+J	J	J	H	H	H	G+J	J	J	H	J	J	H	H	J	J	H	J	G+J	J	H	J	J	J	J	J
	1	IR55423-01	J	J	J	J	J	J	J	J	J	J	J	J	J	J	J	J	J	J	J	J	J	J	-	J	J	J	J	J	J	J
	14	RAM24	J	J	J	J	G	J	J	J	J	J	G	J	J	J	J	J	J	J	J	J	J	J	G+J	J	J	J	J	J	J	J
	51	IR55423-01 x RAM24	H	H	J	H	G+J	J	J	H	H	H	G+J	J	J	H	J	J	H	H	J	J	H	J	G+J	J	H	J	J	J	J	J
	1	IR55423-01	J	J	J	J	J	J	J	J	J	J	J	J	J	J	J	J	J	J	J	J	J	J	J	J	J	J	J	J	J	J
	25	RAM163	J	J	J	J	G	J	J	J	J	J	G	J	J	J	J	J	J	J	J	J	J	J	G+J	J	J	J	J	J	J	J
	52	IR55423-01 x RAM163	H	H	H	H	G+J	J	H	H	H	H	G+J	J	J	H	J	J	H	H	J	J	H	J	G+J	J	H	J	J	J	J	J
	5	IR69502-6-SRN	J	J	J	J	J	J	J	J	J	J	J	J	J	J	J	J	J	J	J	J	J	J	-	J	J	J	J	J	J	J
	19	RAM118	J	J	J	J	G	J	J	J	J	J	G	J	J	J	J	J	J	J	J	J	J	J	G+J	J	J	J	J	J	J	J
	53	IR69502 x RAM118	H	H	J	H	G+J	J	J	H	H	H	G+J	J	J	H	J	J	H	H	J	J	H	J	G+J	J	H	J	J	J	J	J
	5	IR69502-6-SRN	J	J	J	J	J	J	J	J	J	J	J	J	J	J	J	J	J	J	J	J	J	J	-	J	J	J	J	J	J	J
	21	RAM121	J	J	J	J	G	J	J	J	J	J	G	J	J	J	J	J	J	J	J	J	J	J	G+J	J	J	J	J	J	J	J
	54	IR69502 x RAM121	H	H	J	H	G+J	-	J	H	H	H	G+J	J	J	H	J	J	H	H	J	J	H	J	G+J	J	-	J	J	J	J	J
	5	IR69502-6-SRN	J	J	J	J	J	J	J	J	J	J	J	J	J	J	J	J	J	J	J	J	J	J	-	J	J	J	J	J	J	J
	25	RAM163	J	J	J	J	G	J	J	J	J	J	G	J	J	J	J	J	J	J	J	J	J	J	G+J	J	J	J	J	J	J	J
	55	IR69502 x RAM163	H	H	H	H	G+J	J	J	H	H	H	G+J	J	J	H	J	J	H	H	J	J	H	J	G+J	J	H	J	J	J	J	J
japonica-like x glaberrima	2	IR60080-46-A	J	J	J	J	J	J	J	J	H	J	J	J	J	J	J	J	J	J	J	J	J	J	J	J	H	H	H	J	J	H
	26	IG10	J	J	H	J	G+J	-	H	J	J	J	G+J	H	J	J	H	H	H	G+J [2]	H [3]	H	J	H	sMEAR [4]	SMEAR	H	H	H	J	J	H
	56	IR60080-46-A x IG10	J	J	J	J	G	J	J	J	J	J	G	J	J	J	J	J	J	J	J	J	J	J	J		H	H	H	J	H	J
	4	IR68703-AC-24-1	J	J	H	H	J	J	H	J	J	J	G	J	J	J	H	J	J	J	J	J	J	J	J	SMEAR	H	H	H	H	H	H
	27	CG14	J	J	J	J	G	J	-	J	J	J	G+J	H	J	J	J	J	J	G+J	J	J	J	H	J	SMEAR	H	H	H	J	J	J
	57	IR68703-AC-24-1 x CG14	J	J	J	J	J	J	J	J	J	J	J	J	J	J	J	J	J	J	J	J	J	J	J	SMEAR	H	H	H	J	H	J
	2	IR60080-46-A	J	J	J	J	G	J	J	J	J	J	G	J	J	J	J	J	J	J	J	J	J	J	J	SMEAR	J	J	H	J	J	H
	27	CG14	J	J	J	J	J	J	-	J	J	J	G+J	J	J	J	H	J	J	G+J	J	J	J	H	J	SMEAR	H	H	H	H	H	H
	58	IR60080-46-A x CG14	J	J	H	H	J	J	J	H	H	H	G+J	H	J	J	H	J	H	J	J	J	J	H	SMEAR	SMEAR	H	H	H	J	H	H

Supplementary Table 3 Conts.

Subspecies [1]	Entry No.	Description	S04129B	S05064	S05080A	S06001	S07011	S07048	S07050C	S07050A	S07101	S07103	S08066	S08090	S08106	S08107	S09000A	S09026B	S09040B	S09058	S09062B	S09065	S09073	S09075A	S09075B	S09093A	S10001	S10003A	S10013A	S10019	S10026C	S10071	S10072	S11004A	S11006A	S11028	S12011B	S12030	S12109B
indica x glaberrima	3	IR64	I	I	I	I	I	I	I	I	I	I	I	I	I	I	I	I	I	I	I	I	I	I	I	I	I	I	I	I	I	I	I	I	I	I	I	I	I
	15	RAM54	I	I	I	I	J	I	I	J	I	I	J	I	J	J	J	J	I	I	J	J	I	G	I	G	G	I	J	J	J	G	J	I	G	J
	43	IR64 x RAM54	I	I	I	H	I	I	H	I	H	I	I	H	I	H	I	H	H	H	I		G+I	G+I	H	H	H	H	G+I	H	I	G+I	I	H
	3	IR64	I	I	I	I	I	I	I	I	I	I	I	I	I	I	I	I	I	I	I	I	I	I	I	I	I	I	I	I	I	I	I	I	I	I	I	I	I
	16	RAM86	I	I	I	I	J	I	I	J	I	I	J	I	J	J	J	J	I	I	J	J	I	G	I	G	G	I	J	J	J	G	J	I	G	J
	44	IR64 x RAM86	I	I	I	H	I	I	H	I	H	I	I	H	I	H	I	H	H	H	I		G+I	G+I	H	H	H	H	G+I	H	I	G+I	I	H
	3	IR64	I	I	I	I	I	I	I	I	I	I	I	I	I	I	I	I	I	I	I	I	I	I	I	I	I	I	I	I	I	I	I	I	I	I	I	I	I
	17	RAM90	I	I	I	I	I	I	I	I	I	I	J	I	J	J	J	J	I	I	J	J	I	G	I	G	G	I	J	J	J	G	J	I	G	J
	45	IR64 x RAM90	I	I	I	H	-	I	H	I	H	I	I	H	I	H	I	H	H	H	I		G+I	G+I	H	H	H	H	G+I	H	I	G+I	I	H
	3	IR64	I	I	I	I	I	I	I	I	I	I	I	I	I	I	I	I	I	I	I	I	I	I	I	I	I	I	I	I	I	I	I	I	I	I	I	I	I
	20	RAM120	I	I	I	I	J	I	J	J	I	I	J	I	J	J	J	J	I	I	J	J	I	G	I	G	G	I	J	J	J	G	J	I	G	J
	46	IR64 x RAM120	I	I	I	H	I	I	H	I	H	I	I	H	I	H	I	H	H	H	I		G+I	G+I	H	H	H	H	G+I	H	I	G+I	I	H
	3	IR64	I	I	I	I	I	I	I	I	I	I	I	I	I	I	I	I	I	I	I	I	I	I	I	I	I	I	I	I	I	I	I	I	I	I	I	I	I
	23	RAM134	I	I	I	I	I	I	I	I	J	I	I	J	I	I	I	I	I	I	J	J	I	G	I	G	G	I	J	J	J	G	J	I	G	J
	47	IR64 x RAM134	I	I	I	H	I	I	H	I	H	I	I	H	I	H	I	H	H	H	I		G+I	G+I	H	H	H	H	G+I	H	I	G+I	I	H
	3	IR64	I	I	I	I	I	I	I	I	I	I	I	I	I	I	I	I	I	I	I	I	I	I	I	I	I	I	I	I	I	I	I	I	I	I	I	I	I
	22	RAM131	I	I	I	I	J	I	I	I	I	I	J	I	J	J	J	J	I	I	J	J	I	G	I	G	G	I	J	J	J	G	J	I	G	J
	48	IR64 x RAM131	I	I	I	H	I	I	H	I	H	I	I	H	I	H	I	H	H	H	I		G+I	G+I	H	H	H	H	G+I	H	I	G+I	I	H
	6	PSBRC18	I	I	I	I	I	I	I	I	I	I	I	I	I	I	I	I	I	I	I	I	I	I	I	I	I	I	I	I	I	I	I	I	I	I	I	I	I
	18	RAM111	I	I	I	I	J	I	I	I	J	I	I	I	I	I	I	I	J	J	I	I	I	J	J	I	G	I	G	G	I	J	J	J	G	J	I	G	J
	49	PSBRC18 x RAM111	I	I	I	H	I	I	I	H	I	H	-	I	H	I	I	H	I	H	I	H	H	H	I		G+I	G+I	H	H	H	G+I	H	I	G+I
	1	IR55423-01	I	I	I	I	J	I	I	I	-	I	I	I	I	I	I	I	I	I	I	I	I	I	I	I	I	I	I	I	I	I	I	I	I	I	I	I	I
	13	RAM3	I	I	I	I	I	I	I	I	I	I	J	I	J	J	J	J	I	I	J	J	I	G	I	G	G	I	J	J	J	G	J	I	G	J
	50	IR55423-01 x RAM3	I	I	I	H	H	I	I	H	I	H	I	I	H	I	H	I	H	H	H	I		G+I	G+I	H	H	H	H	G+J	I	I	G+I	I	H
	1	IR55423-01	I	I	I	I	J	I	I	I	-	I	I	I	I	I	I	I	I	I	I	I	I	I	I	I	I	I	I	I	I	I	I	I	I	I	I	I	I
	14	RAM24	I	I	I	I	J	I	I	I	I	I	J	I	J	J	J	J	I	I	J	J	I	G	I	G	G	I	J	J	J	G	J	I	G	J
	51	IR55423-01 x RAM24	I	I	I	I	H	I	I	H	I	H	I	I	H	I	H	I	H	H	H	I		G+I	G+I	H	H	H	G+J	H	I	G+I	I	H
	1	IR55423-01	I	I	I	I	I	I	I	I	-	I	I	I	I	I	I	I	I	I	I	I	I	I	I	I	I	I	I	I	I	I	I	I	I	I	I	I	I
	25	RAM163	I	I	I	I	I	I	I	I	I	I	J	I	J	J	J	J	I	I	J	J	I	G	I	G	G	I	J	J	J	G	J	I	G	J
	52	IR55423-01 x RAM163	I	I	I	H	H	I	I	H	I	H	I	I	H	I	H	I	H	H	H	I		G+I	G+I	H	H	H	G+J	H	I	G+I	I	H
	5	IR69502-6-SRN	I	I	I	I	I	I	I	I	I	I	I	I	I	I	I	I	I	I	I	I	I	I	I	I	I	I	I	I	I	I	I	I	I	I	I	I	I
	19	RAM118	I	I	H	J	I	I	I	I	I	I	J	I	J	J	J	J	I	I	J	J	I	G	I	G	G	I	J	J	J	G	J	I	G	J
	53	IR69502 x RAM118	I	I	I	H	H	I	I	H	I	H	I	I	H	I	H	I	H	H	H	I		G+I	G+I	H	H	H	G+I	H	I	G+I	I	H
	5	IR69502-6-SRN	I	I	I	I	I	I	I	I	I	I	I	I	I	I	I	I	I	I	I	I	I	I	I	I	I	I	I	I	I	I	I	I	I	I	I	I	I
	21	RAM121	I	I	I	I	I	I	I	I	I	I	J	I	J	J	J	J	I	I	J	J	I	G	I	G	G	I	J	J	J	G	J	I	G	J
	54	IR69502 x RAM121	I	I	I	H	I	I	I	H	I	H	I	I	H	I	H	H	H	I	H	H	H	I		G+I	G+I	H	H	H	G+I	H	I	G+I	I	H
	5	IR69502-6-SRN	I	I	I	I	I	I	I	I	I	I	I	I	I	I	I	I	I	I	I	I	I	I	I	I	I	I	I	I	I	I	I	I	I	I	I	I	I
	25	RAM163	I	I	I	I	I	I	I	I	I	I	J	I	J	J	J	J	I	I	J	J	I	G	I	G	G	I	J	J	J	G	J	I	G	J
	55	IR69502 x RAM163	I	I	I	H	H	I	I	H	I	H	I	I	H	I	H	I	H	H	H	I		G+I	G+I	H	H	H	G+I	H	I	G+I	I	H
japonica-like x glaberrima	2	IR60080-46-A	I	J	J	J	J	J	J	J	J	J	J	J	J	J	J	J	J	J	J	J	J	J	J	J	J	J	J	I	I	J	J	J	J	J	J	J	J
	26	IG10	-	I	I	I	I	J	I	I	I	I	I	I	I	I	I	I	I	I	I	I	J	J	I	G	I	G	G	I	J	J	J	G	J	I	G	J
	56	IR60080-46-A x IG10	H	H	H	I	J	I	I	I	J	H	J	H	J	G+H	J	J	J	J	H	H	J	J	H	G+J	J	G+J	G+I	J	G+J	I	I	G+J	I	J
	4	IR68703-AC-24-1	I	J	J	J	J	J	J	J	J	J	I	J	J	J	J	J	J	J	J	J	J	J	J	J	J	J	J	I	I	J	J	J	J	J	J	J	J
	27	CG14	I	I	I	I	I	I	I	I	I	I	J	I	J	J	J	J	I	I	J	J	I	G	I	G	G	I	J	J	J	G	J	I	G	J
	57	IR68703-AC-24-1 x CG14	I	H	H	H	J	H	H	J	H	J	I	J	G+H	J	J	J	J	H	H	J	J	H	G+J	H	G+I	H	J	G+J	I	H	G+J	H	J
	2	IR60080-46-A	I	J	J	J	I	J	J	J	J	J	J	J	J	J	J	J	J	J	J	J	J	J	J	J	J	J	J	I	I	J	J	J	J	J	J	J	J
	27	CG14	I	I	I	I	I	I	I	I	I	I	J	I	J	J	J	J	I	I	J	J	I	G	I	G	G	I	J	J	J	G	J	I	G	J
	58	IR60080-46-A x CG14	H	H	H	H	J	H	H	J	H	J	H	J	G+H	J	J	J	J	H	H	H	J	H	G+J	I	G+J	G+J	J	G+J	I	I	G+J	H	J

1) *O.sativa* varieties were classified into *indica* and *japonica*-like varieties based on SPI calculation. (see text). Although some varieties were classified into *japonica*-like group, they have *indica* alleles of some markers.

2) For some markers.hetero alleles or *glaberrima* alleles were amplified in F_1 while parents were not polymorphic.

3) For another markers, indica-japonica hetero alleles were amplified in F1 while parents were not polymorphic. Smear and unclear several bands were amplified by these primers.

Table 3. Successful polymorphic markers between *O. glaberrima* and *O .sativa*.

Subspecies of *O. sativa*[1]	Chromosome													Marker frequency [2] (%)
	1	2	3	4	5	6	7	8	9	10	11	12	Total	
indica	3	2	4	1	0	0	3	1	4	3	1	1	23	34.33
japonica	1	3	5	6	2	1	3	1	5	0	0	1	28	41.79
indica/japonica	1	1	1	0	0	0	0	0	0	2	1	1	7	10.45
-	0	1	1	1	0	0	0	2	1	2	1	0	9	13.43

[1]Two bands from parents represent hetero alleles were successfully amplified when *O. glaberrima* accessions were crossed with corresponding subspecies of *O. sativa*. [2]Total number of markers out of total of 67 subspecies-specific STS markers.

Table 4. Reference markers classifying 30 *O. glaberrima* accessions.

Group	Name	I	J	-	G	H	G+I	G+J	SPI=I/(I+J+H)	Reference markers		Allele change from G1[1]
G1	RAM86	36	24	0	7	0	1	0	60.00			
	RAM90	36	24	0	7	0	1	0	60.00			
	RAM111	36	24	0	7	0	1	0	60.00			
	RAM118	36	24	0	7	0	1	0	60.00			
	RAM120	36	24	0	7	0	1	0	60.00			
	RAM121	36	24	0	7	0	1	0	60.00			
	RAM131	36	24	0	7	0	1	0	60.00			
	RAM134	36	24	0	7	0	1	0	60.00			
	RAM152	36	24	0	7	0	1	0	60.00			
	RAM163	36	24	0	7	0	1	0	60.00			
	IG10	34	24	2	7	0	1	0	58.62			
	CG14	36	24	0	7	0	1	0	60.00			
	CG17	36	24	0	7	0	1	0	60.00			
	TOG6472	35	25	1	6	0	1	0	58.33			
	TOG7235	36	24	1	6	0	1	0	60.00			
G2	RAM3	35	25	0	7	0	1	0	58.33	S08107		I->J
	RAM24	35	25	0	7	0	1	0	58.33			
	RAM54	35	25	0	7	0	1	0	58.33			
G3	Acc.103477	35	25	0	7	0	1	0	58.33	S03041		I->J
G4	CG20	36	25	0	6	0	1	0	59.02	S01160		G->J
G5	TOG5681	35	25	1	5	0	2	0	58.33	S02085		G->G+I
G6	TOG5674	34	25	1	6	0	2	0	57.63	S02085	S02140	G->G+I, I->G
G7	TOG6508	34	25	1	6	0	2	0	57.63	S01160		I->G+I
	TOG6589	34	24	2	6	0	2	0	58.62			
G8	TOG5860	35	23	2	6	1	1	0	59.32	S01054		J->H
	TOG6631	36	24	0	6	1	1	0	59.02			
	TOG7442	36	24	0	6	1	1	0	59.02			
G9	TOG6597	33	22	5	6	1	1	0	58.93	S01054	S01160	J->H, G->G+I
G10	TOG7291	35	24	0	7	1	1	0	58.33	S01054	S02140	J->H, I->G

O. glaberrima accessions in different groups can be distinguished using corresponding reference markers by observation of allele change. For example, 'TOG5674' can be distinguished from 'CG14' by additional *indica* allele of S02085 and specific allele of S02140.

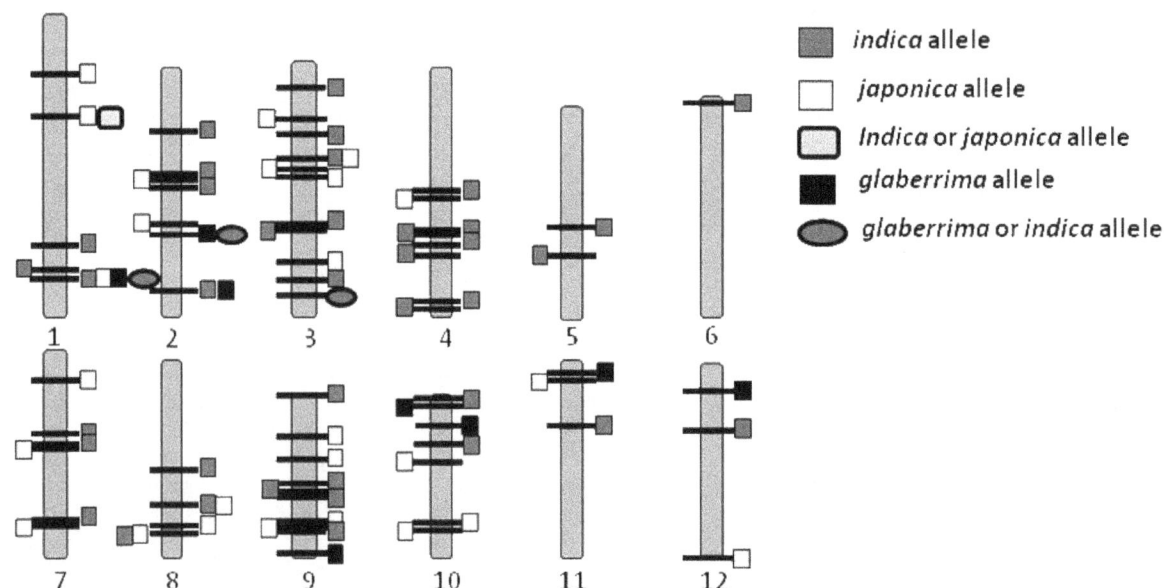

Figure 3. Comparative view of alleles of *O. glaberrima* based on Nipponbare genome.

REFERENCES

Barry MB, Pham JL, Noyer JL, Billot C, Courtois B, Ahmadi N (2007). Genetic diversity of the two cultivated rice species (*O. sativa & O. glaberrima*) in Maritime Guinea. Evidence for interspecific recombination. Euphytica 154: 127-137.

Bi XZ, Xiao YH, Liu WF (1997). Studies on subspecies differentiating protein markers in *Oryza sativa* by two-dimensional polyacrylamide gel electrophoresis. Rice Genet. Newslett., 14: 31-33.

Bimpong IK, Mendoza EMT, Hernandez JE, Mendioro MS, Brar DS (2009). Identification and Mapping of QTLs for Drought Tolerance Introgressed from Oryza *glaberrima* Steud. into Indica Rice (*Oryza. sativa* L). PhD thesis submitted to the University of the Philippines. Los Banos Philippines

Bimpong IK, Carpena AL, Borromeo TH, Mendioro MS, Brar DS (2004). Nematode resistance of backcross derivatives of *Oryza sativa* L crosses with O*ryza glaberrima* Steud. and molecular characterization of introgression. Thesis submitted to the University of the Philippines. Los Banos Philippines

Causse MA, Fulto TM, Cho YG, Ahn SN, Chuncongse J (1994). Saturated molecular map of the rice genome based on an interspecific backcross population. Genetics 138: 1251-1274.

Chin JH, Kim JH, Jiang W, Chu SH, Woo MO, Han L, Brar D, Koh HJ (2007). Identification of Subspecies-specific STS Markers and Their Association with Segregation Distortion in Rice (*Oryza sativa* L.) J. Crop Sci. Biotechnol. 10(3): 175-184.

Chin JH, Kim JH, Kwon SW, Cho YI, Piao ZZ, Han LZ, Koh HJ (2003). Identification of subspecies-specific RAPD markers in rice. Korean J. Breed. 35(2): 102-108.

Cho YC, Shin YS, Ahn SN, Gregorio GB, Kang KH, Brar D, Moon HP (1999). DNA fingerprinting of rice cultivars using AFLP and RAPD markers. Korean J. Crop Sci. 44(1): 26-31.

Enriquez EC, Rosario TL, Brar DS, Mendioro MS, Hernandez JE, Barrion AA (2001). Production of doubled haploids from *Oryza sativa* L. x *O. glaberrima* Steud. and their characterization using microsatellite markers. PhD thesis submitted to the University of the Philippines. Los Banos Philippines

Feltus FA, Wan J, Schulze SR, Estill JC, Jiang N, Paterson AH (2004). An SNP resource for rice genetics and breeding based on subspecies *indica* and *japonica* genome alignments. Genome Res. 14: 1812-1819.

Glaszmann JC (1987). Isozymes and classification of Asian rice varieties. Theor. Appl. Genet. 74: 21-30.

Heuer SM, Meizan KM (2003). Assessing hybrid sterility in *O. glaberrima* x *O. sativa* hybrid progenies by PCR marker analysis and crossing with wide compatibility varieties. Theor. Appl. Genet. 107: 902-909.

Kitampura E (1962). Studies on cytoplasmic sterility of hybrids in distantly related varieties of rice *O. sativa* L. In Fertility of F1 hybrids between strains derived from certain Philippine and Japanese variety crosses and Japanese varieties. Jpn. J. Breed., 12: 81-84.

Koide Y, Onishi K, Kanazawa A, Sano Y (2008). Genetics of speciation in rice. In Rice Biology in the Genomics Era. Edited by Hirano, H.,Sano, Y., Hirai, A., Sasaki, T. Biotechnology in Agriculture and Forestry. Springer-Verlag; pp: 247-259.

Kubo T, Yoshimura A (2005). Epistasis underlying female sterility detected in hybrid breakdown in a *japonica-indica* cross of rice (*Oryza sativa* L.). Theor. Appl. Genet. 110: 346-355.

Li J, Xu P, Deng X, Zhou J, Hu F, Wan J, Tao D (2008). Identifcation of four genes for stable hybrid sterility and an epistatic QTL from a cross between *Oryza sativa* and *Oryza glaberrima*. Euphytica 164: 699-708

Lorieux M, Ndjionjop, N, Ghesquire A (2000). A first interspecific *Oryza sativa* × *Oryza glaberrima* microsatellite-based genetic linkage map. Theor. Appl. Genet. 100: 593-601.

McCouch SR, Teytelman L, Xu Y, Lobos KB, Clare K, Walton M, Fu B, Maghirang R, Li Z, Xing Y, Zhang Q, Kono I, Yano M, Fjellstrom R, DeClerck G, Schneider D, Cartinhour S, Ware D, Stein L (2002). Dvelopment and mapping of 2240 new SSR markers for rice (*Oryza sativa* L.). DNA Res. 9: 199-207.

McNally K, Childs KL, Bohnert R, Davidson RM, Zhao K, Ulat VJ, Zeller G, Clark RM, Hoen DR, Bureau TE, Stokowski R, Ballinger DG,

Frazer KA, Cox DR, Padhukasahasram B, Bustamante CD, Weigel D, Mackill DJ, Bruskiewich RM, Ratsch G, Buell CR, Leung H, Leach JE (2009). Genomewide SNP reveals relationships among landraces and modern varieties of rice. PNAS. 106(30): 12273-12278.

Ohmido N, Fukui K (1995). Cytological studies of African cultivated rice. *Oryza glaberrima*. Theor. Appl. Genet. 91: 212-217.

Park KC, Kim, NH, Cho YS, Kang, KH, Lee, JK, Kim NS (2003). Genetic variations of AA genome *Oryza* species measured by MITE–AFLP. Theor. Appl. Genet. 107: 203-209

Perry DJ, Isabela N, Bousquet J (1999). Sequence-tagged-site (STS) markers of arbitrary genes: the amount and nature of variation revealed in Norway spruce. Heredity 83: 239-248

Morishima H, Hinata K, Oka HI (1963). Comparison of modes of evolution of cultivated forms from two wild rice species, *Oryza breviligulata* and *O. perennis*. Evolution 17: 170-181.

Ni J, Colowit PM, Mackill DJ (2002). Evaluation of genetic diversity in rice subspecies using microsatellite markers. Crop Sci. 42: 601-607.

Neiman M, Linksvayer TA (2006). The conversion of variance and the evolutionary potential of restricted recombination. Heredity, 96: 111-121.

Porteres R (1956). Taxonomie agrobotanique des riz cultives *O. sativa* L. et *O. glaberrima*. S. J. Agric. Trop. Bot. Appl., 3: 341-384; 541-580; 627-700; 821-856.

Panaud O, Chen X, MCcouch SR (1996). Development of microsatellite markers and characterization of simple sequence length polymorphism (SSLP) in rice (Oryza sativa L.). Mol. Gen. Genet. 252: 597-607.

Pham J, Bougerol B (1993). Abnormal segregation in crosses between two cultivated rice species. Heredity, 70: 447-466.

Qian HR, Zhuang JY, Lin HX, Lu J, Zheng KL (1995). Identification of a set of RFLP probes for subspecies differentiation in *Oryza sativa L.* Theor. Appl. Genet. 90: 878-884.

Ren F, Lu BR, Li S, Huang J, Yingguozhu (2003). A Comparative study of genetic relationships among the AA genome Oryza species using RAPD and SSR markers. Theor. Appl. Genet. 108: 113-120.

Robeniol JA, Constantino SV, Resurreccion AP, Villareal CP, Ghareyazie B, Lu BR, Katiyar SK, Menguito CA, Angeles ER, Fu H, Reddy YS, Park W, McCouch SR, Khush GS, Bennett J (1996). Sequence-tagged sites and low-cost DNA markers for rice. : [IRRI] International Rice Research Institute. 1996. Rice genetics III. Proc. 3rd International Rice Genetics Symposium, 16-20 Oct 1995. Manila (Philippines).

Sarla N, Mallikarjuna SPB (2005). *Oryza glaberrima*: A source for the improvement of *Oryza sativa* Cur. Sci. 89: 955-963.

Semon M, Nielsen R, Jones MP, MCcouch SR (2005). The population structure of African cultivated rice *Oryza glaberrima* (Steud.): Evidence for elevated levels of linkage disequilibrium caused by admixture with *O. sativa* and ecological adaptation. Genetic. 169: 1639-1647

Sun CQ, Wang XK, Yoshimura A, Doi K (2002). Genetic differentiation for nuclear, mitochondrial and chloroplast genomes in common wild rice (*Oryza rufipogon* Griff.) and cultivated rice (*Oryza sativa* L.). Theor. Appl. Genet. 104: 1335-1345.

Tanksley SD, Genal MW, Prince JP, de Vicente MC, Bonierbale MW, Broun P, Fulton TM, Giovannoni JJ, Grandillo S, Martin GB, Messeguer R, Miller JC, Miller L, Paterson AH, Pineda O, Roder MS, Wing RA, Wu W, Young ND (1992). High density molecular linkage maps of the tomato and potato genomes. Genetics, 132: 1141-1160.

Temnykh S, Park WD, Ayres N, Cartinhour S, Hauck N, Lipovtiesich L, Cho YG, Ishii T, McCouch SR (2000). Mapping and genome organization of microsatellite sequences in rice (*Oryza*Theor. Appl. Genet. 100: 697-712.

Vaughan DA (1994). The Wild Relatives of Rice: A genetic resources handbook. International Rice Research Institute (IRRI), Manila, Philippines.

Wang CM, Li LH, Zhang XT, Gao Q, Wang RF, An DG (2009). Development and Application of EST-STS Markers Specific to Chromosome 1RS of Secale cereal. Cereal Re. Comm. 37(1): 13-21.

Wang RRC, Li X, Chatterton J (2001). A proposed mechanism for loss of heterozygosity in rice hybrids. Euphytica, 118: 119-126.

12

Genetic variation within and among three invasive *Prosopis juliflora* (Leguminosae) populations in the River Nile State, Sudan

Nada Babiker Hamza

Department of Molecular Biology, Commission for Biotechnology and Genetic Engineering, National Centre for Research, P. O. Box 2404, Khartoum, Sudan.
Nile Basin Research Programme, UNIFOB-Global, University of Bergen, P. O. Box 7800, N-5020 Bergen, Norway.
Email: nada.hamza@gmail.com.

The species of *Prosopis* (Leguminosae) are trees or shrubs well adapted to grow in arid and semi arid regions. In Sudan *Prosopis juliflora* was introduced in 1917. Currently, it has become a noxious weed spreading aggressively in natural and managed habitats. The structure of genetic diversity within and among *P. juliflora* populations infesting three forests in the River Nile State were assessed by RAPD technique. A total of 56 bands were obtained from seven primers. The mean percentage of polymorphic loci over all populations was (55.36%). Kulhuda population had the highest percentage of polymorphic loci (64.29%) and the highest number of private alleles (3). Makabrab population had the lowest percentage of polymorphic loci (46.43%) and two private alleles. Mean expected heterozygosity was (0.218). High genetic differentiation was found among populations (PhiPT = 0.328, P = 0.001). There was a genetic variation of 33% among the populations and within them 67% (AMOVA, P < 0.001). The mean Shannon information index was (I = 0.319, SE = 0.023). UPGMA clustering did not precisely reflect the geographic position of the populations. The results show the current structure of the populations and the similarities between groups of populations, might be due to the recent introduction of the species into Sudan, the limited seed source, the extensive endozoic dispersal seed system and limited pollen dispersal.

Key words: *Prosopis juliflora*, invasive, genetic variation, RAPD, Sudan.

INTRODUCTION

Mesquite trees belong to the family Mimosaceae, Subfamily Mimosoideae, genus *Prosopis* which includes 44 species grouped in 5 sections and 6 series, it occurs worldwide in arid and semiarid regions (Burkart, 1976).

Several *Prosopis* species possess remarkable colonising ability (Bessega et al., 2000a). Raven and Polhill (1981) reported that the dispersion and evolution of the genus *Prosopis* is thought to have taken place approximately 70 million years ago, earlier than the separation of the African and the South-American continents occurred. Two centres of diversity of *Prosopis* occur in the American continent, the Texan-Mexican centre and the Argentinean-Paraguayan centre (Burkart, 1976).

The sections Algarobia have been largely considered

Obligate out-crossers (Simpson, 1977; Simpson et al., 1977). More recently, Bessega et al. (2000b) gave evidence that selfing may occur in some species. They are pollinated by insects and the seeds are endozoically dispersed by herbivores (Burkart, 1976).

Prosopis species have been widely introduced in several countries around the world over the past 150 years for the production of fuel wood, fodder and their ability to grow in the poorest soils and survive in areas where no other trees can survive (Pasiecznik et al., 2001). It is highly recognised for windbreaks, soil binders and sand stabilizers, moreover, provide shelter and food to animals that feed on its nectar, pollen, leaves and fruits (Golubov et al., 2001).

Prosopis juliflora (Swartz) DC is a leguminous, perennial phreatophyte. It is a deciduous, thorny tree (Burkart and Simpson, 1977). It belongs to the section Algarobia (Burkart, 1976). World wide, this dominant woody plant

Abbreviation: PCA, Principal coordinate analysis.

exists in about 45 million ha of grazing lands from sea level up to 1500 m. It is tolerant to very high temperatures (e.g. 48°C) and annual rainfall range of 150 - 750 mm (Darke, 1993; Geilfus, 1994). The roots penetrate to great depths in the soil and can grow in wide range of soils, such as saline, alkaline, sandy and rocky soils (George et al., 2007). P. juliflora is also tolerant to heavy metals (Sinha et al., 2005) and have been proposed to be suitable solution for treatment of soils contaminated with cadmium, chromium and copper (Senthilkumar et al., 2005). Recently, George et al. (2007) investigated the stress-induced genes in P. juliflora through analysis of expressed sequence tags. Their study reveals some insights into the genes responsible from abiotic stress tolerance in P. juliflora as some of the genes in their library known to play a significant role in stress tolerance.

Mesquite (P. juliflora) was introduced into Sudan in 1917 from South Africa and Egypt and planted in Khartoum (Brown and Massey, 1929). Ever since, there had been repeated introductions in several regions of the country.

Introductions in the River Nile State in Sudan were first reported in 1948 to act as shelter belts around Gandatu Agricultural project and continued later on to act as shelterbelts to protect Agricultural Projects from moving sand. In later years, several shelterbelts had been introduced in several villages east and western Nile. More recently (between 1985 - 1996) 23 belts were established around more villages.

Recently, in Sudan, it has been perceived as noxious weed rather than being useful for environmental amelioration. It has invaded diverse habitats, both natural and managed. Over the years this tree species has spread to northern, central and eastern Sudan, with over 90% of the invasion in the Eastern State. Between 1992 and 1996, it was estimated to have spread at a rate of 460 ha per year and by 2006 it covered approximately 230,000 ha of land (Babiker, 2006).

It has become a nuisance in agricultural lands as it forms impenetrable thickets that affect native vegetation community structure development and pastures. It constitutes a threat to biodiversity and affects livelihoods of populations who depend on livestock keeping and subsequent farming as their main sources for income generation (Elhouri, 1986). Consequently, there has been growing calls for its elimination, culminating in a presidential decree in 1996 and followed by several campaigns more recently for its eradication. However, complete eradication of mesquite was not successful due to i) nature of its infestations, ii) longevity of seeds, iii) free of natural enemies (Babiker, 2006) iv) competitive ability (Elsidig et al., 1998) and allelopathy (Mohamed, 2001).

As argued by Ward et al. (2008), there are very few published studies done to understand the genetic diversity and genetic structure within invasive plant populations and for limited number of invasive plants (Ward et al. 2008). No information exists on the genetic diversity of Prosopis species in Sudan based on molecular markers.

The objectives of this study were: (1) to investigate the structure and the distribution of genetic variability within and among P. julflora populations infesting three forests in the River Nile State in Sudan; (2) to find the relationships between those populations based on the RAPDs fragment analysis. The findings should enable us to understand how these populations establish, adapt and expand in their environment.

MATERIALS AND METHODS

Population samplings and plant materials

Three populations of Prosopis growing in three different forests were sampled (Table 1, Figure 1). A total of 45 individuals of Prosopis were collected (15 per forest) in the River Nile State in Sudan. Two populations were collected along the Atbara River and one along the River Nile. From each sampled tree, leaves were collected and dried in silica gel and kept in ventilated room under shade until DNA extraction was done. Plants sampled were at least 100 m from each other, to avoid sampling of ramets of the same vegetative clone. Voucher specimens from each site are kept at the Department of Molecular Biology, Commission for Biotechnology and Genetic Engineering, NCR, Khartoum, Sudan.

DNA extraction

Genomic DNA was extracted from dry leaves of 45 individuals using modified CTAB method (Porebski et al., 1997). In this method the fine powdered plant materials were transferred into 13 ml Falcon tubes containing 6 ml of pre-warmed buffer solution. Tubes containing the samples were then incubated in a water bath at 65°C with gentle shaking for 30 min and left to cool at room temperature for 5 min. Isoamyl alcohol chloroform mixture (1:24) was added to each tube and the phases were mixed gently for 5 min at room temperature to make a homogenous mixture. The cell debris was removed by centrifugation at 5000 rpm for 15 min and the resulted clear aqueous phases (containing DNA) were transferred to new sterile tubes. The step of the chloroform: isoamyl alcohol extraction was repeated twice. The nucleic acids in the aqueous phase were precipitated by adding equal volume of deep cooled isopropanol. The contents were mixed gently and collected by centrifugation at 4000 rpm for 10 min. The formed DNA pellet was washed twice with 70% ethanol and the ethanol was discarded after spinning with flash centrifugation. The remained ethanol was removed by leaving the pellet to dry at room temperature. The pellet was dissolved in TE buffer (10 mM Tris, 1 mM EDTA, pH 8) and stored at -20°C for further use. The extracted DNA samples were observed under UV illumination after staining with ethidium bromide and agarose gel electrophoresis. The purity and the concentrations of the DNA were then spectrophotometrically assessed following Sambrook et al. (1989) method.

RAPD technique

The PCR was carried out in 25 µl final volume using 1 µl of genomic DNA (20 - 40 ng) containing, 2.5 µl of 10X Taq buffer, 1.5 µl MgCl2 (50 mM) , 2.5 µl dNTPs (2 mM/µl), 2 µl random primer (10 pmol/ µl), 0.5 units of Taq DNA polymerase (Vivantis). The mixture was made up to 25 µl by addition of sterilized distilled water. The mixture was amplified in a thermal cycler (Biometra) which was programmed for one cycle of initial denaturation at 94°C for 5 min, 40 cycles of 94°C for 1 min, followed by annealing 36°C for 1 min and ended by extension at 72°C for 1 min followed by a final extension cycle that

Table 1. Geographic information and area of the three forests of the sampled populations of *Prosopis juliflora*.

Forest name	Symbol	Area (ha)	Latitude (N)	Longitude (E)	Location
Makabrab	M	340	17.44854	33.89567	Eastern bank of the River Nile
Um Sayala	S	360.36	17.36925	34.44579	Eastern bank of the Atbara river
Kulhuda	K	840	17.33573	34.36975	Western bank of the Atbara river

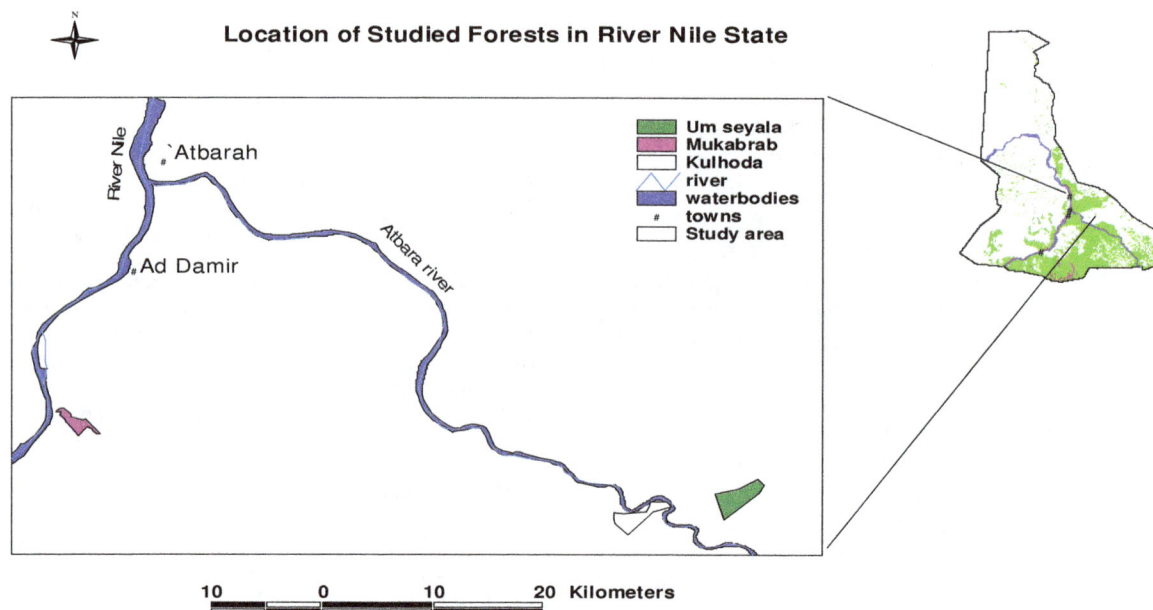

Figure 1. Map of the Nile River State indicating the three forests sites.

performed at 72°C for 7 min. The PCR machine was adjusted to hold the product at 4°C.

Each amplified product was mixed with 3 µl of loading dye (12.5 bromophenol blue, 2 g sucrose) and spun briefly in a micro centrifuge before loading. The PCR products and 1 kp DNA ladder were electrophoresed on 2% agarose gel (stained with EtBr) at 80 volts. The separated fragments were visualized with an ultraviolet (UV) transilluminator. In this study, several RAPD primers from the University of British Columbia (UBC),Vancouver, British Columbia, Canada and from Operon Technologies Inc., Ala- meda, CA; were tested for their reproducibility and cautions were made to avoid unspecific amplification.

Statistical analysis

Analysis of genetic diversity

Each band in the RAPDs profile was treated as an independent locus with two alleles. The numbers of bands produced for each primer were scored manually for presence (1), or absence (0) and a binary matrix was generated and then used for analysis.
The PhiPT (analogue of F_{ST} fixation index) value for genetic variability, the percentage of polymorphism (% P), heterozygosity (he), number of observed alleles (Na), number of effective alleles (Ne) and Shannon's information index (I) were calculated for each population using GenALEx v. 6.1 (Peakall and Smouse, 2006).
With AMOVA, the variance components and their significance

levels for variation among populations and within population using RAPD data of the 45 individuals was obtained. It was conducted in GenALEx v. 6.1. using the PhiPT "P value is calculated as the number of values ≥ observed value (Including Observed Value) ÷ (Number of Permutations + 1)".

Genetic structure

A principal coordinate analysis (PCA) was conducted with GenALEx v. 6.1 (Peakall and Smouse, 2006). This multivariate approach was chosen to complement the cluster analysis information, because cluster analysis is more sensitive to closely related individuals whereas PCA is more informative regarding distances among major groups (Hauser and Crovello, 1982).
Pairwise genetic distances between individuals were calculated by the percentage disagreement method. These data were used in cluster analysis with the unweighted pair-group method using arithmetic averages (UPGMA), in which samples are grouped based on their similarity with the aid of statistical software package STATISTCA- ver.6 (StatSoft, Inc., 2001).

RESULTS

After screening of a series of primers, a total of seven primers (10 mer) that produced strongly amplified polymorphic bands were selected for RAPD-PCR analysis

Table 2. Polymorphism detected by the use of seven random primers on three *Prosopis juliflora* populations.

Name of primer	Sequence of primer (5'-3')	Total number of bands	Number of polymorphic bands	% of polymorphic Bands
A-01	CAGGCCCTTC	7	7	100
B-8	GTCCACACGG	13	10	76.92
B-20	GGACCCTTAC	9	6	66.67
C-2	GTGAGGCGTC	4	4	100
C-10	TGTCTGGGTG	9	8	88.89
UBC-101	GCGGCTGGAG	7	7	100
UBC-122	GTAGACGAGC	7	6	85.71

Table 3. Gene diversity in the three *Prosopis juliflora* populations.

Population	Number of observed alleles (Na)	Number of effective alleles (Ne)	Shannon's information index (I)	Heterozygosity (He)	Polymorphism (%)
M	1.411(0.080)	1.293(0.050)	0.250 (0.039)	0.169 (0.027)	46.43
S	1.411(0.098)	1.438(0.057)	0.344 (0.043)	0.239 (0.030)	55.36
K	1.589(0.080)	1.427(0.051)	0.364 (0.039)	0.247 (0.027)	64.29

Numbers in parenthesis refers to standard error.

Table 4. Summary of the analysis of molecular variance (AMOVA) within and among *Prosopis juliflora* populations.

Source	df	SS	MS	Est.Var	%	Stat	Value	Probability
Among Pops	2	87.733	43.867	2.573	33			
Within Pops	42	221.200	5.267	5.267	67	PhiPT	0.328	0.001

The analysis is based on RAPD phenotypes consisting of 56 band states. Levels of significance are based on 1000 iterations.

Table 5. Summary of Nei pairwise population genetic distance (NeiP), genetic structure (PhiPT) values and their level of significance, estimates of georgraphic distances and Nm values between populations.

Population 1	Population 2	PhiPT	Geographic distance (Km)	Nei genetic distance (GD)
S	K	0.372(P<0.001)	9.180	0.203
S	M	0.260(P<0.001)	58.520	0.131
K	M	0.329(P<0.001)	52.130	0.180

(Table 2). The selected primers generated an appropriate amplification pattern with clear and consistent reproducible bands. A total of 56 bands, were obtained.

The seven informative primers were selected and used to evaluate the degree of polymorphism and genetic relationships within and between all individuals under study. The maximum number of band fragments was produced by the primer B-8 (13 bands) with 76.92% polymorphism while the minimum number of fragments was produced by the primer C-2 (4 bands) with 100% polymorphism. With an average of 6.86 bands per primer. Pattern of RAPD fragments produced by all the highest (0.247, SE = 0.027) (Table 3).

The mean population diversity using the Shannon infor-

mation index (I) was 0.319 (SE = 0.023). Kulhuda population was the most diverse (I = 0.364) and lowest diverse population was Makabrab (I = 0.25) (Table 3).

The analysis of molecular variance (AMOVA) for populations showed significant differentiation (P < 0.001), with 67% of the differentiation attributed to within populations and 33% attributed to among primers is shown in (Table 2).

The highest number of effective alleles was in Umsayala population (1.438, SE = 0.057) and the lowest in Makabrab (1.293, SE = 0.05), with an average 1.386 (SE = 0.031) over all populations. The mean percentage of polymorphic loci over all populations was 55.36 (5.15%). Kulhuda population had the highest percentage

Table 6. Matrix of genetic distances based on percentage disagreements

	S1	S2	S3	S4	S5	S6	S7	S8	S9	S10	S11	S12	S13	S14	S15	K16	K17	K18	K19	K20	K21	K22	K23	K24	K25	K26	K27	K28	K29	K30	M31	M32	M33	M34	M35	M36	M37	M38	M39	M40	M41	M42	M43	M44	M45
S1	0.00																																												
S2	0.11	0.00																																											
S3	0.18	0.07	0.00																																										
S4	0.25	0.25	0.14	0.00																																									
S5	0.32	0.21	0.21	0.11	0.00																																								
S6	0.25	0.21	0.25	0.14	0.18	0.00																																							
S7	0.09	0.13	0.13	0.20	0.23	0.23	0.00																																						
S8	0.13	0.00	0.13	0.20	0.27	0.27	0.14	0.00																																					
S9	0.16	0.36	0.20	0.16	0.23	0.18	0.18	0.11	0.00																																				
S10	0.25	0.18	0.21	0.25	0.25	0.18	0.25	0.20	0.20	0.00																																			
S11	0.18	0.21	0.23	0.20	0.21	0.30	0.27	0.25	0.23	0.29	0.00																																		
S12	0.13	0.00	0.23	0.30	0.30	0.27	0.18	0.14	0.21	0.30	0.20	0.00																																	
S13	0.13	0.09	0.09	0.14	0.27	0.20	0.14	0.07	0.16	0.23	0.21	0.18	0.00																																
S14	0.18	0.14	0.11	0.18	0.21	0.25	0.25	0.18	0.16	0.23	0.21	0.23	0.13	0.00																															
S15	0.18	0.18	0.21	0.21	0.34	0.36	0.36	0.13	0.20	0.27	0.25	0.36	0.13	0.14	0.00																														
K16	0.34	0.27	0.30	0.30	0.34	0.30	0.41	0.43	0.39	0.39	0.32	0.39	0.23	0.30	0.34	0.00																													
K17	0.41	0.38	0.38	0.34	0.41	0.25	0.39	0.41	0.39	0.38	0.38	0.39	0.32	0.27	0.29	0.14	0.00																												
K18	0.34	0.30	0.34	0.30	0.41	0.41	0.36	0.36	0.36	0.30	0.38	0.32	0.32	0.30	0.18	0.29	0.25	0.00																											
K19	0.25	0.21	0.30	0.32	0.32	0.32	0.27	0.23	0.27	0.27	0.41	0.39	0.23	0.32	0.27	0.38	0.39	0.18	0.00																										
K20	0.38	0.34	0.38	0.38	0.41	0.41	0.43	0.39	0.36	0.36	0.39	0.38	0.36	0.36	0.29	0.36	0.32	0.25	0.14	0.00																									
K21	0.27	0.21	0.27	0.34	0.43	0.34	0.36	0.36	0.27	0.29	0.38	0.36	0.23	0.29	0.25	0.18	0.25	0.14	0.05	0.00	0.00																								
K22	0.38	0.30	0.34	0.34	0.32	0.34	0.30	0.30	0.29	0.27	0.38	0.32	0.32	0.30	0.18	0.25	0.30	0.25	0.23	0.23	0.21	0.00																							
K23	0.45	0.38	0.38	0.36	0.34	0.36	0.43	0.48	0.36	0.38	0.45	0.38	0.39	0.34	0.27	0.30	0.30	0.14	0.30	0.30	0.32	0.20	0.00																						
K24	0.29	0.27	0.32	0.36	0.36	0.32	0.36	0.39	0.36	0.25	0.38	0.30	0.38	0.32	0.29	0.30	0.18	0.07	0.16	0.16	0.14	0.13	0.18	0.00																					
K25	0.39	0.36	0.39	0.39	0.43	0.25	0.38	0.38	0.34	0.27	0.34	0.34	0.38	0.36	0.25	0.29	0.25	0.13	0.23	0.20	0.20	0.34	0.36	0.18	0.00																				
K26	0.46	0.43	0.46	0.50	0.43	0.38	0.48	0.41	0.38	0.39	0.46	0.38	0.45	0.30	0.29	0.30	0.32	0.38	0.36	0.36	0.34	0.32	0.38	0.18	0.32	0.00																			
K27	0.30	0.27	0.30	0.34	0.38	0.34	0.34	0.36	0.32	0.34	0.41	0.32	0.32	0.30	0.30	0.38	0.41	0.25	0.30	0.30	0.32	0.21	0.41	0.25	0.27	0.38	0.00																		
K28	0.38	0.38	0.38	0.38	0.41	0.34	0.41	0.39	0.36	0.38	0.38	0.34	0.36	0.30	0.23	0.32	0.36	0.21	0.13	0.16	0.14	0.25	0.18	0.16	0.18	0.21	0.14	0.00																	
K29	0.39	0.38	0.38	0.36	0.43	0.25	0.36	0.36	0.36	0.32	0.32	0.34	0.32	0.34	0.29	0.30	0.39	0.27	0.20	0.20	0.16	0.25	0.23	0.27	0.20	0.29	0.20	0.16	0.00																
K30	0.38	0.30	0.30	0.34	0.38	0.25	0.36	0.36	0.32	0.36	0.39	0.38	0.36	0.30	0.30	0.41	0.38	0.38	0.34	0.38	0.38	0.34	0.45	0.36	0.41	0.50	0.34	0.38	0.32	0.00															
M31	0.27	0.27	0.21	0.30	0.34	0.34	0.18	0.18	0.21	0.25	0.34	0.32	0.14	0.23	0.18	0.38	0.32	0.32	0.30	0.30	0.30	0.21	0.41	0.20	0.21	0.41	0.14	0.21	0.21	0.38	0.00														
M32	0.16	0.20	0.14	0.27	0.27	0.23	0.18	0.16	0.21	0.16	0.16	0.21	0.14	0.14	0.18	0.29	0.39	0.27	0.18	0.25	0.14	0.25	0.29	0.14	0.27	0.34	0.29	0.25	0.16	0.32	0.16	0.00													
M33	0.21	0.21	0.14	0.27	0.32	0.32	0.23	0.16	0.18	0.21	0.32	0.16	0.16	0.16	0.25	0.36	0.25	0.16	0.23	0.18	0.20	0.30	0.30	0.20	0.30	0.36	0.23	0.16	0.14	0.30	0.18	0.09	0.00												
M34	0.21	0.14	0.14	0.25	0.25	0.27	0.23	0.18	0.23	0.25	0.32	0.16	0.16	0.25	0.21	0.30	0.30	0.27	0.18	0.21	0.16	0.34	0.36	0.21	0.36	0.36	0.27	0.21	0.20	0.21	0.18	0.07	0.07	0.00											
M35	0.18	0.18	0.18	0.29	0.36	0.32	0.23	0.16	0.20	0.20	0.29	0.29	0.23	0.15	0.14	0.41	0.30	0.38	0.30	0.30	0.20	0.38	0.29	0.25	0.38	0.41	0.27	0.20	0.18	0.32	0.09	0.14	0.14	0.18	0.00										
M36	0.25	0.25	0.25	0.32	0.34	0.32	0.32	0.27	0.23	0.23	0.36	0.20	0.13	0.25	0.20	0.34	0.41	0.27	0.27	0.27	0.16	0.30	0.39	0.27	0.39	0.38	0.27	0.27	0.21	0.34	0.18	0.07	0.14	0.07	0.18	0.00									
M37	0.25	0.25	0.21	0.27	0.29	0.27	0.27	0.23	0.20	0.20	0.36	0.21	0.13	0.15	0.13	0.36	0.30	0.38	0.30	0.30	0.20	0.30	0.30	0.25	0.36	0.39	0.34	0.23	0.21	0.36	0.16	0.14	0.07	0.09	0.09	0.18	0.00								
M38	0.16	0.16	0.13	0.30	0.38	0.30	0.18	0.16	0.21	0.21	0.29	0.27	0.16	0.18	0.14	0.41	0.29	0.27	0.25	0.25	0.20	0.30	0.41	0.27	0.38	0.38	0.29	0.21	0.21	0.23	0.09	0.05	0.09	0.09	0.09	0.07	0.18	0.00							
M39	0.16	0.16	0.09	0.23	0.30	0.27	0.18	0.14	0.18	0.27	0.30	0.21	0.14	0.14	0.14	0.34	0.32	0.32	0.30	0.25	0.21	0.32	0.38	0.29	0.38	0.34	0.30	0.25	0.23	0.34	0.18	0.09	0.09	0.05	0.05	0.20	0.07	0.20	0.00						
M40	0.25	0.25	0.18	0.32	0.34	0.32	0.25	0.25	0.20	0.32	0.36	0.29	0.07	0.29	0.21	0.34	0.30	0.34	0.21	0.34	0.20	0.38	0.39	0.27	0.39	0.38	0.27	0.16	0.34	0.34	0.20	0.14	0.14	0.14	0.18	0.15	0.18	0.16	0.14	0.00					
M41	0.18	0.18	0.14	0.29	0.34	0.30	0.20	0.18	0.20	0.25	0.36	0.25	0.16	0.16	0.14	0.36	0.32	0.39	0.23	0.32	0.20	0.23	0.36	0.23	0.36	0.36	0.32	0.16	0.21	0.36	0.16	0.09	0.16	0.20	0.21	0.14	0.23	0.16	0.14	0.14	0.00				
M42	0.23	0.20	0.23	0.30	0.34	0.27	0.25	0.23	0.32	0.27	0.34	0.29	0.18	0.27	0.23	0.43	0.45	0.32	0.32	0.38	0.25	0.36	0.43	0.27	0.39	0.38	0.29	0.23	0.27	0.38	0.27	0.16	0.13	0.09	0.09	0.20	0.21	0.36	0.27	0.20	0.16	0.00			
M43	0.16	0.16	0.14	0.27	0.30	0.34	0.18	0.16	0.25	0.25	0.34	0.23	0.14	0.23	0.25	0.38	0.36	0.36	0.25	0.32	0.14	0.25	0.32	0.20	0.32	0.35	0.21	0.14	0.34	0.36	0.14	0.16	0.13	0.13	0.14	0.16	0.14	0.13	0.14	0.20	0.13	0.11	0.00		
M44	0.18	0.14	0.14	0.25	0.30	0.27	0.20	0.14	0.25	0.21	0.29	0.23	0.13	0.13	0.18	0.30	0.38	0.38	0.27	0.34	0.23	0.34	0.36	0.25	0.36	0.36	0.34	0.16	0.34	0.30	0.16	0.14	0.14	0.14	0.18	0.18	0.16	0.18	0.13	0.25	0.16	0.09	0.04	0.00	
M45	0.13	0.13	0.16	0.23	0.27	0.30	0.14	0.11	0.21	0.23	0.30	0.21	0.11	0.20	0.13	0.25	0.32	0.32	0.16	0.21	0.18	0.32	0.29	0.20	0.41	0.30	0.29	0.07	0.29	0.29	0.14	0.13	0.13	0.20	0.13	0.20	0.11	0.11	0.20	0.20	0.20	0.11	0.04	0.05	0.00

of polymorphic loci (64.29%) and the highest number of private alleles (3). Makabrab population had the lowest percentage of polymorphic loci (46.43%) and two private alleles. Mean expected heterozygosity was 0.218 (0.016), with Kulhuda having populations. The PhiPT estimates were high (0.328) and highly significant (P = 0.001) (Table 4).

Principal components analysis of molecular variance performed in GenAlex 6.1, based on individual band pattern. The first two axes explain 56.59% (Figure 3). The fixation index value of PhiPT over all populations was (0.328) and highly significant (P < 0.001). Between populations, the highest PhiPT value (0.372, P < 0.001) was found between Umsayala and Kulhuda and the lowest between Umsayala and Makabrab (0.260, P < 0.001), Table 5. The UPGMA and percent

disagreement values (PDV) were used to estimate the degree of relationships between the individuals analysed based on common amplified fragments. Based on the matrix obtained (Table 6), the mean average percent disagreement values (PDV) for all individuals was 0.25 ranging from 0.04 to 0.46. The dendrogram (Figure 2) shows a trend of clustering of regional populations to an extent. Nevertheless, most of Kulhuda

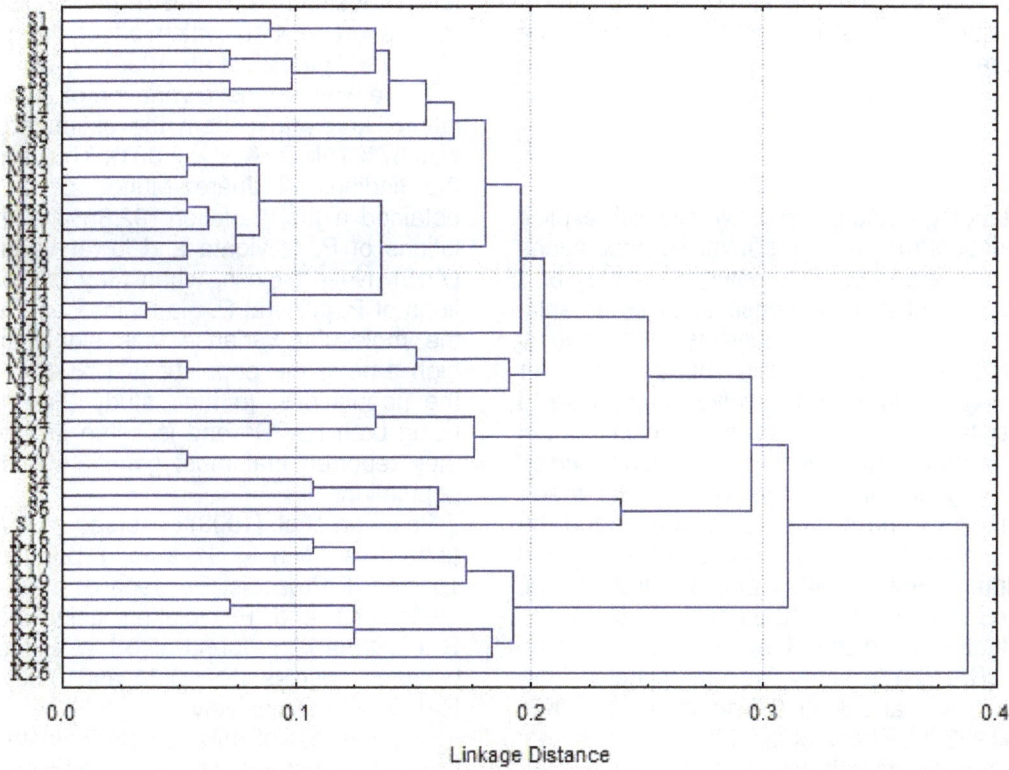

Figure 2. UPGMA tree showing relationships among three populations of *Prosopis juliflora* based on 54 RAPD loci.

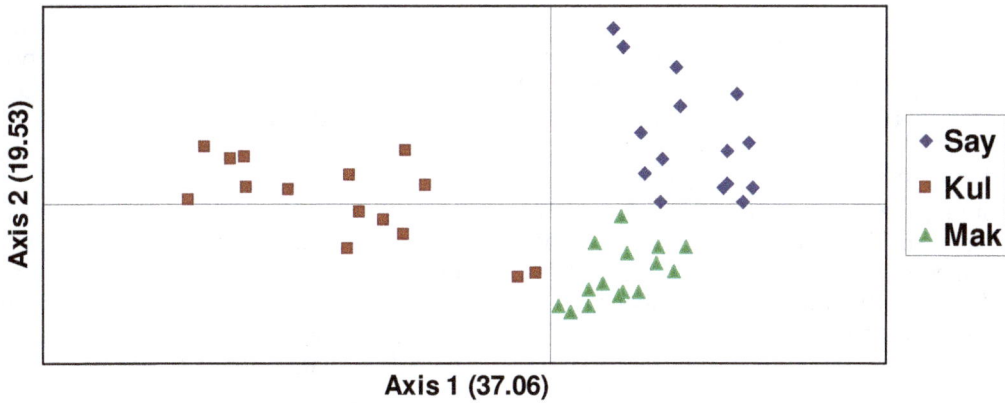

Figure 3. Principal coordinate analysis of the three *Prosopis juliflora* populations based on 56 random amplified polymorphic DNA loci. Note the separation between the Kulhuda individuals and the other two populations.

individuals formed a distant cluster from the other two populations. The pattern that is depicted by the PCA analysis differs from the UPGMA tree. The PCA seems to show three groups that correspond to the three populations, with two individuals of Kulhuda clearly differentiated (Figure 3). In accordance with the results obtained of higher gene diversity values in case of Kulhuda samples, the PCA plot also shows wide scattering when compared to Makabrab, which shows opposite trend both in the PCA plot and the corresponding gene diversity values. The first three coordinates explained 71.7% of the total variation.

Of the three regions, the Makabrab samples show more homogeneity as against the rest two regions (gene diversity measures).

DISCUSSION

A clear understanding of the genetic diversity with explicit analyses of genetic structure of the invasive populations is important; it can help in understanding the history of a particular invasion and their response to environmental changes (Ward, 2006). The mechanisms of their local spread and adaptation and predict the potential for populations of invasive species to evolve in response to evolution of resistance to herbicides or biological control agents based on the diversity levels (Sakai, 2001), which can lead to setting effective control plan for its eradication. The majority of the work done to understand the genetic diversity in the Prosopis genus used the isozyme markers (Saidman 1986, 1990; Saidman and Vilardi, 1987; Keys and Smith, 1994; Saidman et al., 1996, 1998). Later studies, used the Random Amplified Polymorphic DNAs (RAPDs) to study Prosopis species from Africa, South America and Asia (Saidman et al., 1998; Ramirez et al., 1999; Landreas et al., 2006) and to detect inter and Intra-specific genetic variability thanks to their ability to detect the polymorphism (Juárez-Muñoz et al., 2002).

Genetic variation

The He and P values obtained in the present study are similar to those obtained in previous studies on several Prosopis species using both isozymes and RAPDs (Saidman, 1986, 1990; Saidman et al., 1997, 1998, Saidman and Vilardi, 1987; Verga, 1995; Bessega et al., 2000c). In this study, the average number of fragments produced per primer (6.86) using seven primers is less than the one obtained by (Juárez-Muñoz et al., 2002) (8.6) by using five primers on four populations of P. juliflora, Propis laevigata and Propis glandulosa.

Population genetic structure, mating system and seed dispersal

PhiPT and FST are analogous standardizing measures of the degree of genetic differentiation among populations: scores for both measures range from 0 (no differentiation) to 1 (no alleles shared).

In this study, the PhiPT value obtained over all populations was (0.328) and highly significant (P < 0.001), suggesting structuring among the populations. Bessega et al. (2000b) argued that Prosopis species populations are expected to be structured because pollen and seed dispersal in species of Algarobia is limited, causing popu-

lation substructure. Seeds dispersal distances could vary, from short to long which affect the population genetic structure (Hamrick et al., 1992).

There was a genetic variation of 33% among the populations and within them the estimated genetic variation was 67% (AMOVA, P < 0.001). This is in accordance with the findings of Juárez-Muñoz et al. (2002), as they obtained higher variation (92.85%) within the two populations of P. laevigata and lower variation among them (7.15). Nevertheless, when they compared four populations of P. juliflora, P. glandulosa and two of P. laevigata the molecular variance was significantly different and high among the populations (72.48%) and 27.52 within the populations. In their study (Ferreyra et al., 2007), using both RAPDs and isozyme in species of Prosopis, they reported that most genetic variation occurs within populations.

Saidman et al. (1998) observed non-significant levels of gene flow (Nm < 1) among populations of different species of Prosopis. Bessega et al. (2000c) studying P. glandulosa and P. velutina using both isozyme and RAPDS markers, found also low estimates of gene flow between species (Nm=0.39 and 0.60) for isozyme and RAPD loci, respectively.

The results of this study indicate that geographic proximity is not indicative of genetic similarity and hence, is not a guide for understanding the genetic structure of this species.

UPGMA cluster analysis showed a trend of clustering of regional populations. There was considerable overlap among Umsayala and Makabrab populations, which reflects their genetic closeness although they are 58.5 km distant. The main source for seed propagation and distribution in the River Nile State was Gandatu Agricultural project (Abdel Magid T.D, Ex-National Coordinator of Mesquite in the Forests National Corporation, personal communication, 2009).

The introduction of mesquite in the River Nile State in Sudan, prevailing drought, with extensive livestock and animal movement added to decrease in land use and over exploitation of natural vegetation have led to spread of Prosopis into various areas. Browsing animals such as camels, goats, donkeys, cattle and sheep are main agents in spreading it. Furthermore, human beings who collect the pods to feed their animals also contributed.

The fruit pods of mesquite trees are considered as a rich food for domestic animals and human beings in hot dry countries. The pods contain much sugar and a fair amount of protein (Abdel, 1986). Abdel (2001) found that pods contained 26% glucose and 9 to 14% protein and 55% carbohydrates. In Central America the pods are ground into a meal for use in concentrated rations (Laurie, 1974).

It has been reported by Brown and Archer (1987), who studied the dispersal of P. glandulosa var. glandulosa seeds by cattle in a savannah woodland in Texas, that 75% of dung pats of the cattle contained Prosopis seedlings, with an average of 4.2 seedlings per pat and when the cattle were excluded, no establishment of Prosopis was reported. Reynolds (1954) reported a maximum dispersal distance of 50 m by kangaroo rats, whereas livestock and larger animals disperse it to a maximum distance of 4 to 6 km.

After germination, mesquite seedlings grow vigorously. The roots are fast in developing to deeper depth and the un-palatability of the green leaves by animals increase the survival chances of the seedlings, especially in areas that undergo heavy grazing (Mohamed, 2001).

In this case study, pollen dispersal mediated by insects is not expected to be distant as the main vectors of seed dispersal are herbivorous mammals, which are not expected to transport seeds over large distances.

Kulhuda population was genetically differentiated from the other two populations. That might be due to the special microclimate due to closeness of its position to the Atbara River, where, livestock visiting this forest would aid in transporting the seeds from other areas and dispose them in this forest. The moist and fertile soil will aid in the fast germination and establishment of such seeds. Also, in later stages, the selection pressures at this site will be less than in the other two populations.

Although Makabrab forest also lies very close to the eastern bank of the River Nile, it has less number of recently established mesquite individuals (3 - 4 years) than the other forests. The lower genetic diversity found can be explained by the same seed source in the area. Nevertheless, being lower in genetic diversity than the other populations, this population maintain 2 private alleles. Factors increasing inbreeding in Prosopis is being entomophilic where the pollen is usually unable to migrate large distances and having endozoic dispersal seed system (Genisse et al., 1990). Bessega et al. (2000b) study on mating system parameters of seven species of Algorobia, showed that out crossing rate ranges between 0.72 and 1, with an average of 0.85, indicating 15% of selfing may occur in natural populations.

It has been shown in previous studies that the genetic structure of populations affect the efficacy of control of invasives. The control of a population with a genetically homogeneous structure, due to asexual mode of reproduction can be easier with matching a biological control agent to the host genotype, where it is vulnerable to the biological enemies (Van Driesche and Bellows, 1996). However, in sexually reproducing weeds, the greater genetic variation leads to fast adaptive evolution and escape from the biological control agent (Sakai et al., 2001).

Sakai et al. (2001) suggested that if the eradication of invasive populations is impossible, then, setting control strategies to alter the population genetic structure in order to reduce adaptive variation.

Molecular markers have proven to be very useful tools in this study in estimating the genetic variation within and among Prosopis populations. This study area represents limited area where Prosopis cause problems in Sudan. Therefore, substantial genetic differentiation might be expected in this species when studying more populations from different regions.

In our study, high diversity within and similarities between groups of populations, were indicated by the RAPD markers. The recent introduction of the species into Sudan, the limited seed source introduced, the extensive endozoic dispersal seed system and limited pollen dispersal might have shaped the current structure of the populations.

Despite the problem caused by the Prosopis in Sudan, further information on the genetic background of the existing populations is not available. The results of the present investigation constitute the first effort to study the genetic variation within and among some selected populations. Therefore, more research in future is encouraged.

ACKNOWLEDGEMENTS

Thanks to the Forests National Corporation, especially Dr. Abdalla Gafar, Alhag A. Alawad, Abdelhafeez Kamal, Hafiz Habib and Osman Karar for field support, help in sampling and hosting. I am grateful to Habeballa R and Ismail A for help in laboratory work; as well to Dr. Tore Satersdal and Nile Basin Research Program (University of Bergen, Norway) for encouragement and support during the research stay at the University of Bergen.

REFERENCES

Abdel Bari E (1986). The identity of "The Common Mesquite Prosopis spp. Pamphlet No. 1, pp Prosopis project, supported by IDRC. Forest Research Corporation.

Abdel Magid TD (2001). Biodiversity in forests and Forest Products other than Wood. Forest National Corporation Publications (in Arabic). pp. 34-35.

Babiker AG (2006). Problems posed by the introduction of Prosopis spp. In selected countries. Report by: FAO, Rome.

Bessega C, Ferreyra LI, Vilardi JC, Saidman BO (2000a). Unexpected low genetic differentiation among allopatric species of section Algarobia of Prosopis (Leguminosae). Genetica, 109: 255-266

Bessega C, Ferreyra LI, Julio N, Montoya S, Saidman BO, Vilardi JC (2000b). Mating system parameters in species of genus Prosopis (Leguminosae). Hereditas, 132: 19-27

Bessega C, Saidman BO, Vilardi JC (2000c). Isozyme and RAPD studies in Prosopis glandulosa and P. velutina (Leguminosae, Mimosoideae). Gene. Mole. Biol. 23 (3): 639-648

Brown JR, Archer S (1987). Woody plant seed dispersal and gap formation in a North American subtropical savanna woodland: the role of domestic herbivores. Vegetation, 73: 73-80.

Brown AF, Massey RE (1929). Flora of the Sudan Thomas Murby and CO. p. 376.

Burkart A, Simpson BB (1977). Appendix: the genus Prosopis and annotated key to the species of the world. In: Simpson BB (eds) Mesquite: Its biology in two desert scrub ecosystems. US/IBP Synthesis, Series 4. Dowden. Hutchinson Ross Inc. Academia Press. pp. 201-283.

Burkart A (1976). A monograph of the genus Prosopis (Leguminosae, subfam, Mimosoideae). J. Arnold Arboretum, 57: 217-525.

Darke H (1993). Trees for dry lands. International Scientific Publishing, New York. pp. 370.

Elhouri AA (1986). Some aspects of dryland afforestation in the Sudan, with special reference to *Acacia tortilis* (Forsk) hayne, *Acacia seyal* Willd. And *Prosopis chilensis* (Molina) Stuntz. Forest Ecol. Manage., 16: 209-221.

Elsidig NA, Abdelsalam AH, Abdelmagid TD (1998). Socio-Economic, Environmental and Management Aspects of Mesquite in Kassala State (Sudan) Sudanese Social Forestry Society. pp. 96.

Ferreyra LI, Bessega C, Vilardi JC, Saidman BO (2007). Consistency of population genetics parameters estimated from isozyme and RAPDs dataset in species of genus *Prosopis* (Leguminosae, Mimosoideae). Genetica, 131: 217-230.

Geilfus F (1994). El árbol al servicio del agricultor. Guìa de especies. Turrialba, Costa Rica: Enda-Caribe-Centro Agronómico Tropical de Investigación y Enseñanza. 2: 597.

Genisse J, Palacios RA, Hoc PS, Carrizo R, Moffat L, Mom MP, Agullo MA, Picca P, Torregosa S (1990). Observaciones sobre la biologia floral de *Prosopis* (Leguminosae, Mimosoideae). II fases florales y visitants en el distrito Chaqueño Serrano. Darwiniana. 30: 71-85.

George S, Venkataraman G, Parrida A (2007). Identification of stress-induced genes from the drought-tolerant plant Prosopis juliflora (Swartz) DC. through analysis of expressed sequence tags. Genome, 50: 470-478.

Golubov J, Mandujano M, Eguiarte L (2001). The paradox of mesquites (*Prosopis* spp.): invading species or biodiversity enhancers?. Boletin de la Sociedad Botánica de México, 69: 23-30.

Hamrick JL, Godt MJ and Sherman Broyles S (1992). Factors influencing levels of genetic diversity in woody plant species, 6: 95-124.

Hauser LA, Crovello TJ (1982). Numerical analysis of genetic relationships in Thelypodieae (Brassicaceae). Syst. Bot., 7: 249-268.

Keys RN, Smith SE (1994). Mating system parameters and population genetic structure in pioneer populations of *Prosopis velutina* (Leguminosae). Am. J. Bot., 81(8): 1013-1020.

Juárez-Muñoz J, Carrillo-Castañeda G, Arrreguin R, Rubluo A (2002). Inter-and intra-genetic variation of four populations of *Prosopis* using fingerprints. Biodiver.Conserv., 11: 921-930.

Landreas G, Alfonso M, Pasiecznik NM, Harris PJC, Ramirez L (2006). Identification of Prosopis juliflora and Prosopis pallida accessiobs using molecular markers. Biodiv. Conserv. 15: 1829-1844.

Laurie MV (1974). Tree planting in African Savannas, Food and Agriculture Organization of the United Nations, Rome. pp. 72.

Mohamed AA (2001). Some Aspects of Germination, Dormancy and Allelopathy of *Prosopis juliflora* (Mesquite). M. Sc. Thesis, University of Gezira. pp. 69

Peakall R, Smouse PE (2006) Genealex 6: Genetic analysis in excel. Population genetic software for teaching and research. Mol. Ecol. Notes, 6: 288-295.

Pasiecznik NM, Felker P, Harris PJC, Harsh LN, Cruz G, Tewari JC, Cadoret K, Maldonado LJ (2001) The *Prosopis Julifora- Prosopis pallida* Complex: A Monograph. HDRA, Coventry, UK. pp. 162.

Porebski S, Bailey LG, Baum R (1997). Modification of a CTAB DNA extraction protocol for plants containing high polysaccharide and polyphenol components. Plant Mol. Biol. Reporter, 15(1): 8-15.

Ramirez L, de la Vega A, Razkin N, Luna V, Harris PJC (1999). Analysis of the relationships between species of the genus *Prosopis* revealed by the uses of molecular markers. Agronomie, 19: 31-43.

Raven PH, Polhill RM (1981). Biography of the Leguminosae. In: Polhill RM and Raven PH (eds), Advances in Legume Systematics, Royal Botanic Gardens, Kew, UK, 27-34.

Reynolds HG (1954). Some interrelations of the Merriam kangaroo rat to velvet mesquite. J. Range Manage., 7: 176-180.

Saidman BO, Vilardi JC, Montoya S, Dieguez MJ, Hopp HE (1998) Molecular markers: a tool for understanding the relationships among species of *Prosopis* (Leguminosae, Mimosoideae). In: Puri S (eds) Tree Improvement: Applied Research and Technology Transfer. Oxford and IBH, New Delhi. pp. 311-324.

Saidman BO, Bessega C, Ferreyra L, Vilardi JC (1998). Random amplified polymorphic DNA (RAPDs) variation in hybrid swarms and pure populations of genus *Prosopis* (Leguminosae). International Foundation for Science (IFS). Proceedings of the workshop " Recent Advances in Biotechnology for Tree Conservation and Management". Florianópolis. pp. 122-134.

Saidman BO, Montoya S, Vilardi J C, Poggio L (1997). Genetic variability and ploidy level in species of *Prosopis*. Bol. Soc. Argent. Bot., 32: 217-225.

Saidman BO, Vilardi J, Pocovi MI and Acreche N (1996). Genetic divergence among species of the section Strombocarpa, genus *Prosopis* (Leguminosae). J. Genet., 75: 139-149.

Saidman BO (1990) Isozyme studies on hybrid swarms of *Prosopis* caldenia and sympatric species. Silvae Genet., 39: 5-8.

Saidman BO, Vilardi JC (1987) Analysis of the genetic similarities among seven species of *Prosopis* (Leguminosae, Mimosoideae). Theor. Appl. Genet., 75: 109- 116.

Saidman BO (1986) Isoenzymatic studies of alcohol dehydrogenase and glutamate oxaloacetate transaminase in four South American species of *Prosopis* and their natural hybrids. Silvae Genetica., 35:3-10.

Sakai AK, Allendorf FW, Holt JS, Lodge DM, Molofsky J, With KA, Baughman S, Cabin RJ, Cohen JE, Ellstrand NC, McCauley DE, O'Neill PO, Parker IM, Thompson JN, Weller SG (2001). The population biology of invasive species. Ann. Rev. Ecol. Syst., 32: 305-332.

Sambrook J, Fritsch E, Maniatis T (1989). Molecular cloning: a laboratory manual. Second edition. Cold Spring Harbor Laboratory Press, Cold Spring Harbor, New York, USA.

Senthilkumar P, Prince WS, Sivakumar S, Subbhuraam CV (2005). *Prosopis juliflora-* a green solution to decontaminate heavy metal (Cu and Cd) contaminated soils. Chemosphere, 60: 1493-1496. doi:10.1016/j.chemosphere.2005.02.022.PMID: 16054919.

Simpson BB (1977). Breeding system of dominant perennial plants of two disjuncts warm desert ecosystems. Oecologia, 27: 203- 226.

Simpson BB, Neff JL, Moldenke AR (1977). *Prosopis* flowers as a resource, vol.5, pp.84-107. In: Simpson BB. (eds) Mesquite: Its biology in two desert ecosystems, US/IBP. Syntesis Series Dowden Hutchinson and Ross.

Sinha S, Rai UN, Bhatt K, Pandey K, Gupta AK (2005) Flyash-induced oxidative stress and tolerance in *Prosopis juliflora* L. grown on different amended substrates. Environ. Monit. Assess. 102:447-457. doi:10.1007/s 10661-005-6397-4.PMID:15869202.

StatSoft Inc. (2001). STATISTICA (data analysis software system), version 6. http://www.statsoft.com.

Van Driesche RG, Bellows TS Jr. (1996). Biological Control. New York: Chapman and Hall. pp. 539.

Verga AR (1995). Genetische Untersuchungen an *Prosopis* chilensis und *Prosopis flexuosa* (Mimosaceae) im trockenen Chaco Argentiniens. PhD thesis, in Göttingen Research Notes in Forest Genetics, Göttinger Fortgenetische Berichte. Abteilung für Forstgenetik und Forstpflanzenzüchtung der Universität Göttingen, ISSN 0940-7103.

Ward SM, Gaskin JF, Wilson LM (2008). Ecological genetics of plant invasion: what do we know?. Invasive Plant Sci. Manage., 1: 98-109.

Ward S (2006). Genetic analysis of invasive plant populations at different spatial scales. Biol. Invasions, 8: 541- 552.

13

Genetic variability, heritability and correlation in some faba bean genotypes (*Vicia faba* L.) grown in Northwestern Ethiopia

Tafere Mulualem[1]*, Tadesse Dessalegn[2] and Yigzaw Dessalegn[3]

[1]Pawe Agricultural Research Center, Pawe, Ethiopia.
[2]Bahir Dar University, Bahir Dar, Ethiopia.
[3]Amhara Regional Agricultural Research Institute, Bahir Dar, Ethiopia.

A study was conducted at Dabat, Northwestern Ethiopia, during 2010 cropping season. Genotypic and phenotypic coefficient of variation, heritability and correlation coefficients were performed for yield and its contributing parameters in 10 faba bean genotypes. Analysis of variance for traits studied showed significant (P<0.01) differences among the genotypes. Phenotypic coefficient of variation values for most characters was closer than the corresponding genotypic coefficient of variation values showing little environment effect on the expression of these characters. The estimated values of broad-sense heritability were found to be between 27 (stand count at emergence) and 81% (grain yield). Heritability values determined were 72, 67, 65, 46, 44, 53, 58 and 45% for 100 seed weight, biological yield, number of pods per plant, number of pods per node, disease status, days to flowering, days to maturity and plant height, respectively. High heritability indicated that selection based on mean would be successful in improving these traits. Positive and significant correlation coefficients were also obtained between number of pods per node and each of plant height (r = 0.676**), number of pods/plant (r = 0.636**) and number of nodes per plant (r = 0.421*). Pods per plant had a significant positive correlation with plant height (p<0.01) in this study.

Key words: Correlation, faba bean, grain yield, genetic variability, heritability.

INTRODUCTION

Faba bean (*Vicia faba* L.) is one of the main pulse crops grown for seed in Ethiopia. It is widely considered as a good source of protein, starch, cellulose and minerals for humans in developing countries and for animals in industrialized countries (Haciseferogullari et al., 2003). In addition, faba bean is one of the most efficient fixers of the atmospheric nitrogen and, hence, can contribute to sustainability or enhancement of total soil nitrogen fertility through biological N_2-fixation (Lindemann and Glover, 2003).

Successful breeding program depends on the magni-

tude of genetic variation in the population. Moreover, reliable estimates of genetic and environmental variations will be helpful in estimating the heritability ratio and consequently predicted genetic advance from selection. These estimates are useful to initiate such breeding program in order to improve productivity and quality of the crop. The fraction of the phenotypic variation in a trait that is due to genetic differences can be measured as the heritability of the trait. The simplest model for variation in a quantitative trait splits phenotypic variation into variation due to genetic differences between individuals, and variation due to environmental differences. El-Kady and Khalil (1979) estimated the heritability values for some yield components in three faba bean crosses. Bora et al. (1998) stated that a high heritability was followed by high genetic advance for fruiting branches/plant, pods/

*Corresponding author. E-mail: tafere_mulualem@yahoo.com.

plant and seed yield/plant indicating the scope for their improvement through selection. The relationship between seed yield and its components would be of considerable value to breeders for screening breeding materials and selecting donor parents for breeding programs. Some traits of faba bean have a positive as well as a negative correlation. For example, Bond (1966), Lawes (1974) and Shalaby and Katta (1976) reported that yield was highly correlated with number of pods/plant, number of seeds and seed weight/plant in field bean. Poulsen and Knudsen (1980) determined the phenotypic correlation coefficient among weight, number of seeds/pod and seed weight. Positive relationships were obtained between weight of seeds/pod and both seed weight and number of seeds/pod. However, no correlation was found between seed weight and number of seed/pod. Ulukan et al. (2003) found the direct and indirect effects of plant height, pod number/plant and seed number/pod upon biological yield. The total determination coefficient was found to be 0.636 in the model used. A significant and positive correlation was reported between seed yield and plant height, 100-seed weight, seed weight/plant and biological yield, but a negative correlation was determined with maturity date (Alghamdi and Ali, 2004).

The present investigation aimed to determine the variability of traits, provide information on interrelationships of yield with some important yield components and to partition the observed genotypic correlations into their direct and indirect effects. This means that yield has direct and indirect relations with other agronomic traits.

MATERIALS AND METHODS

Experimental procedures

The field experiment was conducted at Dabat in 2010 main cropping season. Dabat is located 12° 59' 3" North latitude and 37° 45' 54" East longitude. It is found in the Amhara National Regional State, North Gondar zone. The area receives average annual rainfall of about 1100 mm, which is sufficient for crop production. The major soil type of the study area is vertisol. The average annual maximum and minimum temperatures are 19.9 and 8.58°C, respectively. Nine faba bean genotypes which are inbreds obtained from Holleta Agricultural Research Center (Degaga, Moti, Gebelcho, Dosha, Wolki, Wayu, Selale, EH99051-3 and Holetta-2 one local check that is, CS20DK) were included in this study. The trial was laid down in randomized complete block design with three replications. Each genotype was planted in four rows of 4 m length by 0.4 m spacing between rows. The distance between replications and plots was 1.5 and 0.6 m, respectively. Diammonium phosphate (DAP) fertilizer was applied at the recommended rate of 100 kg/ha at sowing. Sowing was done by hand drilling at a seed rate of 40 seeds/row or 160 seeds/plot.

Data collection

Data on different agronomic traits were collected on plot and plant basis. Hundred seed weight (g), plant height (cm), number of nodes per plant, number of pods per node and number of pods per plant,

were recorded on plant basis; whereas biological yield per plot (g), grain yield per plot (g), biomass (g), days to flowering, days to maturity, disease resistance and stand count at emergency were estimated on plot basis.

Data analysis

The mean values of the recorded data were subjected to analysis of variance (Gomez and Gomez, 1984). The mean squares were used to estimate genotypic and phenotypic variance according to Sharma (1998). Phenotypic coefficient of variation (PCV) and genotypic coefficient of variation (GCV) were estimated according to the method suggested by Burton and De Vane (1953). Broad sense heritability was calculated as the ratio of genotypic variance to the phenotypic variance according to Falconer and Mackay (1996).

Phenotypic and genotypic correlations between yield and yield related traits were estimated using the method described by Miller et al. (1958). The coefficients of correlations at phenotypic level were tested for their significance by comparing the value of correlation coefficient with tabulated r-value at g-2 degree of freedom. However, the coefficients of correlations at genotypic level were tested for their significance using the formula described by Robertson (1959).

RESULTS AND DISCUSSION

Genetic variability and heritability

In the present study, high phenotypic coefficient of variation (PCV) was observed for 100 seed weight (22.08%), disease resistance (23.98), number of pods per node (33.54%), number of pods per plant (60%), grain yield (67.39%) and biological yield (98.49%). Medium PCV were observed for plant height (16%), stand count at emergency (12.21%) and number of nodes per plant (20%), but the remaining traits showed low PCV (Table 1). The result more or less agreed with that reported by Swarup and Changle (1962).

As reported previously by other investigators like Bond (1966), Omar et al. (1970), Mahmoud et al. (1986) and Abul-Naas et al. (1989), the genetic variance components in traits such as seed yield, number of pods per plant, 100 seed weight and plant height, played an important role in the total variation (Table 1).

High genotypic coefficients of variation (GCV) were observed for biomass yield (80.95%), grain yield (60.85%), number of pods per plant (48.73%) and number of pods per node (22.78%). Moderate genotypic coefficients of variation were estimated for plant height, number of nodes per plant and 100 seed weight. These results were also reported by several authors such as Abul-Naas et al. (1989), El-Hosary and Nawar (1984) and Mahmoud et al. (1986). Low GCV was observed for days to flowering (3.33%), days to maturity (0.85%) and stand count at emergency (6.35%) (Table 1). High GCV value of characters suggested the possibility of improving these traits through selection. Similarly, El-Hosary and Nawar (1984) estimated different levels of GCV in faba bean. Moreover, the differences between PCV and GCV were

Table 1. Variances, coefficient of variations and heritability.

Trait	$\delta^2{}_p$	$\delta^2{}_g$	H%	PCV	GCV
Days to flowering	7.12	3.81	53.51	4.54	3.33
Days to maturity	2.47	1.44	58.29	1.12	0.85
Stand count at emergency	108.06	29.25	27.06	12.21	6.35
Plant height	193.68	88.77	45.83	16.0	10.83
Pods per plant	33.28	21.84	65.62	60.0	48.73
Pods per node	0.52	0.24	46.15	33.54	22.78
Nodes per plant	11.96	3.32	27.76	20.06	10.56
100 seed weight	213.59	154.28	72.23	22.08	18.76
Chocolate spot	39.35	17.41	44.25	23.98	15.95
Biomass yield	496.94	335.72	67.56	98.49	80.95
Seed yield	56.47	46.04	81.53	67.39	60.85

$\delta^2{}_p$, Phenotypic variance; $\delta^2{}_g$, genotypic variance; H, heritability; PCV, phenotypic coefficient of variation; GCV, genotypic coefficient of variation.

very narrow which indicated the importance of genetic variance in the inheritance of the studied characters.

Heritability (H) in broad sense estimates were generally high for most studied traits which ranged from 27.06% for stand count at emergency to 81.53% for grain yield. The highest estimate of broad sense heritability (H) was recorded by grain yield, 100 seed weight, biological yield, number of pods per plant, number of pods per node, disease status, days to flowering, days to maturity and plant height, with heritability of 81.53, 72.23, 67.56, 65.62, 46.15, 44.25, 53.51, 58.29 and 45.83%, respectively (Table 1). Hence, these traits can be assumed as mainly determined by their genetic constitution. Stand count at harvest (27.06%) showed medium heritability including number of pods per plant (27.76%) which makes selection for these traits difficult because environmental effect is more evident than genetic effect. However, Dixit et al. (1970) reported that high heritability and GCV were not always associated with high genetic advance. Meanwhile, Swarup and Changle (1962) reported that both heritability ratio and GCV% gave the best picture for the expected genetic advance. Those traits that showed high and moderate heritability are found to have high GCV value than traits that showed low heritability. Selection for these traits is relatively easy because most of the variation is genetic rather than environmental. On the other hand, traits with high PCV have less heritability which means variation for these traits is more of environmental than genetic and it is not advisable to select for these traits. Dabholkar (1992) explained that whenever values are stated for heritability of a character, it refers to a particular population under particular environmental conditions. He classified heritability estimates as low (5 to 10%), medium (10 to 30%) and high (>30%). Accordingly, all the agronomic characters considered for analysis showed high heritability, constituting high breeding value which has

more additive genetic effects which is important for crop improvement.

Correlation coefficients

Positive and significant correlation coefficients were obtained between number of pods per node and each of plant height (r = 0.676**), number of pods/plant (r = 0.636**) and number of nodes per plant (r = 0.421*). In the present study, pods per plant had a significant positive correlation with plant height (p<0.01) but, according to Alan and Geren (2007) and Ulukan et al. (2003), such pair of characters showed a low level of significance (p<0.05) and non-significance, respectively for faba bean as correlation. Seed yield was strongly correlated with number of nodes per plant, number of pods per node, number of pods per plant and plant height with the value of 0.484**, 0.56**, 0.634** and 0.649**, respectively (Table 2). This result shows the yield of plant is determined by these traits. There was a significant correlation between biological yield and plant height as it was also reported by Ulukan et al. (2003). Negative and significant correlations were observed between 100-seed weight and number of pods per plant (r = -0.530**) as well as number of pods per node (r = -0.418*). These results are in disagreement with those obtained by Bond (1966), Lawes (1974), Shalaby and Katta (1976), Poulsen and Kundesn (1980), Ulukan et al. (2003) and Alghamdi and Ali (2004).

Bianco et al. (1979) found positive relationships between yield and plant height, number of branches and pods/plant, number of seeds/pod and 1000-seed weight, whereas, seed yield was negatively correlated with flowering date and the lowest node bearing pods. These findings indicate that selection for each or both of number of pods, nodes and biomass would be accompanied by high yielding ability under such conditions.

Table 2. Correlation coefficients of the main traits of faba bean genotypes.

	DF	DM	NPP	PPN	PPP	PH	HSW
DF							
DM	0.719**						
NPP	0.011	0.1009					
PPN	0.427*	0.382*	0.432*				
PPP	0.1064	0.195	0.410*	0.477**			
PH	0.2619	0.1971	0.421*	0.676**	0.636**		
HSW	0.0694	-0.268	-0.223	-0.418*	-0.530**	-0.305	

DF, Days to flowering; DM, days to maturity; NPP, nodes per plant; PPN, pods per node; PPP, pod per plant; PH, plant height; HSW, hundred seed weight; * and ** significant at 0.05 and 0.01 level of probability, respectively.

Conclusion

The present study illustrated the existence of wide ranges of variations for most of the traits among faba bean genotypes and opportunities of the genetic gain through selection or hybridization. Phenotypic and genotypic correlation analysis showed the positive correlation of grain yield with important agro-morphological characters. Hence, improving one or more of the traits could result in high grain yield for faba bean. Presence of genetic variability and heritability estimates would be helpful to the breeder to estimate genetic advance and to predict percentage of genetic advance in the population(s) under study. Success of genetic improvement is attributed to the magnitude and nature of variability present for a specific character. Accordingly, all the agronomic characters considered for analysis showed high heritability, constituting high breeding value which has more additive genetic effects which is important for crop improvement.

ACKNOWLEDGMENTS

Our deep gratitude and acknowledgement goes to Local Seed Business (LSB) project later changed to ISSD, for providing the research fund.

REFERENCES

Abul-Naas AA, Abdel-Barry AA, Rady, S, El-Shawaf IS, El-Hosary AA (1989). Breeding studies in yield and its components in faba bean (*Vicia faba* L.). Egypt Agron. 14(1-2):117-149.

Alghamdi SS, Ali Kh A (2004). Performance of several newly bred faba bean lines. Egypt. J. Plant Breed. 8:189-200.

Bianco VV, Damato G, Miccolis V, Polignano G, Poerceddu E, Scipa G (1979). Variation in a collection of Vicia faba L. and correlations among agronomically important characters. In: Some current research on *Vicia faba* in Western Europe. Ed. D.A. Bond, G.T. Scarascia-Mugnozza and M.K. Poulsen. Pub. EEC, EUR 6244:125-143.

Bond DA (1966). Yield and components of yield in diallel crosses between inbred lines of winter beans (*Vicia faba* L.). J. Agric. Sci. Camb. 57:352-336.

Bora GC, Gupta SN, Tomer YS, Singh S (1998). Genetic variability, correlation and path analysis in faba bean (*Vicia faba*). Indian J. Agric. Sci. 68(4):212-214.

Burton GW, de Vane EW (1953). Estimating heritability in tall Fescue (*Festuca arundinacea*) from replicated clonal material. Agron. J. 45:478-481.

Dabholkar AR (1992). Elements of biometrical genetics. Concept Publishing Company, (Eds.), New Dehli. p. 431.

El-Hosary AA, Nawar AA (1984). Gene effects in field beans (vicia faba L.) earliness and maturity. Egypt J. Genet. Cytol. 13:109-119.

El-Kady MAK, Khalil SA, (1979). Behaviour of seed yield components in cross between broad bean cultivars and selection for superior yield. Egypt. J. Agron. 4:159-170.

Falconer DS, Mackay TFC (1996). Introduction to quantitative genetics. 4th Ed. Longman, New York. p. 580.

Gomez KA, Gomez AA (1984). Statistical procedures for agricultural research. 2nd Ed. John Willey and Sons, New York.

Haciseferogullari H, Gezer I, Bahtiyarca Y, Menges HO (2003). Determination of some chemical and physical properties of Sakiz faba bean (vicia faba L. Var. major).J. Food Eng. 60:475-479

Lawes DA (1974). Field beans: improving yield and reliability. Span 17:21-23.

Lindemann C, Glover R (2003). Nitrogen fixation by legumes. New Mexico State University, Mexico. http://www.cahe.nmsu.edu/pubs/_a/a-129.pdf.

Mahmoud SA, Al-Ayobi D (1986). Combining ability for yield and its components in broad bean (*Vicia faba* L.) Ann. Agri. Sci. 24(4):1923-1936.

Miller PA, Williams C, Robinson HF, Comstock RE, (1958). Estimates of genotypic and Environmental variances and co-variances in upland cotton and their implications in selection. Agric. J. 50:126-131.

Omar Abdel-Aziz, Selim AKA, Hassanein SH, Abdel-Hafiz SM (1970). Mode of inheritance of protein content and seed weight in broad bean (*Vicia faba* L.) Ain Shams Univ. Res. Bull. 626:1-10

Poulsen MH, Knudsen JN (1980). Breeding for many small seeds/pod in Vicia faba L. FABIS Newsletter 2:26-28.

Robertson A (1959). The sampling variance of the genetic correlation coefficient. Biometrics 15:469-485.

Shalaby TA, Katta YS (1976). Path coefficient analysis of seed yield and some agronomic characters in field beans (*Vicia faba* L.). J. Agric. Res. Tanta Univ. 2(2):70-79.

Sharma JR (1998). Statistical and biometrical techniques in plant breeding. New Age International Publication, New Delhi. p. 432.

Swarup V, Changle DS (1962). Studies on genetic variability in Sorghum phenotypic variation and its inheritable component in some important quantitative characters contributing towards yield. Ind. J. Genet. 22:31-36

Ulukan H, Culer M, Keskin S (2003). A path coefficient analysis of some yield and yield components in faba bean (Vicia faba L.) genotypes. Pak. J. Biol. Sci. 6(23):1951-1955.

14

Genetic polymorphism of kappa-casein gene in indigenous Eastern Africa goat populations

SK. Kiplagat[1], M. Agaba[3], IS. Kosgey[2], M. Okeyo[3], D. Indetie[4], O. Hanotte[5] and MK. Limo[1]*

[1]Department of Biochemistry and Molecular Biology, Egerton University, Kenya.
[2]Department of Animal Sciences, Egerton University, Kenya.
[3]International Livestock Research Institute (ILRI), Nairobi, Kenya.
[4]Kenya Agricultural Research Institute (KARI), National Beef Research Centre-Lanet, Kenya.
[5]School of Biology, University of Nottingham, UK.

Indigenous goat breeds kept by majority of smallholder rural farmers in Eastern Africa are adapted to the local environment. These goats are critical for nutrition and income of their keepers. Milk production per doe is extremely variable. The variation in milk yield in goats is due to varied management practices and variability in genetic make-up of the animals. The variation in kappa-casein gene and the distribution frequencies of its variants amongst indigenous Eastern Africa goat populations were investigated. A 458 base pairs sequence in exon 4 of 296 goat samples were amplified, sequenced and variation analyzed. Nine point mutations corresponding to base transitions were identified. Three sites were synonymous substitutions while the other six mutations were non-synonymous. All the amino acid substitutions were conservative. Analysis of the association of the mutations yielded nine haplotypes. The occurrence of these haplotypes in ten goat populations indicated that only one haplotype occurred at a rather high frequency. The prevalent κ-casein variant was CSN3*B with frequencies ranging from 0.750 to 0.953. The second most common allele was CSN3*A. Further studies on other casein loci are necessary to establish associations of all the casein mutations and the effects of the haplotypes to milk production traits.

Key words: Goat, indigenous, k-casein, polymorphism.

INTRODUCTION

Indigenous goats are naturally highly adapted to hot environments and can withstand recurrent drought in Eastern Africa better than cattle. Goats are important to the subsistence, economic and social livelihoods of a large human population in the region. They are especially important to women, children and the aged who are often the vulnerable members of the society in terms of under-nutrition and poverty (Kosgey, 2004). Goat milk exceeds cow milk in monounsaturated, polyunsaturated fatty acids and medium chain triglycerides in which all are known to be beneficial for human health, especially for cardio-vascular conditions (Haenlein, 1992). However, milk pro-

duction per doe is relatively low and extremely variable. The variation in milk yield being due to manage-ment practices like disease control, feeding and housing and variability in genetic make-up of the goats (Azevedo et al., 1994). This provides an opportunity for improve-ment of goat milk production by changes in management practices and adoption of genetic improvement technologies.

The association of genetic polymorphism with milk production and composition has stimulated interest in using genetic polymorphism of casein genes in molecular-marker assisted selection (MAS) to improve milk productivity in farm animals (Kumar et al., 2006). Casein is made up of many components; the main ones are α_{s1}-casein, α_{s2}-casein, β-casein and κ-casein (Walstra, 1999). Kappa-casein plays an important role in the formation, stabilization and aggregation of the casein

*Corresponding author. E-mail: mklimoh@yahoo.com.

micelles thus altering the manufacturing properties and digestibility of milk (Jann, 2004). The casein proteins are encoded in a locus that comprises four casein genes; the evolutionary related calcium - sensitive casein encoding genes (α_{s1}, α_{s2} and β) and the functionally related κ-casein gene (Rijnkels et al., 1997). Kappa-casein gene (CSN3) sequence and the promoter region of the gene have been reported (Coll et al., 1993, Coll et al., 1995). The goat κ-casein mRNA contains an open reading frame of 579 bp coding for 21 amino acids for signal peptide and 171 amino acids of mature protein, the coding sequence for mature protein is contained in exon three (9 amino acids) and exon four (Yahyaoui et al., 2003).

Studies that have examined the CSN3 in cattle have indicated an association of some κ-casein alleles and milk yield, composition and quality. Studies by Bovenhuis et al. (1992) which involved 305 days milk production records of 10,151 first lactation cows observed that k-casein genotypes had a significant effect on milk production. Kappa-Casein AA cows produced 173 kg of milk more than κ-casein BB cows. Furthermore, κ-casein genotypes had a highly significant effect on protein content; the κ-casein BB cows produced milk with a 0.8% higher protein content than that of the AA cows. Generally, majority of researchers believe that the k-casein B variant is associated with higher fat, protein and casein in the milk and has a significant influence on cheese making properties of milk and superior rennet co-agulation properties in comparison to AA or AB variants (Gangaraj et al., 2008). The genotypes BB and AB are used in artificial insemination programs to obtain a greater increase of the frequency of these alleles in cattle populations of commercial interest (Otaviano et al., 2005).

Caseins are rapidly evolving gene family, presumably due to the minimal structural requirements for functioning (Bonsing and Mackinlay, 1987). Whereas CSN3 is considered to be monomorphic in sheep (Moioli et al. 1998), however, recent studies on goat CSN3 showed that the gene is highly polymorphic (Caroli et al., 2001; Yahyaoui et al., 2001; Angiolillo et al., 2002; Yahyaoui et al., 2003; Chessa et al., 2003; Jann et al., 2004; Reale et al., 2005; Prinzenberg et al., 2005). A total of 16 poly-morphic sites have been identified in the domesticated goat, of which 13 are protein variants and 3 are silent mutations involving a total of 15 polymorphic sites in CSN3 exon 4 (Prinzenberg et al., 2005). Recent studies emphasised, that the analyzed breeds show differences in the occurrence and frequency of the alleles, the allele distribution reflecting the geographic origin (Moioli, 2007). The distribution of such alleles has been influenced either by selection pressure for milk production or, more likely, by genetic drift (Yahyaoui et al., 2003).

The purpose of the current study was to analyze the genetic polymorphism of κ-casein gene exon 4 by sequencing. And further determine the distribution frequencies of the variants in one exotic and nine indigenous goat popula-tions found in five countries in Eastern Africa. This study

serves as the bases for the design of an association studies that will provide information on breeding of goats that have better milk production traits.

MATERIALS AND METHODS

Sample collection and DNA extraction

A total of nine indigenous goat populations from five Eastern Africa countries were considered for the study. In addition, one exotic breed was also analysed. Small East Africa goat (n = 35) Kenya (Baringo district), Small East Africa goat (n = 27) Kenya (Samburu district), Long-Earned Somali (n = 27) Ethiopia, Maasai (n = 30) Tanzania, Keffa (n = 32) Ethiopia, Afar (n = 33) Ethiopia, Shilluk (n = 16) Sudan, Short-Eared Somali (n = 35) Somalia, Long-Eared Somali (n = 28) Somalia, Toggenberg (n = 33) Germany.

Genomic DNA was isolated from the whole blood samples using the method described by Sambrook et al. (1989) with minor modifications. After checking the quantity and quality of the DNA using a NanoDrop ND-1000 spectrophotometer (NanoDrop Technologies, Wilmington, DE), the DNA was diluted to a final concentration of 50 ngμl $^{-1}$ in water and stored at 4°C, until the time for use.

DNA amplification and purification

A 458 bp fragment of goat κ-casein exon 4 was amplified by polymerase chain reaction (PCR) from gDNA (genomic DNA) samples using primers CSN3_1A (TATGTGCTGAGTAGGTATCC) and CNS3_1B (TTGTCCTCTTTGATGTCTCC) which were designed relative to the caprine κ-casein cDNA (Coll et al., 1993). The PCR was performed in a 15 μl final volume containing primer CNS3_1A (20 pM; 0.15 μl), primer CNS3_1 B (20 pM; 0.15 μl), ABgene Reddy mix (7.5 μl), water (5.7 μl) and gDNA (1.5). Thermal cycling conditions was: 94°C for 4 min, 5 cycles of 94°C for 15 s, 59°C for 1 min and 70°C for 30 s, followed by 30 cycles of 94°C for 15 s, 54°C for 30 s, 70°C for 30 s, with a final extension at 72°C for 10 s. To confirm amplifi-cation, PCR products were resolved by agarose gel electrophoresis and visualized by gel red staining. PCR products were purified using the QIAquick PCR purification kit (QIAGEN GmbH, Germany).

DNA sequencing reactions

Purified fragments were sequenced to determine the precise nucleotide sequence. Direct sequencing of CSN3 exon 4 was per-formed using the same primers used for amplification that is primers CSN3_1A and CSN3_1B. Sequencing was done using the BigDye® Terminator version 3.1 Cycle Sequencing Kit (Applied Bio-systems). The purified products from the BigDye® Terminator were electrophoresed on an ABI 3730 XL automated capillary DNA sequencer (Applied Biosystems) for one hour and the resulting electropherograms were analyzed.

Sequence data analysis

BioEdit (Ibis Biosciences) was used for the viewing and editing of the DNA sequences. Multiple alignments of the sequences were performed using clustalX program (Thompson et al., 1997). The polymorphic sites were confirmed by visual examination of the electropherograms. The translation of DNA sequence to amino acid sequence was done using ExPASy - Translate tool. Haplotypes were inferred using Clark's algorithm (Clark, 1990) and their fre-

Table 1. Kappa-casein gene haplotypes among indigenous Eastern Africa goats.

Nucleotide position	Protein position	Haplotypes								
		A	B	D	L	M	N	O	P	Q
245	43	T (Tyr)		C (Tyr)		C (Tyr)	C (Tyr)			C (Tyr)
247	44	A (Gln)		G (Arg)						
309	65	G (Val)			A (Ile)					A (Ile)
384	90	G (Asp)				A (Asn)	A (Asn)			
471	119	G (Val)	A (Ile)	A (Ile)	A (Ile)	A (Ile)	A (Ile)		A (Ile)	
545	143	C (Asn)						T (Asn)		
550	145	T (Val)				C (Ala)	C (Ala)			
584	156	G (Ala)					A (Ala)		A (Ala)	
591	159	T (Ser)		C (Pro)	C (Pro)		C (Pro)			C(Pro)

Nucleotide positions are compared with GenBank accession No. X60763.

frequencies determined by direct counting.

RESULTS

Variation in κ-casein gene

CSN3 exon 4 sequences from 296 goats were analyzed for variations. Nine polymorphic sites were identified (Table 1). All were point mutations; corresponding to base transitions, where T and G are substituted by C and A, respectively. Most of the polymorphic sites were heterozygous, except at position 471 which was homozygous for genotype AA in 90% of the goats (266) and heterozygous for genotype AG in the rest.

Analysis of the deduced amino acid sequences showed that three sites were synonymous substitutions corresponding to amino acid residues tyrosine (at position 245), asparagine (545) and alanine (584). The other six mutations were non-synonymous producing codon changes of glutamine to arginine (at position 247), valine to isoleucine (309 and 471), aspartate to asparagine (384), valine to alanine (550) and serine to proline (591) (Table 1). All the amino acid substitutions are conservative.

Kappa-casein haplotypes

Nine haplotypes were inferred using Clark's algorithm from the 296 animals (592 chromosomes) sampled. Table 1 shows the haplotype alleles at the variable sites. The analysis of the κ-casein haplotypes in the ten goat populations indicated that only one haplotype occur at a rather high frequency. The prevalent κ-casein variant was CSN3*B with frequencies ranging from 0.750 (Shilluk) to 0.953 (Keffa). The second most common allele was CSN3*A. This allele was inconsistent in that it was not found in three goat populations, while in the rest, the frequencies ranged from low value of 0.014 (Short Eared goat) to a relatively high value of 0.250 (Shilluk). The haplotype frequency in the populations under study is shown in Table 2.

DISCUSSION

Kappa-casein gene variation

K-casein gene exon 4 from 296 goats collected from ten populations, obtained from five Eastern African countries were analyzed for sequence variations. A total of nine SNPs were identified. Seven SNPs had been identified previously in other goat populations. Yahyaoui et al. (2001), first identified polymorphism in Spanish and French breeds in the positions 245 (nucleotide change from T to C), 309 (G to A), 471 (G to A) and in the position 591 (T to C). Polymorphism in the position 247 (A to G) was first identified by Caroli et al. (2001) in Italian goat breeds. While polymorphism in the positions 384 (G to A) and 550 (T to C) was previously identified by Prinzenberg et al. (2005). These SNPs have also been described in various goat populations outside Eastern Africa (Yahyaoui et al., 2003; Angiolillo et al., 2002; Jann et al., 2004). The high number of similarity in polymorphic sites amongst most of the populations in different parts of the world is suggestive of similarity in evolutionary processes undergone by these populations. Two new SNPs identified at nucleotide positions 545 (C/T) and 584 (G/A), occurred in two (Small East Africa goat-Baringo district and Maasai) and three (Long Eared Somali, Short Eared Somali- Somalia and Long Eared Somali-Somalia) goat populations, respectively. The two new SNPs are neutral mutations that are of no consequence on κ-casein protein structure and function. However, they can be used as markers if they are established to be in linkage disequilibrium with alleles associated with milk production traits. Most of the genotypes at polymorphic sites were heterozygous. However, polymorphic site 471 was homozygous for genotype AA in 90% of the samples analyzed, agreeing with earlier observation (Mercier et al., 1976) which postulated isoleucine to be the predominant amino acid at the corresponding amino acid sequence.

Three of the sites are silent mutations corresponding to

Table 2. Haplotype frequencies for the kappa-casein locus in ten goat populations analyzed.

Population	n.	A	B	D	L	M	N	O	P	Q
SEAgB	35	0.029	0.913	-	-	0.029	-	0.029	-	-
SEAgS	27	0.019	0.944	-	-	0.037	-	-	-	-
LES	27	0.037	0.926	-	-	0.019	-	-	0.019	-
MS	30	0.033	0.933	-	-	-	-	0.033	-	-
KF	32	-	0.953	-	0.016	-	-	-	-	0.031
AF	33	-	0.939	0.045	-	-	-	-	-	0.015
SH	16	0.250	0.750	-	-	-	-	-	-	-
TG	33	0.121	0.879	-	-	-	-	-	-	-
SES-S	35	0.014	0.914	-	0.029	-	0.014	-	0.014	0.014
LES-S	28	-	0.946	-	-	0.036	-	-	0.018	-

SEAgB, Small East Africa goat (Baringo district); Small East Africa goat (Samburu district); LES, Long Eared Somali; MS, Maasai; KF, Keffa; AF, Afar; SH, Shilluk; TG, Toggenberg; SES-S, Short Eared Somali-Somalia; LES-S, Long Eared Somali-Somalia. n., Sample size. Haplotype frequencies calculated using chromosome numbers (= 2n).

amino acid residues tyrosine (at position 245), asparagine (545) and alanine (584). The other six mutations were non-synonymous producing codon changes of glutamine to arginine (at position 247), valine to isoleucine (309 and 471), aspartate to asparagine (384), valine to alanine (550), serine to proline (591). Individual κ-casein molecules have been speculated to cross-link into disulfide-bonded polymers with a structure such that the hydrophilic tails project into the milk serum and the hydrophobic regions attach to the micelle core containing calcium-sensitive caseins (Phadungath, 2005). All the non-synonymous single nucleotide polymorphisms caused conservative amino acid substitutions. Thus, the mutations are of less consequence in the function of the κ-casein protein of stabilization of the micelles.

Kappa-casein haplotypes in ten goat populations

A total of nine haplotypes were inferred using Clarks' algorithm (Clark, 1990). Nomenclature developed by Prinzenberg et al. (2005) was adopted in naming previously identified haplotypes, while the new haplotypes were named in alphabetical order following the lastly named haplotype. Out of the nine haplotypes identified, five had previously been found in other goat breeds (Yahyaoui et al., 2001; Caroli et al., 2001; Angiolillo et al., 2002; Yahyaoui et al., 2003; Jann et al., 2004; Prinzenberg et al., 2005).

The analysis of κ-casein variation in ten goat populations indicated that only one haplotype occurred at a rather high frequency. The most prevalent CSN3 haplotype was CSN3*B with frequencies ranging from 0.750 (Shilluk) to 0.953 (Keffa). The haplotype is most likely to be associated with production trait(s) which has/have being selected for, by the local communities, over many generations. This haplotype appears to be almost fixed in these populations. The second most common allele was CSN3*A. This allele was inconsistent in that it was not found in three goat populations (Keffa, Afar and Long Eared Somali-Somalia); this inconsistency might be due to low number of samples analyzed and low haplotype frequency in most of the populations. The frequencies in the rest of the populations ranged from low value of 0.014 (Short Eared Somali-Somalia) to a relatively high value of 0.250 (Shilluk). These findings were in agreement with earlier report by Prinzenberg et al. (2005) in goat breeds from Europe, Africa and Asian part of Turkey, their most common variant was also CSN3*B with frequencies ranging from 0.260 (Hair goat) to 0.674 (Angora goat). Their second most common allele was also CSN3*A, with frequencies ranging from 0.151 (Angora goat) to 0.414 (Borno goat). Two variants were found in only one population each. The haplotype CSN3*D was only found in Afar with a relatively low frequency of 0.045 while variant CSN3*N appeared in Short Eared goat population with a low frequency of 0.014. The alleles CSN3*L and CSN3*O occurred in low frequencies in two goat populations; CSN3*L was found in Short Eared Somali population (Somalia) with a relatively low frequency of 0.029 and in Keffa goat population with a rather very low frequency of 0.016. Whereas, allele CSN3*O appeared in Maasai population with a frequency of 0.033 and in Small East Africa goat (Baringo district) the frequency was 0.029. Haplotype CSN3*M was the most widely distributed variant occurring in four goat populations(Small East Africa goat-Baringo district, Small East Africa goat-Samburu district, Long eared Somali, and Long Eared Somali-Somalia). Based on the high frequency (approximately 90%) of haplotype CSN3*B in nine out of the ten goat populations, it appears that the populations are rather similar at this locus. The relatively low sample size of Shilluk goat population might be the cause of its slight variations in haplotype frequencies. However, due to independent calculation of haplotype frequencies in each population samples using chromosome numbers, the low sample size of Shilluk population does not affect the haplotype frequencies in other goat populations.

The greatest number of haplotypes at CSN3 (six) occurred in Short Eared Somali-Somalia, CSN3*B was at high frequency followed by CSN3*L, while the rest occurred at equal frequency: CSN3*A, CSN3*N, CSN3*P and CSN3*Q. Four alleles were found in Small East Africa goat (Baringo district) and Long Eared Somali. Shilluk and Toggenberg were almost monomorphic at this locus, with allele CSN3*A occurring at a low frequency of 0.250 and 0.121, respectively. In future, it is necessary to determine nucleotide variations within regions of the CSN3 not analyzed in this study, these includes the promoter region, exon 1, 2 and 3 and introns and variation in other casein loci.

Conclusion

CSN3 is a highly polymorphic locus with a total of seventeen SNPs identified in various goat populations globally as at now. Out of the total SNPs, only nine were identified in Eastern Africa goat populations, two of which had not been previously identified in other goat populations. Six out of the nine SNPs are non-synonymous mutations, leading to conservative amino acid substitutions. With the inclusion of the newly inferred haplotypes, the total number of κ-casein haplotypes have rapidly increased from sixteen to twenty. In Eastern Africa goat populations, nine haplotypes were inferred, four of which had never been identified in other goat populations. CSN3*B remain the most common haplotype amongst the majority of breeds in various geographical locations, including Eastern Africa goat populations. On considering the frequencies of the CSN3 haplotypes in the goat populations studied, it appears that variation in this gene may not influence milk yield; therefore, further studies on other casein loci are necessary to establish associations of all the casein mutations and the effects of such haplotypes on milk production traits.

ACKNOWLEDGEMENTS

We would like to thank ASARECA-AARNET, ILRI and Egerton University for financial support.

REFERENCES

Angiolillo A, Yahyaoui MH, Sanchez A, Pilla F, Folch JM (2002). Characterization of new genetic variant in the caprine k-casein gene. J. Dairy Sci. 85: 2679-2680.

Azevedo J, Mascarenhas M, Valentim R, Almeida J, Silva S, Pires S, Teixeira M (1994). Preservação e valorização dos ovinos da raça Churra da Te rr a Q u e n t e. Relatório Final do Projecto PAN I da Associação Nacional de Criadores de Ovinos Churra da Terra Quente, Torre de Moncorvo, Portugal.

Bonsing J, Mackinlay AG (1987). Recent studies on nucleotide sequences encoding the caseins. J. Dairy Res. 54: 447-61.

Bovenhuis H, van Arendonk JAM, Korver S (1992). Associations between milk protein polymorphisms and milk production traits. J. Dairy Sci. 75: 2549-2559.

Caroli A, Jann O, Budelli E, Bolla P, Ja¨ger S, Erhardt G (2001). Genetic polymorphism of goat κ-casein (CSN3) in different breeds and characterization at DNA level. Anim. Genet. 32: 226–230.

Chessa S, Budelli E, Gutscher K, Caroli A, Erhardt G (2003). Short communication: simultaneous identification of five k-casein (CSN3) alleles in domestic goat by Polymerase Chain Reaction-Single Strand Conformation Polymorphism. J. Dairy Sci. 86: 3726-3729.

Clark GA (1990). Inference of haplotypes from PCR-amplified samples of diploid populations. Mol. Bio. Evol. 7: 111-122.

Coll A, Folch JM, Sanchez A (1993). Nucleotide sequence of the goat Kappa casein cDNA. J. Anim. Sci. 71-2833.

Coll A, Folch JM, Sanchez A (1995). A structural features of the 5' flanking region of the caprine kappa casein gene. J. Dairy Sci. 78: 973-977.

Gangaraj DR, Shetty S, Govindaiah MG, Nagaraja CS, Byregowda SM and Jayashankar MR (2008). Molecular characterization of kappa-casein gene in Buffaloes. Sci. Asia 34: 435–439.

Haenlein GFW (1992). Role of goat meat and milk in human nutrition. In: Proceedings of the Fifth International Conference on Goats, vol. II, part II. Indian Council of Agricultural Research Publishers, New Delhi, India, pp. 575–580.

Jann OC, Prinzenberg FM, Luikart G, Caroli A, Erhardt G (2004). High polymorphism in the Kappa-casein (CSN3) gene from wild and domestic caprine species revealed by DNA sequencing. J. Dairy Res. 71: 188-195.

Kosgey IS (2004). Breeding objectives and Breeding Strategies for Small Ruminants in the Tropics. Ph.D. Thesis, Wageningen University, The Netherlands, 272p.

Kumar D, Gupta N, Ahlawat SPS, Satyanarayana R, Sunder S, Gupta SC (2006). Single strand confirmation polymorphism (SSCP) detection in exon I of the α - lactalbumin gene of Indian Jamunapri milk goats (Capra hircus). Genet. Mol. Bio. 29: 271-74.

Mercier JC, Chobert JM (1976). Comparative study of the amino acid sequences of the caseinomacropeptides from seven species. Fed. Eur. Biol. Soc. Lett. 72-208.

Moioli B, D'Andrea M, Pilla F (2007). Candidate genes affecting sheep and goat milk quality. Small Rumin. Res. 68: 179–192.

Moioli B, Pilla F, Tripaldi C (1998). Detection of milk protein genetic polymorphism in order to improve dairy traits in sheep and goats: a review. Small Rumin. Res. 27: 185-195.

Otaviano AR, Tonhati H, Sena JAD, Munoz MFC (2005). Kappa-casein gene study with molecular markers in female buffaloes (Bubalus bubalis). Genet. Mol. Bio. 28: 237–41.

Phadungath C (2005). Casein micelle structure: a concise review. Songklanakarin J. Sci. Tech. 27: 201-212.

Prinzenberg EM, Gutscher K, Chessa S, Caroli A, Erhardt G (2005). Caprine kappa-Casein (CSN3) Polymorphism: New Developments in Molecular Knowledge J. Dairy Sci. 88: 1490-1498.

Reale S, Yahyaoui MH, Folch JM, Sanchez A, Angiolillo AF (2005). Short communication: Genetic polymorphism of the κ-casein (CSN3) gene in goats reared in Southern Italy. Ital. J. Anim. Sci. 4: 97-101.

Rijnkels M, Kooiman PM, Boer HA, Pieper FR (1997). Organization of the bovine casein gene locus. Mammalian Genome 8: 148-152.

Sambrook J, Fritsch EF, Marriatis T (1989). Molecular cloning: A laboratory manual 2nd edition. Cold spring harbour, cold spring laboratory press, N.Y., USA, 224p.

Thompson JD, Gibson TJ, Plewniak F, Jeanmougin F, Higgins DG (1997). "The CLUSTAL_X windows interface: flexible strategies for multiple sequence alignment aided by quality analysis tools". Nucl. Acids Res. 25: 4876-4882.

Walstra P (1999). Casein sub-micelles: do they exist? Inter. Dairy J. 9: 189-192.

Yahyaoui MH, Angiolillo A, Pilla F, Sanchez A, Folch JM (2003). Characterization and genotyping of the caprine kappa casein variants. J. Dairy Sci. 86: 2715-2720.

Yahyaoui MH, Coll A, Sanchez A, Folch JM (2001). Genetic polymorphism of the caprine kappa casein. J. Dairy Res. 68: 209-216.

15

Molecular characterization of cotton using simple sequence repeat (*SSR*) markers and application of genetic analysis

Ambreen Ijaz[1], Sadia Ali[1], Usman Ijaz[2], Smiullah[2] and Tayyaba Shaheen[1]

[1]Department of Bioinformatics and Biotechnology, GC University, Faisalabad, Pakistan.
[2]Ayub Agriculture Research Institute, Faisalabad, Pakistan.

Cotton is grown worldwide for the production of fiber and an important oil seed crop. Genetic diversity and correlation between varieties are of great importance for cotton breeding. To find out polymorphism and genome changeability, DNA markers are most comprehensively used. The current work was conducted to study the genetic relatedness among 20 cotton genotypes using simple sequence repeats markers using 31 simple sequence repeats (*SSRs*) belonging to BNL series. These 31 DNA markers amplified fragments of 80 to 340 bp in size through SSR profiling. Seven markers out of 31 were found to be polymorphic. A total of 41 loci were amplified out of these, 17 loci were informative showing 41% polymorphism. On average, 1.3 loci per primer were amplified. A dendrogram was constructed using pair group method of arithmetic means (UPGMA) method comprising eight main groups that is, A, B, C, D, E, F, G and H. Genotypes MNH-147 and GOHAR-87 emerged as genetically most similar with a value of 97% followed by 97% similarity between the genotypes CIM-443 and FVH-53. The genotypes CIM-1100 and BH-136 were found most divergent showing 58% genetic similarity. Dissimilarity coefficient of the generated information obtained on genetic relatedness would be supportive in further breeding of cotton, the selection of parents for crossing and will also be helpful in widening the genetic base of breeding materials.

Key words: Genetic diversity, simple sequence repeat (SSR).

INTRODUCTION

The cotton genus (*Gossypium*) have almost 50 species, disseminated in arid to semi-arid regions of the tropic and subtropics (Wendel and Cronn, 2003; Rahman et al., 2012). Cotton is a vital source of seed oil and protein meal. Cotton (*Gossypium* spp.), a major cash crop in the world, is a foremost natural fiber and is cultivated in more than 50 countries. China, India, USA, Pakistan and Uzbekistan are the foremost cotton growing countries of the world (Khadi et al., 2010; Rahman et al., 2012). Cotton uniqueness is evident as the four separate species were cultivated autonomously (Brubaker et al.,

1999) for the particular single-celled trichomes, or fibers, that occur on the epidermis of the seeds. This process implicated four species, two from Africa-Asia, namely *Gossypium arboretum* and *Gossypium herbaceum* and two from Americas, *Gossypium hirsutum* and *Gossypium barbadense*, (Wendel and Cronn, 2003). Several thousand years back, the uniqueness of cotton fibers discovered by native people made them helpful for textiles and other applications. Consequently, cotton cultivation became increasingly renowned, such that over the millennia, cotton became firmly established as the world's

Table 1. Name and origin of cotton (*Gossypium hirsutum* L.) genotypes.

S/N	Genotype	Origin of genotype	S/N	Genotype	Origin of genotype
1	GP-319	Unknown	11	SLS-1	Unknown
2	CIM-70	Cotton research institute, Multan	12	S-14	Cotton research institute, Multan
3	S-12	Cotton research institute, Multan	13	KARAS-HS	Unknown
4	FH-87	Cotton research institute, Faisalabad	14	FH-634	Cotton research institute, Faisalabad
5	GOHAR-87	Cotton research station, Bahawapur	15	CIM-448	Cotton research institute, Multan
6	RH-1	Cotton research station, Rahim Yar Khan	16	MNH-339	Unknown
7	CIM-109	Cotton research institute, Multan	17	CIM-1100	Cotton research institute, Multan
8	MNH-147	Cotton research institute, Multan	18	FVH-53	Cotton research station, Vehari
9	CIM-240	Cotton research institute, Multan	19	CIM-443	Cotton research institute, Multan
10	BH-136	Unknown	20	CIM-446	Cotton research institute, Multan

most important fiber crop. The main cotton varieties grown worldwide are Egyptian, American Pima, Sea Island, Asiatic and Upland (Wendel and Cronn, 2003; Rahman et al., 2012).

Development of modern cultivars in upland cotton is limited due to narrow genetic base (Zhang et al., 2008; Carpenter, 2011). The lack of genetic diversity in cotton is concerned with the slowing of progress in developing new cotton cultivars with improved yield and quality potential, as well as stress resistance (Esmail et al., 2008). Genetic diversity of cotton cultivars can be effectively evaluated by molecular markers and the study provides useful information on the selection of parents in the development of cotton cultivars and hybrids as well (Wu et al., 2006; Ullah et al., 2012). Advances in DNA marker technology were the keys to the emergence of genomics. Simple sequence repeats are short, tandemly repetitive DNA sequences that comprises two to six nucleotide core units (Litt and Lutty, 1989; Dongre and Parkhi, 2005). A PCR-based co-dominant, multi allelic, genetic marker system can be based on these sequences (Preetha and Raveendren, 2008; Yonca *et al.*, 2011).

This study was primarily aimed to show the genetic diversity in a set of cotton genotypes with the following specific objectives: (i) To check polymorphism among cotton genotypes through SSR marker system; (ii) To determine distance and relatedness among cotton genotypes using DNA markers; (iii) To construct the dendrogram for determination of phylogenetic relationships.

MATERIALS AND METHODS

Plant material

The present study was conducted at Biotechnology Laboratory, Agricultural Biotechnology Research Institute, AARI Faisalabad. For this study, 31 SSR markers were surveyed on 20 cotton genotypes. Plant material for this study was collected from different sources mentioned in Table 1. After tagging of samples, leaves were

washed with distilled water and transported in liquid nitrogen to the ABRI, Faisalabad and were kept in -80°C freezer to avoiding degradation of leaf samples. DNA was extracted by CTAB method from the fresh leaves (Iqbal et al., 1997). Then quantification of DNA was done using software (NanoDrop-1000 3.3.1). 50 ng DNA was used to check the quality of DNA by running on 0.8% agarose gel and then absorbance ratio was calculated at 260/280 on spectrophotometer.

Microsatellite markers and amplifications

In the present research, 31 SSR markers of BNL series were employed on 20 cotton genotypes. Amplification carried out in 20 μl of reaction mixture consisting of 2 μl DNA at a concentration of 30 ng/μl DNA template, 2 μl of 10× PCR buffer, 4.5 μl of 0.2 mMdNTPs, 2 μl of 50 mM $MgCl_2$ and 1.5 μl of 30 ng/μl of each forward and reverse primer. The amplification was done in Eppendrof thermocycler consisting of initial denaturation step of 5 min at 94°C afterward 35 cycles of 30 s at 94°C for DNA denaturing, 30 s for annealing at 55°C, and 1 min for DNA extention at 72°C with a final extention for 10 min at 72°C.

The amplified fragments were separated electrophoretically using a denaturing 3% agarose gel and also by polyacrylamide gel. The ethidium bromide (0.5 mg/mL) was used to stain the gel for about 30 min. A 100 bp DNA ladder was spotted on first well with an amount of 5 μl as a fragment length standard and in the remaining wells, about 10 μl of each sample was loaded and photographed under ultraviolet light.

Fragment length was determined graphically by comparison with the DNA ladder. After taking the snap, data was scored as 0 for dearth of band and 1 for presence of band. Genetic similarity coefficients were calculated based on Nei and Lei (1979) coefficient using numerical taxonomy multivariate analysis system (NTSYSpc) version 2.2 software package. The unweighted pair group method of arithmetic means (UPGMA) was used to construct the dendrogram.

RESULTS

Polymorphism revealed by SSR

The 20 cotton genotypes were amplified by 31 BNL series of SSR markers. Out of these 31 SSR primers, seven were found informative revealing an average poly-

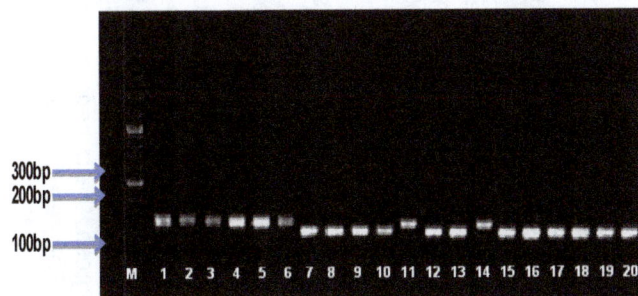

Figure 1. DNA fingerprints of 20 cotton genotypes amplified with SSR primer BNL-193. Amplicons were resolved on agarose gel stained with ethidium bromide. Here 2 loci appeared, one with 130 bp (CIM-109, MNH-147, CIM-240, BH-136, S-14, KARAS-SH, CIM-448, MNH-339, CIM-1100, FVH-53, CIM-443, CIM-446) and other with 140 bp (GP-319, CIM-70, S-12, FH-87, GOHAR-87, RH-1, SLS-1, FH-637). M stands for DNA size Marker.

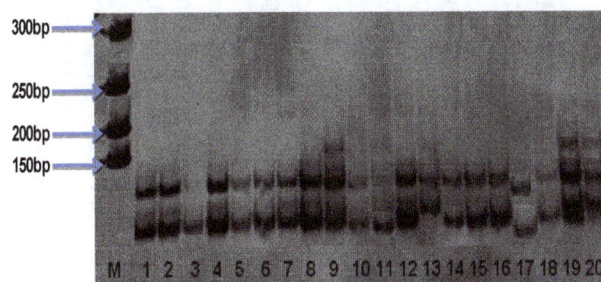

Figure 2. DNA fingerprints of 20 cotton genotypes amplified with SSR primer BNL-150. Polyacrylamide gel was used to resolve the amplicons and then further stained by ethidium bromide. M stands for DNA size marker.

Table 2. Amplification statistics of 7 polymorphic SSR primers employed on 20 cotton genotypes.

S/N	Primer name	Informative band size (bp)
		180
1	BNL-226	190
	BNL-226	200
2	BNL-150	130
	BNL-150	160
	BNL-530	180
3	BNL-530	190
	BNL-530	200
4	BNL-193	130
	BNL-193	140
5	BNL-1064	280
	BNL-1064	300
	BNL-1061	120
6	BNL-1061	130
	BNL-1061	140
7	BNL-1047	125
	BNL-1047	140

morphism rate of 41%. The 17 loci were found informative from a total of 41 amplified ones on an average of 1.3 loci per primer. Molecular size informative fragments of all primers ranged from 80 to 340 bp. The polymorphic SSR markers were BNL-193 (Figure 1), BNL-226, BNL-530, BNL-1047, BNL-1064, BNL-1061 and BNL-150 (Figure 2) listed in Table 2.

Nei and Lei (1979) coefficient method was used to generate similarity matrix (Table 3). 42 to 97% similarity was speckled between genotypes through pair wise combinations and 97% similarity has been observed between genotypes MNH-147 and GOHAR-87. The genotypes CIM-443 and FVH-53 showed 97% similarity. CIM-448 and FH-634 also showed low dissimilarity of 5% whereas only 8% diversity was found between (MNH-147, RH-1), (MNH-147, CIM-109).

The genotypes CIM-1100 and BH-136 were found most dissimilar with the dissimilarity coefficient of 58% followed by CIM-1100 and CIM-70 with the dissimilarity coefficient of 53%. The average similarity among 20 cotton genotypes was 79%.

A dendrogram was constructed for 20 cotton genotypes using UPGMA method (Liu et al., 2005) by data generated of 41 SSR loci (Figure 3). The ensuing dendrogram grouped the genotypes into eight main clusters namely, A, B, C, D, E, F, G and H. Group A comprised of three genotypes including GP-319, CIM-70 and FH-87. The group A was further subdivided into two subgroups A1 and A2. The A1 subgroup contains two genotypes, that is, GP-319 and CIM-70 whereas, FH-87 was in group A2. Group B was the largest one comprising 10 genotypes (GOHAR-87, MNH-147, RH-1, CIM-109, CIM-240, S-14, FH-634, CIM-448, FVH-53 and CIM-443). Group B is further subdivided in two groups, that is, B1 and B2. B1 includes GOHAR-87, MNH-147, RH-1 and CIM-109, whereas CIM-240, S-14, FH-634, CIM-448, FVH-53 and CIM-443 clustered in B2 group. Group E had two genotypes, that is, SLS-1 and KARAS-HS. Group C, D, F, G and H contains single genotype each, that is, CIM-446, MNH-339, S-12, BH-136 and CIM-1100, respectively. The similarity between genotypes varied with a range of 55 to 98% (Figure 3).

DISCUSSION

A number of DNA markers have been used to study the extent of genetic variation in a number of crops and differ mainly in their principles and engender varying amounts of data. The present study was designed to explore genetic diversity among the cotton genotypes using SSR markers.

In the present study, 31 primers were used to check polymorphism among 20 cotton genotypes. Out of 31

Table 3. Average estimates of genetic similarity between 20 cotton genotypes using 31 SSR primers.

	GP-319	CIM-70	S-12	FH-87	GO-HAR	RH-1	CIM-109	MNH-147	CIM-240	BH-136	SLS-1	S-14	KAR-AS	FH-634	CIM-448	MNH-339	CIM-1100	FVH-53	CIM-443	CIM-446
GP-319	1																			
CIM-70	0.87	1																		
S-12	0.80	0.72	1																	
FH-87	0.87	0.80	0.77	1																
GOHAR	0.82	0.80	0.67	0.80	1															
RH-1	0.82	0.75	0.72	0.85	0.90	1														
CIM-109	0.82	0.80	0.77	0.80	0.90	0.90	1													
MNH-147	0.85	0.82	0.70	0.82	0.97	0.92	0.92	1												
CIM-240	0.75	0.77	0.65	0.72	0.87	0.82	0.87	0.90	1											
BH-136	0.57	0.65	0.57	0.65	0.70	0.65	0.70	0.72	0.82	1										
SLS-1	0.67	0.70	0.57	0.70	0.70	0.65	0.65	0.67	0.72	0.65	1									
S-14	0.85	0.82	0.70	0.82	0.87	0.82	0.82	0.90	0.90	0.72	0.77	1								
KARAS	0.80	0.77	0.70	0.72	0.72	0.67	0.72	0.75	0.75	0.62	0.82	0.85	1							
FH-634	0.90	0.82	0.75	0.82	0.82	0.77	0.77	0.85	0.75	0.57	0.67	0.85	0.75	1						
CIM-448	0.85	0.82	0.70	0.82	0.87	0.82	0.82	0.90	0.80	0.62	0.67	0.90	0.75	0.95	1					
MNH-339	0.72	0.75	0.62	0.70	0.75	0.70	0.70	0.77	0.72	0.60	0.60	0.77	0.62	0.77	0.82	1				
CIM-1100	0.55	0.47	0.65	0.57	0.57	0.57	0.52	0.55	0.45	0.42	0.57	0.55	0.50	0.60	0.60	0.47	1			
FVH-53	0.82	0.80	0.72	0.85	0.85	0.85	0.85	0.87	0.77	0.60	0.70	0.87	0.72	0.87	0.92	0.85	0.62	1		
CIM-443	0.85	0.82	0.70	0.82	0.87	0.82	0.82	0.90	0.80	0.62	0.67	0.90	0.75	0.90	0.95	0.87	0.60	0.97	1	
CIM-446	0.80	0.82	0.65	0.77	0.82	0.77	0.77	0.85	0.80	0.62	0.77	0.85	0.80	0.80	0.85	0.77	0.50	0.82	0.85	1

primers 7 were polymorphic. From the final 41 amplified loci, 17 were informative with the total percentage of 41%. Dendrogram constructed using data of 31 SSR loci by NTSYspc 2.2 software pakage and based on UPGMA method (Table 3). The eight main groups, that is, A, B, C, D, E, F, G and H of genotypes have been obtained from the resulting dendrogram. Genotypes GP-319, CIM-70 and FH-87 present in group A, even though they were collected from different research stations but their genetic makeup seemed to be similar. Cluster B was the major one comprising of 10 genotypes (GOHAR-87, MNH-147, RH-1, CIM-109, CIM-240, S-14, FH-634, CIM-448, FVH-53 and CIM-443). Dendrogram showed that in group B, the genotypes (GOHAR-87, MNH-147), (FVH-53, CIM-443) had greatest similarity of 98%. Cluster E having two genotypes SLS-1 and KARAS-HS were most similar to each other. Group C, D, F, G, H, contained single genotype each, CIM-446, MNH-339, S-12, BH-136, CIM-1100, respectively resulting to most diverse genotypes.

The 41% polymorphism of current results showed reasonable genetic diversity among cotton genotypes and the results were according to Abdalla et al. (2001), Van Esbroeck et al. (1998) and Sapkal et al. (2011) reported the limited genetic variations in cotton. However, Guang and Xiong-Ming (2006) reported higher polymorphism (80%) among the cotton genotypes.

According to Li et al. (2008), the genetic diversity is higher in cotton genotypes through SSRs. According to Rehmanet al. (2009), the cotton germplasm has tapered genetic base therefore to evolve the varieties with promising characters, it is strongly required to expand the genetic base of cotton germplasm.

The results of present study are according to the results of Wendel and Brubaker (1993), Tatineniet al. (1996). Similarly, Iqbal et al. (1997) concluded that the cultivated upland cotton G. hirsutumhad narrow genetic diversity. The present results are also in accordance with the findings of Lacape et al. (2003), Iqbal et al. (2001) and Dongreet al. (2007) who found lack of genetic diversity in cultivated cotton genotypes. Narrow genetic base in cotton resulted in the presence of low genetic diversity (Liu et al., 2000).

For the economic and quality traits, extensive selection is one of the major reasons for low genetic diversity in modern crop cultivars. Dissimilar to crop plants, tree species, specifically propagated by vegetative means do not undergo hybridization and extensive selection thus during the course of progress, the genetic resources are more conserved. The result obtained on genetic

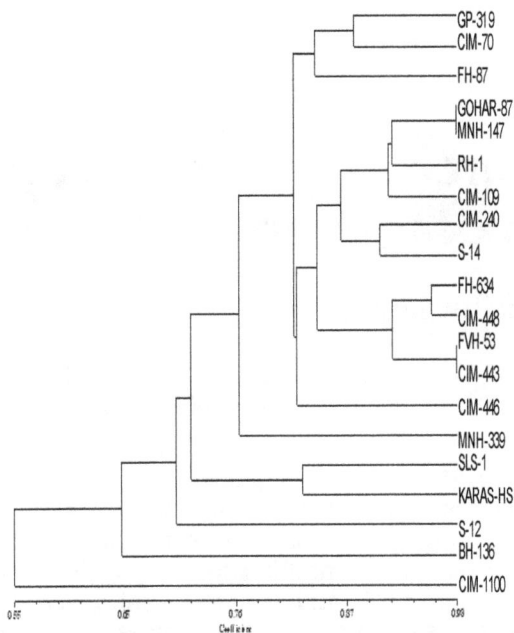

Figure 3. Dendrogram constructed for 20 cotton genotypes based on genetic distances using SSR primers.

relatedness will be helpful in further breeding of cotton cultivars, the selection of parents for crossing and will also be helpful in widening the genetic bases of breeding materials.

REFERENCES

Abdalla M, Reddy OUK, Ei-Zik KM, Pepper AE (2001). Genetic diversity and relationships of diploid and tetraploid cotton. Theore. Appl. Genet. 6:222-229

Brubaker CL, Paterson AH, Wendel JF (1999). Comparative genetic mapping of allotetraploid cotton and its diploid progenitors. Genome, 42:184-203.

Candida HC, Bertini DM, Schuster I, Sediyama T, Barrosi EGD, Moreira MA (2006). Characterization and genetic diversity analysis of cotton cultivars using microsatellites. Genet. Mol. Biol. 29: 321-329.

Carpenter J (2011).. Impacts of GM crops on biodiversity. *GM Crops* 2:1-17

Dongre AB, Bhandarkar M, Banerjee S (2007). Genetic diversity in tetraploid and diploid cotton (Gossypiumspp.) using ISSR and microsatellite DNA markers. Intl. J. Biotechnol. 6:349-353.

Dongre A, Parkhi V (2005). Identification of cotton hybrid through the combination of PCR based RAPD, ISSR and microsatellite markers. J. Plant Biochem. Biot. 14:53-55.

Esmail RE, Zhang JF, Abdel-Hamid AM(2008). Genetic diversity in elite cotton germplasm lines using field performance and RAPD markers. World. J. Agric. Sci. 4(3):369-375.

Guang C, Xiong-Ming D (2006). Genetic diversity of source germplasm of upland cotton in China as determined by SSR marker analysis. Acta. Genetica. Sinica. 33(8):733-745

Iqbal MJ, Aziz N, Saeed NA, Zafar Y (1997). Genetic diversity evaluation of some elite cotton varieties by RAPD analysis. Theore. Appl. Genet. 94:139-144.

Iqbal MJ, Reddy OUK, El-Zik KM, Pepper AE (2001). A genetic bottleneck in the `evolution under domestication' of upland cotton

Gossypiumhirsutum L. examined using DNA fingerprinting. Theore. Appl. Genet. 103: 547-554

Khadi BM, Santhy V, Yadav MS (2010). Cotton: An Introduction. Biotechnol. Agric. Forest. 65:1-14

Lacape JM, Nguyen TB, Thibivilliers S, Courtois B, Bojinov BM, Cantrell RG, Burr B, Hau B (2003). A combined RFLP SSR-AFLP map of tetraploide cotton based on a GossypiumhirsutumxGossypiumbarbadensebackcross population. Genome 46:612-626

Li ZX, Wang Y, Zhang G, Zhang L, Wu JC, Ma Z (2008). Assessment of genetic diversity in glandless cotton germplasm resources by using agronomic traits and molecular markers. Frontiers of Agriculture in China 2:245-252

Litt M, Luty JA (1989). A hypervariable Microsatellite revealed by in vitro amplification of a dinucleotide repeat within the cardiac muscle actin gene. Am. J. Hum. Genet. 44:397-40

Liu D, Guo X, Lin Z, Nie Y, X. Zhan (2005). Genetic diversity of asian cotton (Gossypiumarboreum L.) in china evaluated by Microsatellite analysis. Genet. Res. Crop. Evol. 53:1145-1152

Liu S, Cantrell RG, Mccarty JC, Stewart MJ (2000). Simple Sequence Repeat based assessment of genetic diversity in cotton race stock accessions. Crop Sci. 40:1459-1469

Nie N, Lei W (1979). Mathematiical model for studying genetic variation in terms of restriction endonuclease. Proc. Natl. Acad. Sci. 76:5269-5273

Preetha S, Raveendren TS (2008). Molecular marker technology in cotton. Biotechnol. Mol. Biol. 3: 032-045

Rahman M, Shaheen T, Tabbasam N, Iqbal MA, Zafar Y,Paterson AH (2012). Cotton genetic resources. A review. Agron. Sustain. Dev. 32:419-432

Rehman OU, Khan SH, Sajjad M (2009). Genetic diversity among cotton (Gossypiumhirsutum L.) genotypes existing in Pakistan. Am. Eurasian. J. Sustain. Agric. 3:816-823

Sapkal DR, Sutar SR, Thakre PB, Patil BR, Paterson AH, Waghmare VN (2011). Genetic diversity analysis of maintainer and restorer accessions in upland cotton (Gossypiumhirsutum L.) J. Plant Biochem. Biotechnol. 20(1):20-28

Tatineni V, Cantrell RG, Davis DD (1996). Genetic diversity in elite cotton germplasm determined by morphological characteristics and RAPDs. Crop. Sci. 36:186-192.

Ullah I, Iram A, Iqbal MZ, Nawaz M, Hasniand SM, Jamil S (2012). Genetic diversity analysis of Bt cotton genotypes in Pakistan using simple sequence repeat markers. Genet. Mol. Res. 11 (1): 597-605

Van EGA, Bowman DT, Calhoun DS, May OL (1998). Changes in the genetic diversity of cotton in the USA from 1970 to 1995. Crop. Sci. 38:33-37

Wendel JF, Brubaker CI (1993). RFLP diversity in Gossypiumhirsutum L. and new insights into the domestication of cotton. Am. J. Bot., 80: 71

Wendel JF, Cronn RC (2003). Polyploidy and the evolutionary history of cotton. Adv. Agron. 78:139-186

Wu Y, Machado AC, White RG, Llewellyn DJ, Dennis ES (2006). Expression profiling identifies genes expressed early during lint fibre initiation in cotton. Plant. Cell Physiol. 47:107-127

Yonca S, Col B, Burun B (2011). Genetic diversity and identification of some Turkish cotton genotypes (Gossypiumhirsutum L.) by RAPD-PCR analysis. Turk J. Biol. 36:143-150

Zhang Y, Lin Z, Xia Q, Zhang M, Zhang X (2008). Characteristics and analysis of simple sequence repeats in the cotton genome based on a linkage map constructed from a BC1 population between Gossypiumhirsutum and G. barbadense. Genome 51:534-546

16

Effect of genotype, explant source and medium on *in vitro* regeneration of tomato

Praveen Mamidala[1,2] and Rama Swamy Nanna[1]*

[1]Department of Biotechnology, Kakatiya University, Warangal, A.P- 506009, India.
[2]Department of Biotechnology, Vaagdevi College of Engineering, Warangal, 506009, India.

For the present investigations, five cultivars including PKM-1, Moneymaker, Microtom, Micro-MsK and White Cherry of tomato (*Lycopersicon esculentum* Mill.) were selected. The explants cotyledon, hypocotyl and leaf of all the cultivars of tomato were cultured on MS medium fortified with different concentrations of Benzyl amino purine (BAP) (1.0 to 3.0 mg/L) and 0.1 mg/L indole-3-acetic acid (IAA). Adventitious shoot buds were induced at the cut ends of the explants after three weeks of culture. Adventitious shoots were induced from hypocotyl, cotyledon and leaf explants on MS medium fortified with 1.0 mg/l BAP + 0.1mg/l IAA (Medium A); 2.0 mg/l BAP+0.1 mg/l IAA (Medium B) and 3.0 mg/L BAP + 0.1 mg/L IAA (Medium C). Of the five genotypes tested, PKM-1 showed highest number of shoots per explant followed by Micro-MSK, Microtom and Money maker on MS medium augmented with different concentrations of growth regulators used (A, B & C). The lowest regeneration efficiency was observed from all the explants used in White Cherry. More number of adventitious shoots was induced from leaf explants compared to cotyledon and hypocotyl explants in all the genotypes of tomato studied. Among the three media tested, medium'B' showed superiority in cotyledon and leaf explants compared to medium 'A'. Whereas medium 'C' had shown poor response on regeneration ability of all the explants tested irrespective of genotypes. *In vitro* rooting was achieved on MS medium augmented with 0.1 mg/L NAA (Naphthalene acetic acid) in all the genotypes. Thus the plant regeneration was found to be influenced by the genotype, explant and type of medium.

Key words: Tomato, cotyledon, hypocotyl, leaf, regeneration, plantlet formation.

INTRODUCTION

Tomato (*Lycopersicon esculentum* Mill.) is a highly nutritive vegetable crop. Though it is a temperate crop plant, it is extensively cultivated in the tropical and subtropical regions of the world round the year. It ranks third among vegetable crops (only next to potato and sweet potato) with an annual production of 283 million metric tonnes in the year 2009 (FAO statistical database, 2011). Tomato is rich in vitamin A and C and fibre, and is also cholesterol-free (Hobson and Davies, 1971). Tomato contains approximately 20 to 50 mg of lycopene/100 g of fruit weight (Kalloo, 1991) which is the most powerful antioxidant in the carotenoid family and it protects humans from free radicals that degrade many parts of the body and is also known to prevent cancer (Block et al.,

1992; Gerster, 1997).

Due to its importance as vegetable and medicinal value, many laboratories are doing research on tomato to manipulate the nutrient quality through transformation studies. Intravarietal differences in in vitro tomato regeneration from various explants are highly pronounced (Ohki et al., 1978; Frankenberger et al., 1981a, b; Kurtz and Lineberger, 1983; Plastira and Perdikaris, 1997). Specific plant growth regulator requirements for each genotype make the tissue culture task in tomato quite difficult and researchers need to deal with each genotype individually (Bhatia et al., 2004). Successful coupling of a regeneration system and gene transfer procedures will assist in addressing both basic and applied research issues (Bhatia et al., 2005). For genetic transformation, the prerequisite is to develop regeneration protocol. The plant regeneration in tomato is genotype, explant and media dependent. A good in vitro regeneration system is

*Corresponding author. E-mail: swamynr.dr@gmail.com.

essential for an effective genetic transformation for commercial applications. Since genotype, explants and media influence the efficiency of regeneration system, we have undertaken the present studies to evaluate genetically diverse cultivars of *L. esculentum* including PKM-1, Money Maker, Microtom, Micro-MSK and White Cherry for plant regeneration from hypocotyls, cotyledon and leaf explants.

MATERIALS AND METHODS

Seeds of tomato (*L. esculentum*) cvs. PKM-1, White Cherry, Moneymaker, Microtom and Micro-MSK were soaked under running tap water for 24 h. These seeds were sterilized with a mixture 10% NaOCl and 2% SDS (10% w/v) for 3 min. Later the sterilized seeds were washed thrice with sterile distilled water and were dried on sterile tissue paper. The sterilized seeds were germinated aseptically on MS basal medium (Murashige and Skoog, 1962) solidified with 0.8% (w/v) agar (Sigma Chemical Co, USA) in 250 ml Erlenmeyer cnical flasks (ca.50 mlmedium/flask). Cotyledon (0.6 to 0.8 cm²) and hypocotyl (0.5 to 0.8 cm long) explants from *in vitro* grown seedling (8 to 10 days old) were excised, whereas the leaf explants (1.0 cm²) were used from three week old axenic seedlings.

Culture media and culture conditions

The hypocotyl, cotyledon and leaf explants were cultured on MS basal medium supplemented with 1.0 mg/L BAP + 0.1 mg/L IAA (Medium A); 2.0 mg/L BAP + 0.1 mg/L IAA (Medium B) and 3.0 mg/L BAP + 0.1 mg/L IAA (Medium C). The media were adjusted to pH 5.8 either with 0.1 N HCl or 0.1 N NaOH before addition of 0.8% (w/v) agar and autoclaved at 121°C under 15 lbs for 15 to 20 min. The media were dispensed into different petriplates and 15 explants were inoculated in each petriplate. All the cultures were incubated at 25°C under 16/8 h (dark/light) photoperiod with light intensity 40 to 50 μmol m⁻²s⁻¹ provided by cool white fluorescent lights (Phillips Ltd). Explants were sub-cultured on the same fresh medium for every four weeks for further proliferation.

In vitro rooting and acclimatization

Micro-shoots (2 to 3 nodes) developed from hypocotyl; cotyledon and leaf explants were excised and transferred on to MS medium fortified with 1.0 mg/L NAA. *In vitro* rooted plantlets were taken out from the culture tubes and washed with sterile distilled water to remove the remains of medium. These were transferred to plastic cups containing sterile vermiculite and covered with polythene bags to maintain the RH (70 to 80%) and later shifted to green house. The acclimatized plantlets were transferred into field.

Data analysis

Data of 15 explants were recorded after four weeks of culture. Each experiment was repeated at least twice with similar results and the data of one representative experiment presented here are the mean value alone and pooled from three explants, five genotypes and three media (A, B and C). Trifactorial experiments were carried out to analyze the effect of genotype, explant and medium and their interactions singly and in combination, on *in vitro* shoot regeneration (number of shoots/explant). Results were analyzed by using ANOVA. Fischer's LSD test (Least Significant Difference) was applied to show statistical significance of difference among the means

Cytological studies

Root tips of about 100 plants regenerated from different explants (pooled from all media tested) were randomly selected developed from three different media were fixed in aceto-alcohol (1:3) mixture. They were treated with 1 N HCl (1 to 2 min) followed by 2% aceto-orcein for about 2 h and then squashed with a drop of 45% acetic acid and observed under a Nikon compound research microscope.

RESULTS AND DISCUSSION

Shoot buds induction and proliferation were found after 10 to 12 days of culture from cut ends of the explants. Data on the average number of shoots formed per explant (hypocotyl, cotyledon and leaf) on different regeneration media; including mediums A, B and C are summarized in Table 1. The cultivar PKM-1 showed the highest number of shoots per explant of all the genotypes tested in the present investigations. Similarly cotyledon explants have responded well on all the media used in contrast to hypocotyl and leaf explants of all the cultivars.

Adventitious shoot buds induction was found to be higher in mediums B and C compared to medium A in all genotypes tested irrespective of explants used. The cv PKM-1 showed the maximum frequency number of adventitious shoots formation in mediums A and B from leaf explants. The cultivar MicroMSK showed better regeneration ability in comparison to Microtom. Money Maker exhibited the maximum frequency of adventitious shoots formation when leaf and cotyledon explants were cultured in mediums A and B. The lowest frequency number of multiple shoots formation per explant was observed from all the explants cultured in White Cherry compared to rest of the genotypes used (Figure 1).

Medium B showed more number of shoots formation in all the explants cultured. Leaf explants of all the genotypes have responded well and produced more number of shoots. However cotyledon explants showed superiority over hypocotyls (Figure 2). It is interesting to note that leaf explants showed efficient regeneration on medium B which suggests that application of leaf as an explant source can minimize the variation brought by the genotypes (Figure 3). Evaluation of three media for shoot regeneration (pooled from 5 genotypes) revealed that medium B is superior to the other combinations of media (Figure 4). When explant and medium interaction for optimal shoot regeneration evaluated, it was observed that leaf explants regenerated maximum on mediums B and C, whereas cotyledon and hypocotyl explants on mediums A and B (Figure 5). Similarly, the literature showed that the combination of IAA + BAP for shoot regeneration from different explants of tomato was found to be more effective (Duzyman et al., 1994; Selvi and Khader, 1993; Villiers et al., 1993; Chandel and Katiyar, 2000; Gunay and Rao, 1980; Kurtz and Lineberger, 1983; Chen et al., 1999). However, Pongtongkam et al. (1993), Plastira and Perdikaris (1997) and Venkatachalam et al. (2000) found that BA alone was more effective than IAA +

Table 1. Comparison of the average number of shoot formed per explant on MS medium supplemented with BAP 1.0 mg/L + 0.1 mg/L IAA (A), BAP 2.0 mg/L + 1 mg/L IAA (B) and BAP 3.0 mg/L + 0.1 mg/L IAA (C) in five genotypes of *L. esculentum.*

Genotype	Medium	Explant		
		Hypocotyl	Cotyledon	Leaf
PKM-1	A	5.4[a]	13.3[b]	14.6[b]
	B	4.8[a]	8.1[b]	8.3[a]
	C	4.3[b]	6.2[a]	6.7[b]
Money Maker	A	4.3[b]	5.5[b]	7.6[b]
	B	4.8[a]	7.2[a]	10.3[a]
	C	NR	5.4[a]	7.0[a]
Micro-Tom	A	2.4[b]	4.7[b]	9.3[b]
	B	2.5[NS]	7.2[b]	14.1[a]
	C	2.4[NS]	3.2[a]	6.5[a]
Micro-Msk	A	4.1[b]	9.0[a]	11.1[a]
	B	4.9[a]	7.7[b]	14.2[b]
	C	4.4[b]	9.4[a]	9.8[a]
White Cherry	A	NR	4.2[b]	7.1[b]
	B	4.6[b]	6.4[a]	9.6[b]
	C	NR	4.0[b]	5.3[a]

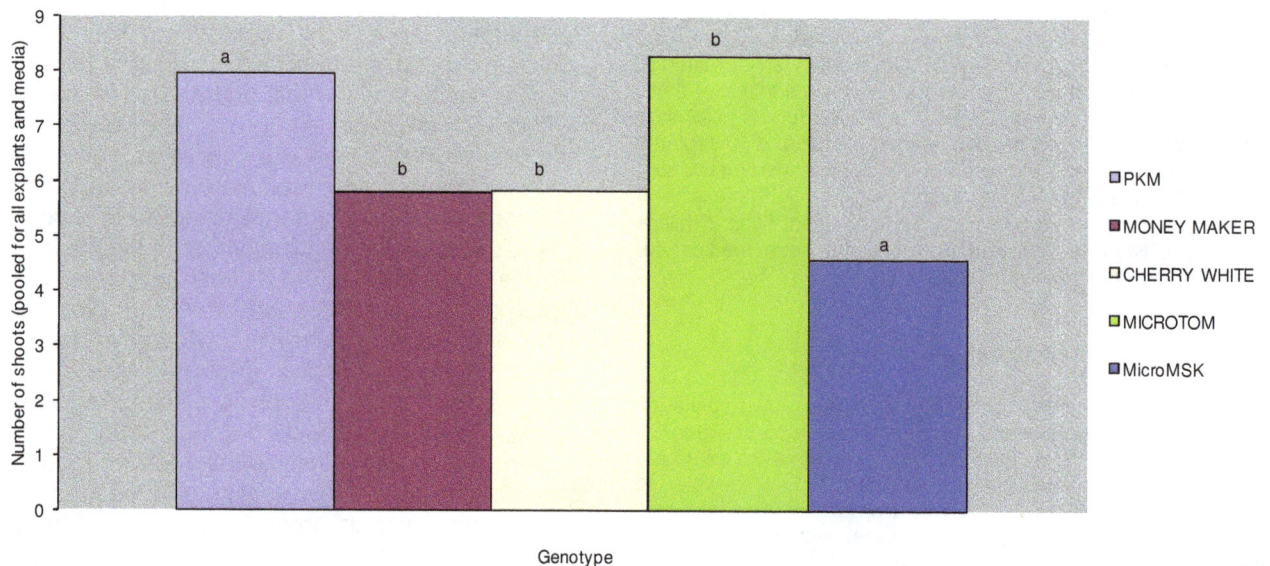

Figure 1. Number of shoots (pooled for three explants and three media) regenerated from *L. esculentum* after 4 weeks of culture. Bars with different letters represent significantly different means (p<0.01) using Fischer's LSD procedure.

BA combination for shoot regeneration from leaf, cotyledon and hypocotyl explants. According to our present investigation on five genotypes, we suggest that the growth regulator concentration varies from genotype to genotype and genotypic specific optimal medium concentration is required.

The present protocols are useful for genetic transformation studies in all these five genotypes of tomato.

In vitro rooting and plantlet establishment

Adventitious shoots developed from different explants were excised (3 to 4 cm) and shifted on to root induction medium (MS + 0.1 mg/L NAA). Profuse rooting was observed within two weeks of incubation in all the cultures tested. The *in vitro* rooted plantlets were taken out from the culture tubes and washed to remove adhered

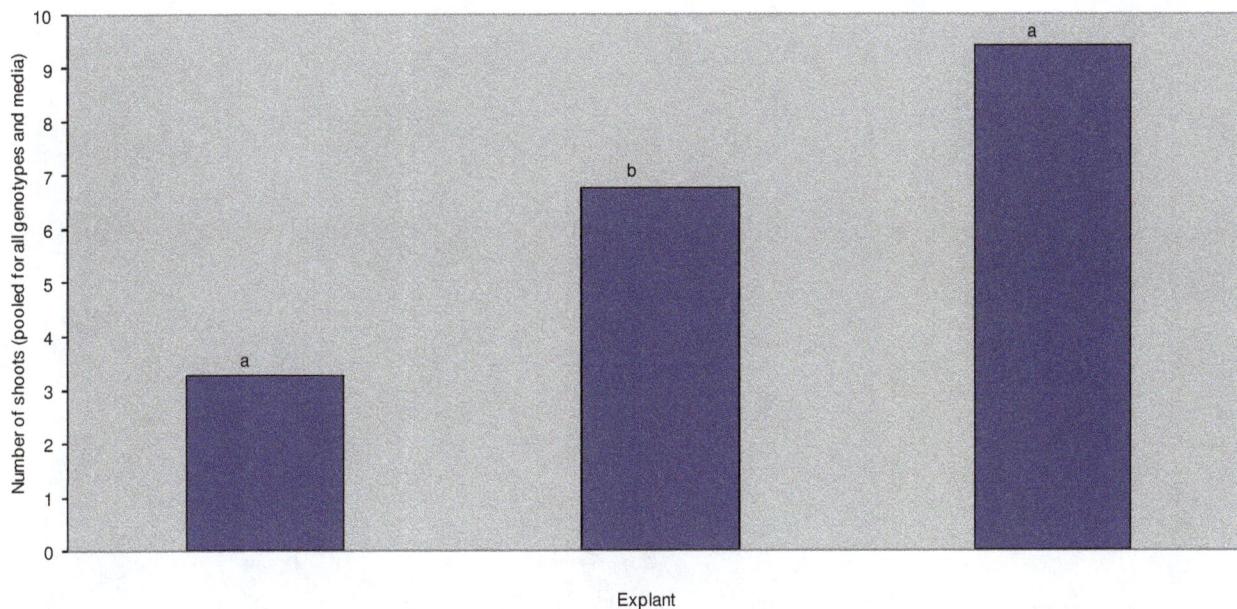

Figure 2. Number of shoots (pooled for five genotypes and three media) obtained from hypocotyl, cotyledon and leaf explants of *L. esculentum* scored after 4 weeks of culture. Bars with different letters represent significantly different means (p<0.01) using Fischer's LSD procedure.

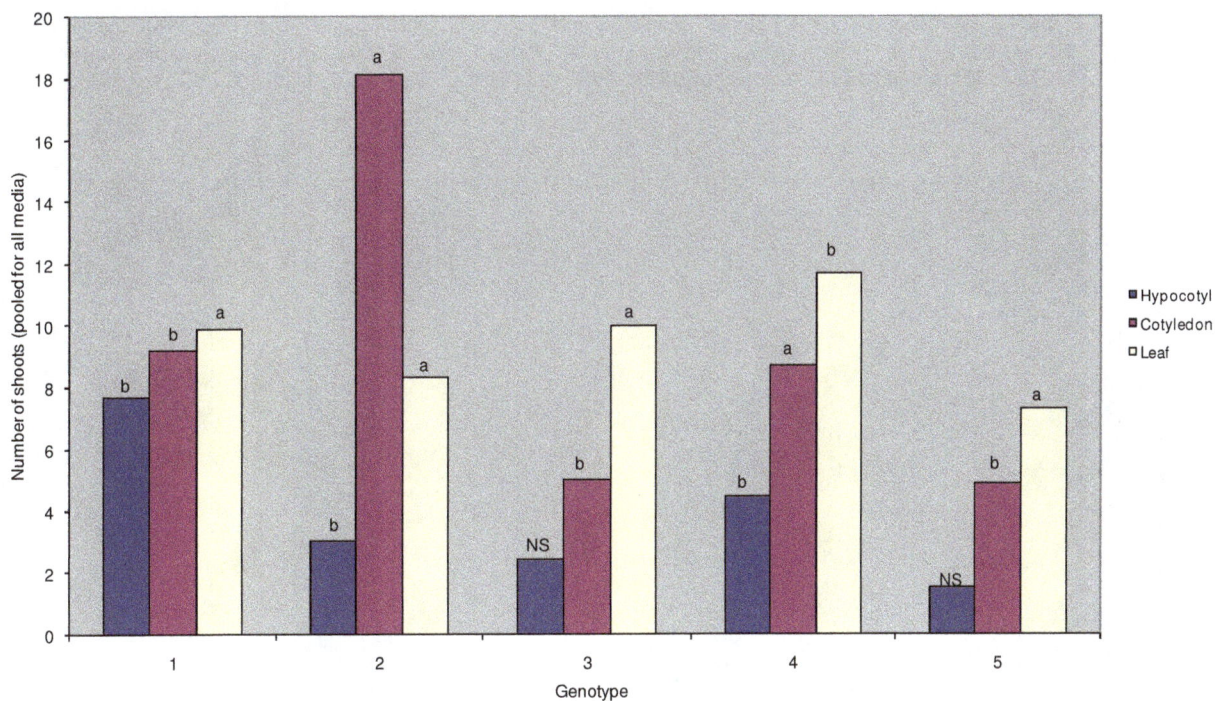

Figure 3. Number of shoots (pooled for three media) obtained from hypocotyl, cotyledon and leaf explants of *L. esculentum* after 4 weeks of culture. Bars with different letters represent significantly different means (p<0.01) using Fischer's LSD procedure.

agar and traces of medium. Plantlets were then transferred to plastic cups containing sterile vermiculite. The hardening was done for 4 weeks in a growth chamber and later moved into green house and transferred to field. The survival percentage of plantlets after transplantation was 72% and all the plants developed *in vitro* were found to be morphologically similar to parental plants. Cytological studies showed that the plants regenerated from various

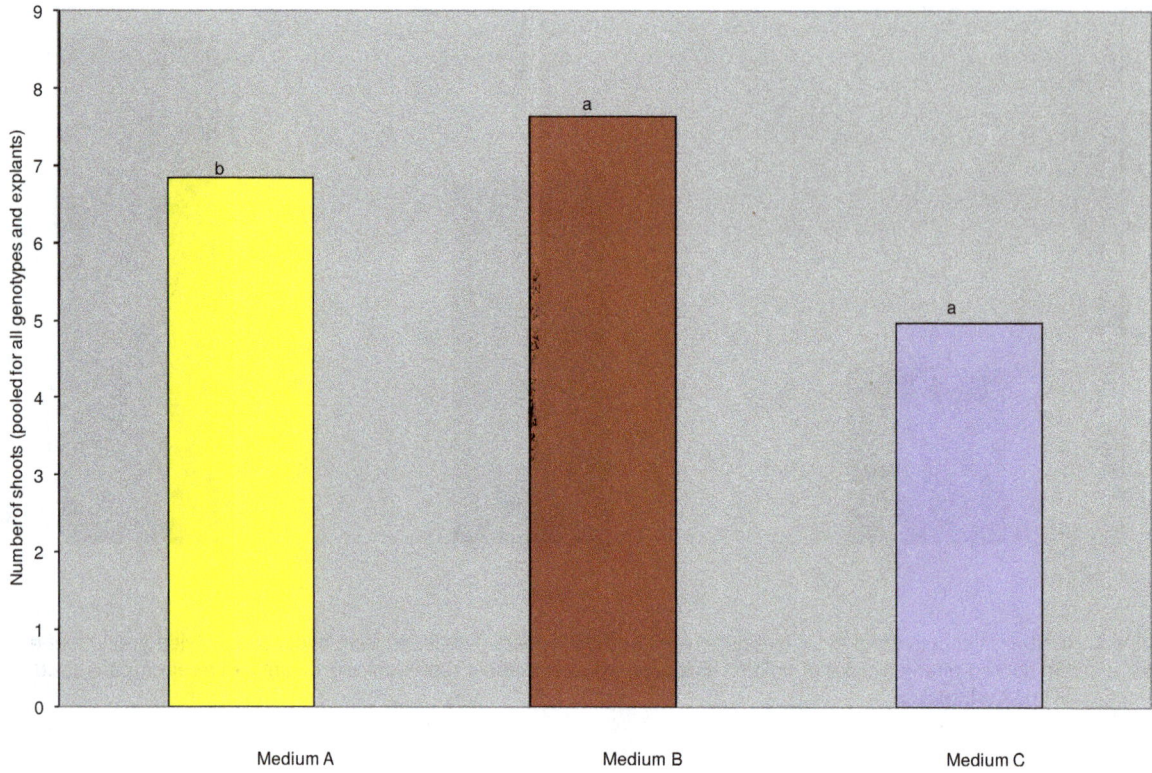

Figure 4. Number of shoots (pooled for seven genotypes and three explants) induced by 1.0 mg/L BAP + 0.1 mg/L IAA (A); 2.0 mg/L BAP + 0.1 mg/L IAA (B), and 3.0 mg/L BAP + 0.1 mg/L IAA(C) media scored after 4 weeks of culture. Bars with different letters represent significantly different means (p<0.01) using Fischer's LSD procedure.

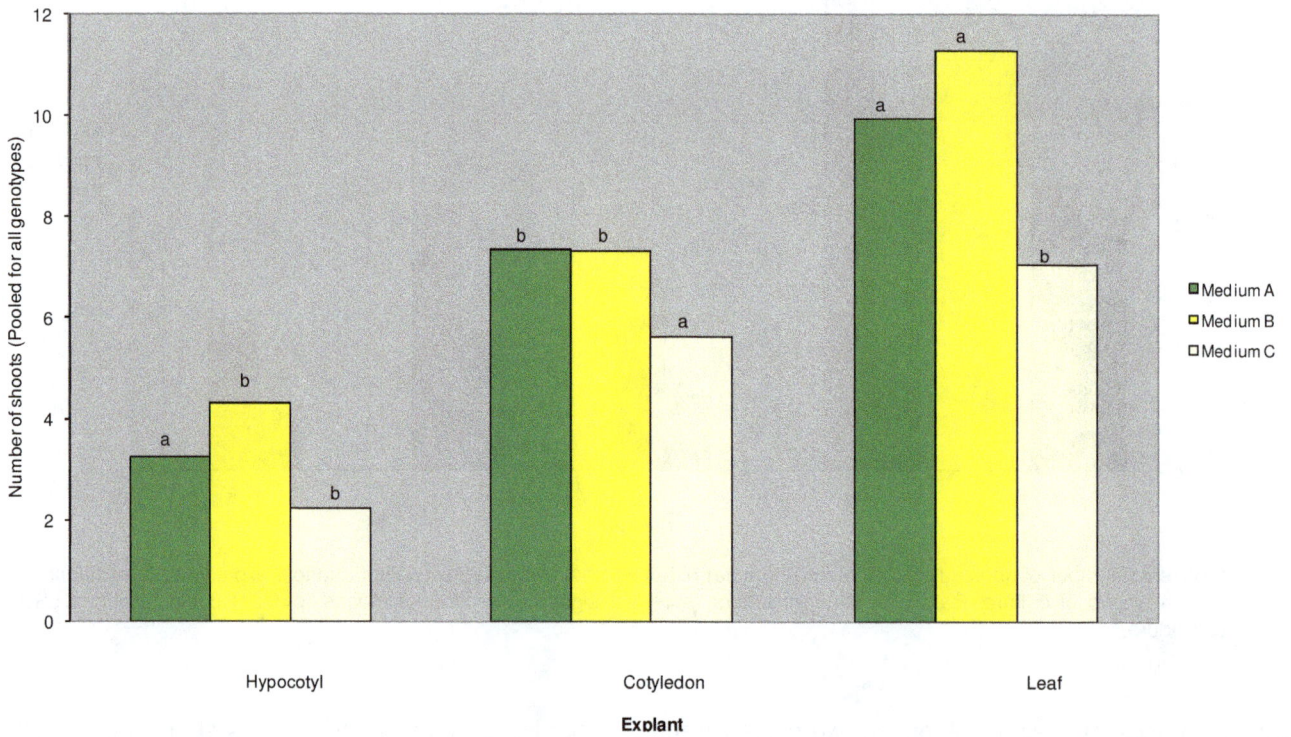

Figure 5. Number of shoots (pooled for five genotypes) induced by 1.0 mg/L BAP + 0.1 mg/L IAA (A); 2.0 mg/L BAP + 0.1 mg/L IAA (B), and 3.0 mg/L BAP + 0.1 mg/L IAA(C) media in hypocotyl, cotyledon and leaf explants after 4 weeks of culture. Bars with different letters represent significantly different means (p<0.01) using Fischer's LSD procedure.

tissue cultures were normal diploids ($2n = 2x = 24$) and without any mitotic aberrations. Similar results were observed in recovery of multiple shoots also from shoot tip explants by Novak and Maskova (1979) in tomato and reported no abnormalities in chromosomes of root tips or no irregularities at meiosis, fruit set, or in seed variability. Thus the present investigation showed that leaf and cotyledon explants were found to be better for regeneration, which is a prerequisite for *Agrobacterium* mediated genetic transformation for the genetic improvement of these important cultivars.

ACKNOWLEDGEMENTS

MP gratefully acknowledges Third World Academy of Sciences (TWAS) Italy, and Chinese Academy of Sciences (CAS), China for financial assistance in the form of CAS-TWAS Postdoctoral Fellowship 2008. Thanks are due to Prof Lazaro E P Peres. Depto.de Ciencias Biologicas, LCB, Universidade de Sau Paulo, Piracicaba, Brazil for providing Microtom and Micro-MSK seeds.

REFERENCES

Block GB, Patterson B, Subar A, (1992). Fruit, vegetables and cancer prevention: a review of the epidemiological evidence. Nutr. Cancer,18: 1-29.

Bhatia P, Nanjappa A, Tissa S, David M (2004). Tissue Culture studies in tomato (*Lycoperiscon esculentum*) Plant Cell, Tiss. Org. Cult., 78: 1-21.

Bhatia P, Ashwath N, Midmore DJ (2005). Effects of Genotype, Explant Orientation, And Wounding On Shoot Regeneration In Tomato *In Vitro* Cell. Dev. Biol. Plant, 41: 457-464.

Chandel G, Katiyar SK (2000). Organogenesis and somatic embryogenesis in tomato (*Lycopersicon esculantum* Mill.). Adv. Plant Sci., 13: 11-17

Chen H, Zhang J, Zhuang T, Zhou G (1999). Studies on optimum hormone levels for tomato plant regeneration from hypocotyls explants cultured *in vitro*. Acta Agric. Shanghai, 15: 26-29.

Duzyman E, Tanrisever A, Gunver G (1994). Comparative studies on regeneration of different tissues of tomato *in vitro*. Acta. Horticult., 34: 235-242.

FAO Statistical Database (2011). FAOSTAT Agriculture data, URL: http:// http://faostat.fao.org/site/567, date of access 26 February 2011.

Frankenberger EA, Hasegawa PM, Tigchelaar E (1981a). Influence of environment and developmental state on the shoot forming capacity of tomato genotypes. Z. Pflanzenphysiol., 102: 221-232.

Frankenberger EA, Hasegawa PM, Tigchelaar EC (1981b). Diallel analysis of shoot-forming capacity among selected tomato genotypes. Z. Pflanzenphysiol., 102: 233-242.

Gerster H (1997). The potential role of lycopene for human health. J. Am. College Nutr. 16: 109-126.

Gunay AL, Rao PS (1980). *In vitro* propagation of hybrid tomato plants (*Lycopersicon esculentum* L.) using hypocotyl and cotyledon explants. Ann. Bot., 45: 205-207.

Hobson G, Davies J (1971). The Tomato. In: Hulme A (eds) The Biochemistry of Fruits and Their Products. Academic Press, New York, London Introduction for regeneration: pp. 337-482

Kalloo G (1991). Genetic Improvement of Tomato. Kalloo G (eds) monographs of Theoretical and Applied Genetics Springer-verlag, Berlin, Heidelberg, New York: pp. 1-9

Kurtz SM Lineberger RD (1983) Genotypic differences in morphogenic capacity of cultured leaf explants of tomato. J. Am.Soc. Hort. Sci., 108: 710-714.

Murashige T, Skoog F (1962). A revised medium for rapid growth and bioassays with tobacco tissue cultures. Physiol. Plant, 15: 473-497.

Novak FJ, Maskova I (1979). Apical shoot tip culture of tomato. Sci. Horticult. 10: 337-344.

Ohki S, Bigot C, Mousseau J (1978). Analysis of shoot forming capacity *in vitro* in two lines of tomato (*Lycoperiscon esculentum* Mill.) and their hybrids. Plant Cell Physiol., 19: 27-42.

Plastira VA, Perdikaris AK (1997). Effect of genotype and explant type in regeneration frequency of tomato *in vitro*. Acta Horticult., pp. 231-234.

Pongtongkam P, Ratisoontorn P, Suputtitada S, Piyachoknagul S, Ngernsiri L, Thonpan A (1993). Tomato propagation by tissue culture. Kasetsart J. 27: 269-277.

Selvi DT, Khader MA (1993). *In vitro* morphogenetic capacity of tomato (*Lycopersicon esculentum* Mill.) var. PKM.1. South Indian Horticult., 41: 251-258.

Villiers RPD, Vuuren RJV, Ferreira DI, Staden JV (1993). Regeneration of adventitious buds from leaf discs of *Lycopersicon esculentum* cv. Rodade: optimization of culture medium and growth conditions. J. S. African Soc. Hort. Sci., 3: 24-27.

Venkatachalam P, Geetha N, Priya P, Rajaseger G, Jayabalan N (2000). High frequency plantlet regeneration from hypocotyls explants of tomato (*Lycopersicon esculentum* Mill.) via organogenesis. Plant Cell Biotechnol. Mol. Biol., 1: 95-100.

17

The Genetic variants of IL1RAPL2 gene associated with non-specific mental retardation in Chinese children

Zhang Kejin[1], He Bo[1], Gong Pingyuan[1], Gao Xiaocai[1,2], Zheng Zijian[1,2], Huang Shaoping[3] and Zhang Fuchang[1,2]*

[1]Key Laboratory of Resource Biology and Biotechnology in Western China (Northwest University) Ministry of Education, College of Life Science, Institute of Population and Health, Northwest University, Xi'an 710069, China.
[2]Institute of Applied Psychology, College of Public Management, Northwest University, Xi'an 710069, China.
[3]The 2nd Affiliated Hospital, Xi'an Jiaotong University, Xi'an 710004, China.

The present study investigated the association between the genetic variants of IL1RAPL2 gene and Non-syndromic mental retardation (NSMR) in the children of QinBa region of China. Five common SNPs (rs5962434, rs5916817, rs3764765, and rs5962298 and rs9887672) of IL1RAPL2 were chosen and examined their individual genotype frequencies using the conventional polymerase chain reaction single strand conformation polymorphism (PCR-SSCP) method, and evaluated the association between these genetic polymorphisms and NSMR with the suitable bio-statistic software. Two SNPs (rs5962298 and rs9887672), whose alleles and genotypes distribution showed a significant differences between the control and NSMR groups (alleles: $p = 0.020$ and 0.017; genotypes: $p = 0.025$ and 0.053, respectively). Furthermore, the different gender effect was found out, when stratified the data set by the sex. Taken together, we provided substantial evidence that IL1RAPL2 conferred a NSMR susceptibility to children of Qinba region in China, and further work should been done.

Key words: Non-syndromic mental retardation, molecular genetics, association analysis, genetic variants.

INTRODUCTION

Mental retardation (MR) is a heterogeneous disorder, characterized by an IQ score of 70 or lower and significant limitation in the social and practical adaptive skills and leads to problems with self care, communi-cation and school activities. MR affects 1 to 3% of children in different countries and regions (Mclaren and Bryson, 1987), and is divided into "syndromic" (SMR; the additional sdymorphic, neurological, and/or metabolic features accompany the mental deficit) and "non-syndromic" (NSMR; the cognitive impairment is the sole definable clinical feature) forms. Although the underlying causes of MR are likely to be extremely heterogeneous, a strong genetic component (even genetic etiologies) has

been found in approximately two thirds of MR cases (Curry et al., 1997).

Recently, most of research works focused on the X chromosome, and try to map, identify and clone the MR candidate genes in this chromosome. So far, there were 34 candidate MR genes which were identified and cloned in X chromosome: 17 of them are NSMR-specific genes, and the other 17 genes can cause both NSMR and SMR, according to the update of Greenwood Genetic Center in July 2009 (Stevenson et al., 2000). Moreover, linkage data for 56 X-linked families have been mapped (log score >2), but based on these data, no NSMR gene has been cloned.

Carrie et al. (1999) identified a mutation encode sequence (NCBI accession number: 300206.0001) in a small family with X-linked NSMR. The sequence encodes a 696 amino acid protein with a homology to interleukin-1 (IL-1) receptor accessory protein (Carrie et al., 1999). Following in Carrie's work, several other reports also have verified the association between IL1RAPL1 and NSMR (Bahi et al., 2003; Tabolacci et al., 2006). Jin et al.

---

*Corresponding author. E-mail: zhfch@nwu.edu.cn.

Abbreviation: NSMR, nonspecific mental retardation; Co, controls.

(2000) found a closely related homolog, which was designated as IL1RAPL2 gene (Jin et al., 2000). The proteins encoded by IL1RAPL1 and IL1RAPL2 share 65% sequence identity as well as a C-terminal sequence absent in other members of the IL1 receptor family. Thus, like IL1RAPL1 and IL1RAPL2 may be the mutation locus causing NSMR (Jin et al., 2000; Sana et al., 2000).

This study was aimed to study the relationship between IL1RAPL2 and NSMR by examining the genetic polymorphisms of five SNPs in IL1RAPL2 gene in a random population. We performed a case–control study to investigate the association between IL1RAPL2 and NSMR in the Han Chinese population.

MATERIALS AND METHODS

Subjects

In total, there were 118 (male/female = 56/62) MR, 116 (male/female = 52/64) borderline, and 322 (male/female = 167/155) controls in our study. All subjects were randomly collected with the requirement of case-control analysis from the Qinba region of Shaanxi province, northwest China. In addition, after the protocol had been fully explained, which was approved by the Ethical Committee of the National Human Genome Center, standard informed consents were obtained from the participants and their guardians.

Intelligence evaluation

The intelligence of each participant was screened with the Chinese Wechsler Young Children Scale of Intelligence (C-WYCSI) (Gong and Dai, 1992) for 4 to 5 year old children and the Chinese Wechsler Intelligence Scale for Children (C-WISC) (Gong and Cai, 1993) for 6 to 14 year old children. The social disability (SD) scores were assessed with the adaptive scale for infants and Children revised by Zuo et al. (1998). To further evaluate the children with an IQ < 85 and SD score no higher than 9, a subsequent clinical examination was carried out by a group of neurologists, pediatricians and gynecologists. We defined the tested IQ < 70 and SD score ≤ 8 as MR and IQ of 70 to 79 and SD score of 9 as a borderline form of MR. The definition and diagnosis criteria of MR and borderline were based on the classification of mental disorders 2nd revision (CCMD-2-R) and the classification of mental and behavioral disorders from the WHO (Cao, 1995; Who, 1992). The cases affected by trachoma, infection, trauma, dystrophy, toxicosis, cerebral palsy, birth complications or other specific clinical causes were excluded from the analysis.

All subjects were of Chinese Han population and were randomly collected. Standard informed consents were obtained from all participating subjects according to the protocol reviewed and approved by the Ethical Committee of the National Human Genome Center of China.

Variants identification and genotyping

Genomic DNAs were extracted from peripheral blood mononuclear cells using a modified phenol chloroform extraction method (Joseph and Davi, 2001) and were stored at -20°C for genotyping. Five SNPs with minor allele frequency over 5% in the IL1RAPL2 region were selected, from the dbSNP (http://www.ncbi.nlm.nih.gov/SNP), they were rs5962434, rs5916817, rs3764765, rs5962298 and

rs9887672, the information of them were displayed in Table 1. The SNP rs5962434 (A/C) is in the 3'-UTR, while rs5916817 (C/T) in intron 2, and rs3764765 (C/T) in exon 9, rs5962298 (A/G), rs9887672 (C/T) are both in intron 9. The last three SNPs are in the domain which encodes the novel C-terminal sequences.

The polymerase chain reaction combined with single strand conformation polymorphism (PCR-SSCP) method and sequencing analysis were used to genotyping the five SNPs. Polymerase chain reactions were carried out in 96-well micro titer plates with each well containing a ten microliter reaction mixture of 20 mM $(NH_4)_2$ SO_4, 5 mM Tris-HCl (pH 8.8 at 25), 0.01% Tween 20, 0.2 mM dNTPs, 2.5 mM $MgCl_2$, 0.3 uM of primer, 20ng DNA and 1U Taq DNA Polymerase (Fermentas International Inc. Ontario, Canada). PCR protocol included an initial 2 min at 95°C, 29 cycles at 94°C for 30 s, annealing temperature for 30 s (the annealing temperature needed for each pair of primers are listed in Table 1), 72°C for 30 s and a final extension period of 2 min at 72°C using the Mastercycler gradient 5531 PCR System (Eppendorf Inc. Hamburg, Germany). PCR products were resolved by SSCP analysis, and verified 20 to 30 samples per SNP by ABI 3700 DNA Sequencer (ABI Inc. Foster, USA) randomly.

PCR products were denatured by using the following buffer (95% formamide, 0.025% xylene-cyanole, 0.025% bromophenol blue, 10 mM EDTA, pH 8.0) at 97°C for 5 min, and snap-chilled on ice for at least 10 min. Samples were then loaded onto 11% nondenaturing polyacrylamide gels (29:1 acrylamide to bisacrylamide) containing 0.5×TBE (0.045M Tris-borate, 0.001M EDTA, pH8.0). Electrophoresis was kept at a constant temperature (4°C) with a cooling unit and was carried out in a vertical unit at 11 V/cm for 15 h. The individual's genotype was confirmed by silver-staining and gel-imaging with Bio-Rad imaging instrument. Randomly selected PCR products that exhibited different migration patterns on gels were selected for sequencing to confirm their polymorphisms and genotypes.

Statistical analysis

Demographic data (including the age and gender) and the allele and genotype frequencies for five SNPs were calculated with SPSS (SPSS Inc., Chicago, IL, USA). Hardy-Weinberg equilibrium (HWE) tests were performed for a single SNP with the method described by Guo et al. (2002). Differences in the genotype and allele distributions between MR and control groups were assessed by the Monte Carlo method with CLUMP 2.3 (10,000 simulations) (Stephens et al., 2001). The haplotypes combined by target SNPs were estimated by PHASE2.1 software package (Stephens and Donnelly, 2003; Stephens et al., 2001). The subjects' haplotypes with a confidence level of ≥ 95% were included. The pair wise linkage disequilibrium (LD), the frequencies of haplotypes and the fit of haplotypes to HWE were estimated by Cube X 2007 (Gaunt et al., 2007), a cubic exact solution online-program rather than an iterative approach (http://www.oege.org/software/cubex/). In this study, the p-values were two-tailed and a difference was considered statically significant when $p < 0.05$. The statistical power analysis was performed with the G*Power program (Faul et al., 2007).

RESULTS

HWE test was carried out on the genotype distribution of five SNPs. The distribution was consistent with HWE no matter in NSMR group or control group.

Table 2 shows the comparison of each SNP allele and genotype frequency distribution among NSMR, borderline and control groups. The allele and genotype frequency distributions of rs5962298 and rs9887672 were significantly different between NSMR and control groups. The alleles frequency of X^A and X^T alleles of them were significantly higher in NSMR than that of control (p = 0.020; p = 0.025, respectively). The genotype frequency distribution also was significant different between the two groups (p = 0.017). Subsequently, the comparison were also performed for girls and boys, respectively; and the distribution differences of rs5962298 and rs9887672 were more significant in girls (allele: p = 0.004, p = 0.019; genotype: p = 0.021, p = 0.014), but vanished in boys. Moreover, no differences were detected between borderline and control groups.

The haplotypes combined by rs5962298 and rs9887672 were analyzed between NSMR and control groups in girls (Table 3). But no significant difference was found in girls. Therefore, the LD and haplotypes fit to HWE were calculated by the Cubex 2007 program. It shows that rs5962298 and rs9887672 did not have a strong LD (D' = -0.671, r^2 = 0.250), and their combined haplotypes did not fit to HWE, although their single SNPs were compatible with HWE.

Moreover, a power calculation was performed by the G*Power program, based on the Cohen's method (Faul et al., 2007). When an effective size index of 0.25 (corresponding to 'weak to moderate' gene effect) was used, the sample size in our study revealed > 95% power for detecting significant associations (α < 0.05).

DISCUSSION

Qin-Ba Mountain region of Shaanxi province is a relatively isolated area in northwest China. The prevalence of MR (3.19%) in the region is much higher than that in most areas of China (1.07%); and if the number of children in the low range of normal IQ (borderline of MR) is included, the prevalence will increase to 8.23% (Zhang et al., 2004). Moreover, family clustering is obviously presented in two counties where several families have multiple affected members in one or more generations. The epidemiological survey reveals that the heritability of MR is as high as 70.23%, when the MR cases affected by trachoma, infection, trauma, toxicity, cerebral palsy and birth complications are excluded (Zhang et al., 2005, 2006). This result suggests that genetic factors play an important role in the etiology of MR in this region.

IL1RAPL1 gene is a form candidate NSMR gene, reported by Bahi et al. (2003), Carrie et al. (1999) and Tabolacci et al. (2006). Both IL1RAPL1 and IL1RAPL2 belong to the IL-1 receptor family comprised of eight members involved in the response to IL-1 and IL-18. They share the same gene structure, the same intron-exon organization and a high similarity at the protein level (Jin et al., 2000; Sana et al., 2000). IL1RAPL1 differs from IL1RAPL2 by the presence of a 150 amino acid C-terminal domain encoded by the sequence region of exon 10 and exon 11, and this domain is not present in any other members of IL-1R family. Furthermore, the novel C-terminal domain shows an important biological role, and it interacts with the neuronal calcium sensor-1 protein (NCS-1). Therefore, researchers have focused on the function of the C-terminal domain. Carrie et al. (199) and Tabolacci et al. (2006) confirmed that the polymorphism in this region (exon 10 and exon 11) affects the interaction between NCS-1 and the C-terminal domain, causing NSMR presumably through the same mechanism with IL1RAPL2 (Gambino et al., 2007).

Our study selected five SNPs of IL1RAPL2, and three of them were used to mark the polymorphism condition of the C-terminal domain. We performed a case-control study to investigate the relationship between IL1RAPL2 and NSMR within the children of Qinba region of China. The single-locus association analysis showed that the SNPs rs5962298 and rs9887672 were close to the region encoding the C-terminal domain of IL1RAPL2. We found significant differences in the allele and genotype distributions between the NSMR and control groups (rs5962298, allele: p = 0.020 and genotype: p = 0.017; rs9887672 allele: p = 0.025 and genotype: p = 0.053, respectively). Furthermore, these differences were even more significant in girls, but disappeared in boys. Surprisingly, when we examined the distribution of haplotypes combined by rs5962298 and rs9887672, no significant difference was found regardless of individual or global haplotype test. The Cubex 2007 program indicated that there was no strong LD between rs5962298 and rs9887672 (D' = 0.671 and r^2 = 0.250), and the combined haplotypes also did not fit to HWE (X^2=17.76, df = 5 P = 0.003) although each of them was consistent with HWE. This result implied that there were one or more recombination hotspots between these two SNPs. They just represented two different regions of IL1RAPL2, respectively. Therefore, the haplotype analysis may not be a suitable method. But our study still showed the potential association between the genetic variants of IL1RAPL2 and NSMR.

Our results can be explained at two fold: First, both SNPs (rs5962298 and rs9887672) are located in the ninth intron of IL1RAPL2, close to the region encoding the C-terminal domain. The variant of this special domain is essential for the protein's biological functions, which may cause individual NSMR. Our study provided some evidence for this assumption. Second, in additional to these two SNPs, there were other mutations related to girls' NSMR in Qinba region of China. The inconsistent results between the single-site analysis and the haplotype analysis indicated that there were more than one mutation nearby rs5962298 and rs9887672 and they may influence the function of IL1RAPL2 and the association with NSMR as well. Given properly selected genetic markers, consistent results would be expected. Meanwhile, the effect of the gender difference should be

Table 1. The primers information and PCR condition of five SNPs of *IL1RAPL2*.

SNPs	Primer sequences (5'->3')	Product size	Tm (°C)	Allele
rs5962434	Forward: CTGGGTGGAGCAGCGAGG Reversed: GTATCGGGGATTCTTGCGG	187 bp	61.5	A/C
rs5916817	Forward: CCCAAGAAGTGAAAGTGTAT Reversed: TAGGGGAGAAGAAGTAATC	342 bp	58.0	C/T
rs3764765	Forward: CCCAAGTGTTGCTGATTT Reversed: TGCCATAGCCCTTTACCC	139 bp	59.0	C/T
rs5962298	Forward: ATTGCTGTCAAATCCCTCC Reversed: ATTGCTTGTTTGTAGATGGC	125 bp	61.0	A/G
rs9887672	Forward: ACCTTTCAGTTTTTCAGTC Reversed: CAAGAAGCCATTTGTGTT	252 bp	61.0	C/T

Table 2. The alleles and genotyps frequencies of five SNPs in NSMR and control groups.

SNPs	Allele (%)		P value (d.f.=1)	OR (95%CI)	Genotype (%)					P value (d.f.=3)
rs5962434	X^A	X^C			X^AX^A	X^AX^C	X^CX^C	X^AY	X^CY	
NSMR	84 (61.3)	53 (38.7)	0.131	1.25 (0.932–1.697)	18 (37.5)	20 (41.7)	10 (20.8)	28 (68.3)	13 (31.7)	0.211
Co.	221 (53.9)	189 (46.1)			33 (24.4)	75 (55.6)	27 (20.0)	80 (57.1)	60 (42.9)	
rs5916817	X^C	X^T			X^CX^C	X^CX^T	X^TX^T	X^CY	X^TY	
NSMR	94 (58.8)	66 (41.3)	0.948	0.99 (0.913–1.103)	20 (37.0)	21 (38.9)	13 (24.1)	33 (63.5)	19 (36.5)	0.617
Co.	271 (59.0)	188 (41.0)			57 (37.5)	70 (46.1)	25 (16.4)	87 (56.1)	68 (43.9)	
rs3764765	X^C	X^T			X^CX^C	X^CX^T	X^TX^T	X^CY	X^TY	
NSMR	47 (28.8)	116 (71.2)	0.165	1.23 (0.923–1.635)	4 (6.8)	31 (53.4)	23 (39.6)	8 (17.0)	39 (83.0)	0.282
Co.	107 (23.4)	351 (76.6)			8 (5.3)	61 (40.1)	83 (54.6)	30 (19.5)	124 (80.5)	
rs5962298	X^A	X^G			X^AX^A	X^AX^G	X^GX^G	X^AY	X^GY	
NSMR	59 (33.5)	117 (66.5)	0.020[a]	1.37 (1.056–1.775)	5 (8.2)	34 (55.7)	22 (36.1)	15 (27.8)	39 (72.2)	0.017[a]
Co.	113 (24.4)	350 (75.6)			14 (9.1)	49 (31.8)	91 (59.1)	36 (23.2)	119 (76.8)	
rs9887672	X^C	X^T			X^CX^C	X^CX^T	X^TX^T	X^CY	X^TY	

Table 2. Contd.

NSMR	135 (75.8)	43(24.2)	0.025[a]	0.72 (0.542-0.948)	33 (54.1)	25 (41.0)	3 (4.9)	44 (78.6)	12 (21.4)	0.053
Co.	386 (83.5)	76 (16.5)			109 (73.6)	33 (22.3)	6 (4.1)	135 (81.3)	31 (18.7)	

[a]Bold font indicates significant associated statistic.

Table 3. The frequencies of haplotypes combine by rs5962298 (A/G) - rs9887672 (C/T) in NSMR and controls in girls.

	The frequencies of haplotypes				
	rs5962298 - rs9887672[a]				Global P value
	X^{A-C}	X^{A-T}	X^{G-C}	X^{G-T}	
NSMR	0.180	0.156	0.631	0.033	0.313
Co.	0.115	0.135	0.712	0.034	
Individual P value	0.129	0.506	0.111	0.872	

[a]The haplotypes which frequence lower than 0.05 were excluded in global p value was calculated.

controlled.

In summary, our study has shown that the genetic variants of the region encoding the C-terminal domain of IL1RAPL2 were associated with NSMR children in Qinba region of China, but further efforts should be made to examine this locus, especially controlling the gender-difference effect and using larger sample sizes.

ACKNOWLEDGEMENTS

We sincerely thank all the participants and researchers in this study. This work was supported by 'the Tenth Five-Year Plan' National Tackle Problem Item (No.2001BA901A49), Special Prophase Project on Basic Research of The National Department of Science and Technology (2007CB516702), Shaanxi Province Science Fund (SJ08C236) and Education Foundation of Shaanxi ((No. 11JK0620).

REFERENCES

Bahi N, Friocourt G, Carrie A, Graham Me, Weiss Jl, Chafey P, Fauchereau F, Burgoyne Rdand Chelly J (2003). Il1 Receptor Accessory Protein Like, a Protein Involved in X-Linked Mental Retardation, Interacts with Neuronal Calcium Sensor-1 and Regulates Exocytosis. Hum. Mol. Genet. 12: 1415-1425.

Cao (1995). Chinese Classification of Mental Disorders Southeast University Press, Nanking.

Carrie A, Jun L, Bienvenu T, Vinet Mc, Mcdonell N, Couvert P, Zemni R, Cardona A, Van Buggenhout G, Frints S, Hamel B, Moraine C, Ropers HH, Strom T, Howell GR, Whittaker A, Ross MT, Kahn A, Fryns J, Beldjord C, Marynen P, Chelly J (1999). A New Member of the Il-1 Receptor Family Highly Expressed in Hippocampus and Involved in X-Linked Mental Retardation. Nat Genet, 2): 25-31.

Curry CJ, Stevenson RE, Aughton D, Byrne J, Carey JC, Cassidy S, Cunniff C, Graham JM, Jones MC, Kaback MM, Moeschler J, Schaefer GB, Schwartz S, Tarleton J, Opitz J (1997). Evaluation of Mental Retardation: Recommendations of a Consensus Conference: Am. Coll. Med. Genet., 72: 468-477.

Faul F, Erdfelder E, Lang AG, Buchner A (2007). G*Power 3: A Flexible Statistical Power Analysis Program for the Social, Behavioral, and Biomedical Sciences. Behav. Res. Meth., 39: 175-191.

Gambino F, Pavlowsky A, Begle A, Dupont Jl, Bahi N, Courjaret R, Gardette R, Hadjkacem H, Skala H, Poulain B, Chelly J, Vitale Nand Humeau Y (2007). Il1-Receptor Accessory Protein-Like 1 (Il1rapl1), a Protein Involved in Cognitive Functions, Regulates N-Type Ca2+-Channel and Neurite Elongation. Proc. Natl. Acad. Sci. U S A., 104: 9063-9068.

Gaunt TR, Rodriguez S, Day IN (2007). Cubic Exact Solutions for the Estimation of Pairwise Haplotype Frequencies: Implications for Linkage Disequilibrium Analyses and a Web Tool 'Cubex'. BMC Bioinformatics, 8: 428.

Gong YX, Cai TS (1993). Wechsler Intelligence Scale for Children, Chinese Revision (C-Wisc). Hunan: Map Press, Changsha.

Gong YX, Dai XY (1992). Chinses-Wechsler Yong Children Scale of Intelligence (Cwycsi). Hunan: Map Press, Changsha.

Guo Y, Guo C, Sui Q (2002). The Basis of Medical Genetics. Shandong University press, Ji Nan.

Jin H, Gardner RJ, Viswesvaraiah R, Muntoni F, Roberts RQ (2000). Two Novel Members of the Interleukin-1 Receptor Gene Family, One Deleted in Xp22.1-Xp21.3 Mental Retardation. Eur. J. Hum. Genet., 8: 87-94.

Joseph S, Davi WS (2001). Molecular Cloning: A Labortory Manual (Third Editor). Cold Spring Harbor Laboratory Press, New York.

Mclaren J, Bryson SE (1987). Review of Recent

Epidemiological Studies of Mental Retardation: Prevalence, Associated Disorders, and Etiology. Am. J. Ment. Retard., 92: 243-254.

Sana TR, Debets R, Timans JC, Bazan JF, Kastelein RA (2000). Computational Identification, Cloning, and Characterization of Il-1r9, a Novel Interleukin-1 Receptor-Like Gene Encoded over an Unusually Large Interval of Human Chromosome Xq22.2-Q22.3. Genomics, 69: 252-262.

Stephens M, Donnelly P (2003). A Comparison of Bayesian Methods for Haplotype Reconstruction from Population Genotype Data. Am. J. Hum. Genet., 73: 1162-1169.

Stephens M, Smith NJ, Donnelly P (2001). A New Statistical Method for Haplotype Reconstruction from Population Data. Am. J. Hum. Genet., 68: 978-989.

Stevenson RE, Schwartz CE, Schroer RJ (2000). X-Linked Mental Retardation, pp. Oxford University Press.

Tabolacci E, Pomponi MG, Pietrobono R, Terracciano A, Chiurazzi P, Neri G (2006). A Truncating Mutation in the Il1rapl1 Gene Is Responsible for X-Linked Mental Retardation in the Mrx21 Family. Am. J. Med. Genet. A., 140: 482-487.

WHO (1992). The Ico-10 Classification of Mental and Behavioural Disorders. WHO, Geneva.

Zhang FC, Li RI, Gao XC, Zheng ZJ, Huang SP, Song HY, Xi H, Li F (2004). An Investigation of Mental Retarded Children Aged 0 - 14 in Qinba Mountain Areas. J. Fourth Mil. Med. Univ., 25: 511-514.

Zhang KJ, Zhang FC, Zheng ZJ, Gong PY, He G, Nan YP, Zhang R, Ms J (2005). An Analysis of Inheriting Type of Non-Causing Mental Retarded Children in Zhashui Experimental Station. J. North W. Univ. (Nat. Sci. Ed.), 35: 597-600.

Zhang SM, Zhang FC, Gong PY, Zhang KJ, Yhang XZ, He G, Li N (2006). The Discussion of Inheriting Type of Mental Retardation in Ankang Experimental Station. J. North W. Univ. (Nat. Sci. Ed.), 36: 97-100.

Zuo QH, Zhang ZX Wu LZ (1998). Adaptive Scale of Infant and Children. Med. Univ. China, Beijing, China, 36: 97-100.

18

The frequency of Y chromosome microdeletions in infertile men from Chennai, a South East Indian population and the effect of smoking, drinking alcohol and chemical exposure on their frequencies

V. G. Abilash[1], Radha Saraswathy[1] and K. M. Marimuthu[2]

[1]Division of Biomolecules and Genetics, School of Bio Sciences and Technology, VIT University, Vellore 632014, Tamilnadu, India.
[2]Emeritus Professor, Dept of Genetics, University of Madras, Chennai 600113, Tamilnadu, India.

The aim of the study were to estimate the frequency of Y chromosome microdeletion in infertile men from a new geographical ethnic region, Chennai, South East India, to explore the effect of smoking, alcohol drinking, chemical exposure and cellular chromosomal aberration on the frequency of infertility in 34 azoospermia and 55 oligospermia patients. The frequency of Y chromosome microdeletion was estimated using 12 STS markers and the chromosomal aberrations were estimated in leukocyte cultures. In azoospermia the frequency of microdeletions in AZFa, AZFb, AZFc and AZFd were 27, 4, 56 and 13% respectively. In oligospermia they were 33, 7, 48 and 12% in the same order. These frequencies of Y chromosome microdeletion are significantly higher than that of European population. The chromosome aberrations per cell in azoospermia and oligospermia were higher than that of the control at the level of $p > 0.001$. The percentage of microdeletion observed in unexposed azoospermia had 15%, azoospermia smokers 22%, azoospermia smokers and alcoholics 25%; whereas the unexposed oligospermia had 7%, oligospermia smokers 12%, oligospermia smokers and alcoholics 37%. It seems that the etiology of male infertility may differ between ethnic populations and smoking, alcohol drinking and chemical exposure may have deleterious effect on human fertility.

Key words: Y chromosome microdeletion, sequence-tagged site (STS), chemical exposure, chromosomal aberrations, ethnic region.

INTRODUCTION

Infertility, defined as the inability to conceive after 12 months of unprotected intercourse, affects 10 – 15% of all couples (Mosher and Pratt, 1991). In roughly half of the cases, a male factor is identified, while an occult male factor may be involved in 15–24% percent of cases in which no etiology is uncovered ("unexplained" infertility) (Skakkebaek et al., 1994; Templeton and Penney, 1982). A variety of occupational exposures have been linked to impaired male fertility (Sheiner et al., 2003). However, studies have been limited by inadequate sample sizes, inappropriate study designs, and/or selection bias

(Lahdetie, 1995; Bonde et al., 1996; Cohn et al., 2002). Hence well defined further study with a good sample size of 34 azoospermia and 55 oligospermia patients is planned.

The AZF (azoospermia factor) region Yq11 contains genes vital for spermatogenesis. Vogt et al. (1996), Affara et al. (1999) subdivided this region into AZFa, AZFb and AZFc. Deletions within these sub-regions cause various spermatogenic and infertility phenotypes (Affara and Mitchell, 2000). Molecular studies have shown that microdeletions at Yq11 may represent the etiological factor in as many as 10 - 15% of cases with idiopathic azoospermia or severe oligozoospermia (Reijo et al., 1995; Vogt et al., 1996). The microdeletion events appear in three critical regions of the long arm of the Y chromosome, initially considered non overlapping, called

---

*Corresponding author. E-mail: r_saraswathy@yahoo.com.

Azoospermia factor (AZFa, AZFb and AZFc) (Vogt et al., 1996). Around 71% of men with Y chromosomal microdeletions and severe oligozoospermia or idiopathic azoo-spermia were found to have AZFc deletions compared with 13% with AZFa and 31% with AZFb deletions (Ferlin et al., 1999; Reynolds and Cooke, 2005). It is planned to estimate the frequency of various Y chromosome micro-deletion in 34 azoospermia and 55 oligospermia patients selected from a new geographical and ethnic population from Chennai and its surroundings in South India.

To what extent there is a genetic contribution is unclear. It has been reported that in a certain ethnic group, men with a particular haplotype (II) have a lower sperm concentration compared with men with haplotypes (III) and (IV) and, the frequency of haplotype (II) is more common in azoospermic men compared with normal men (Kuroki et al., 1999). Based on this, it appears that the genetic contribution towards male fertility on account of a de-creased sperm concentration might be significant in some ethnic groups (Seshagiri, 2001). Hence it is planned to study the frequency and pattern of infertility in a new ethnic population at Chennai in South India.

In recent years, there has been increasing concern about the possible deleterious effects of sperm quality. Amongst other factors, cigarette smoke is one factor which can theoretically influence male fertility in several ways. Unfortunately, the results of several studies in this area have been contradictory. In some studies, no relation could be demonstrated between tobacco consumption and sperm quality whilst in other reports, an association has been described between smoking and low sperm count, a relatively higher proportion of abnormal spermatozoa and reduced sperm motility (Oldereid et al., 1989). Hence it requires further study to get a clear picture.

There appears to be a worldwide concern over decreasing human sperm concentration but this has been highly controversial (Seshagiri, 2001). Decreasing sperm counts are attributed to the deleterious effects of environmental contamination by heavy metals and chemical exposure during working in chemical industry (Mehta and Anandkumar, 1997; Benoff et al., 2000; Sharpe, 2000). Hence it is planned to study the effect of chemical exposure to understand the degree of deleterious effect on sperm count.

Microdeletion of the Y chromosomes that remove associated fertility genes have received attention of late (Chandley, 1998). Structural abnormalities do lead to phenoltypic male reproductive disorder or may predispose to severe congenital abnormality when gametes are formed (Diemer and Desjardins, 1999). A study is planned to estimate and detect the frequency of chromosomal aberrations in all the patients studied.

It may be concluded that though it was known that azoos-permia and severe oligospermia cause infertility in human yet it is further decided to detect in this study the impact of certain life style like smoking, drinking alcohol, exposure to some toxic chemicals and environmental induced chromosomal aberrations on the severity of infertility infertility.

MATERIALS AND METHODS

34 azoospermic and 55 oligospermic men, selected from the Andrology Department, Stanley Medical College and Hospitals, Chennai, India, were included in the present study. The age groups of azoospermic men ranged from 24 - 38 years and oligospermic men age ranged from 16 - 37 years. With the help of an experienced urologist at Stanley Hospital, a detailed case history and clinical examination of every patient were carried out. The life style habit and chemical exposure of the probands were recorded, including smoking habit, alcohol drinking and exposure to toxic chemicals.

Semen analysis is routinely performed on the male partner of couple coming for infertility treatment. Hundred random fertile Indian men were included in this study as control. Blood samples from each azoospermic, oligospermic and control men were collected by the Physicians with the written consent. For this study ethical clearance was obtained from the Madras Medical College, Chennai, India.

Cytogenetic analysis

2 ml of intravenous blood was collected from every patient and control by using sodium heparin coated vaccutainer. The cytogenetic studies were carried out to find out the karyotype, frequency and type of chromosomal aberrations. Chromosome preparations were obtained from PHA-stimulated peripheral blood lymphocytes by using modified method of Hungerford, (1965). At least fifty well spread metaphase plates were scored by direct microscopic analysis. Scoring of chromosomal aberrations including chromatid and chromosomal breaks and deletions were carried out and recorded from well spread and stained cells under oil immersion objective lens (100×) of the light microscope. Well spread metaphases were photographed under oil immersion objective lens (100X) of Leica DM2000 microscope with Metasystems camera and the photomicrographs of banded spreads were karyotyped using automatic IKaros software (Metasystems). The karyotype was described according to the International System for Human Cytogenetic Nomenclature (Shaffer and Tommerup, 2005).

Molecular analysis

The molecular study was carried out in all the patients and control sample to make a thorough analysis to detect the Y chromosome microdeletion in AZF region.

DNA extraction and quantification

9 ml of intravenous blood was collected from all the patients and control sample by using EDTA coated vaccutainer. The genomic DNA was extracted from peripheral blood by using modified method of Lotery et al. (2000). Qualitative analysis of DNA was carried out by 0.8% Agarose gel electrophoresis and quantification of DNA by using Biophotometer (Eppendorf). Dilutions of DNA were made up to 10 ng/µl concentration by using TE buffer, pH 8.0. The 10 ng/µl of concentrated DNA solution was checked on 0.8% agarose gel.

Polymerase chain reaction analysis

The polymerase chain reaction (PCR) based studies for

microdeletion on azoospermic, oliogospermic and control men were carried out using STS markers on the long arm of Y chromosome. Screening for AZF region was done using 12 STS markers. The AZFa region was analyzed with sY82 and sY84. The AZFb region was analyzed with sY164. The AZFc region was analyzed with sY158, sY160, sY240, sY254, sY255, sY277 and CDY. The AZFd region was analyzed with sY145 and sY152. The lyophilized primers were ordered and received from the company (Ist Base Pvt. Ltd, Singapore). Polymerase chain reaction consisted of 10µl PCR reaction mixture and included 1.0 µl PCR buffer (10×), 1.0 µl MgCl2 (25 mM), 0.8 µl deoxynucleotide tri-phosphates (10 mM), 0.5 pM of each primer, 1unit of Taq Polymerase (1 unit/ul) and 20 ng of genomic DNA. All the reagents for PCR were purchased from Vivantis Technologies (Malaysia). Each marker was amplified separately in a 0.2 mL thin wall tube using an Eppendorf thermal cycler with a female negative control sample. PCR conditions used for STS markers were as follows: initial denaturation (95°C for 5 min), subsequent denaturations (94°C for 1min) and extension (72°C for 1 min) were the same for all the samples. Different annealing temperatures that were used for different STS markers were as follows: 60°C for 15 s for sY82, sY254 and sY277; 60°C for 1 min for sY158, sY160, sY240, and sY255; 57°C for 30 s for sY84, Chromodomain Y(CDY) and sY145; 58°C for 30 s for sY164 and sY152. The PCR products were separated by eletrophoresis on 2% agarose gel. A 100 bp DNA ladder was loaded with PCR products to estimate band size. The gel was stained with ethidium bromide and visualized under UV transilluminator and photographed.

RESULTS AND DISCUSSION

Cytogenetic analysis

Spermatogenesis is a complex process and it is subject to the influence of many genes. Genetic factors involved in male infertility are manifested as chromosomal disorders, monogenic disorders, multi-factorial disorders and endocrine disorders of genetic origin (Diemer and Desjardins, 1999; Egozcue et al., 2000). Chromosomal abnormalities are common in infertile men, for example, Klinefelter syndrome (Egozcue et al., 2000). Besides numerical abnormalities, structural abnormalities also lead to phenotypic male reproductive disorder or may predispose to severe congenital abnormality when gametes are formed (Diemer and Desjardins, 1999).

In the present study no chromosomal abnormality was observed. G banded metaphase analyses of all 34 cases of azoospermia revealed 46, XY normal karyotype. The frequency of naturally occurring chromosome aberrations per cell in the patient was 0.18 and in the control it was 0.014. In the patients the frequency of chromosomal aberrations was significantly higher than that of the control at the level of p > 0.001.

In the same manner in all 55 cases of oligospermia revealed 46, XY normal karyotype. The frequency of chromosome aberrations per cell in the patient was 0.16 and in the control it was 0.014. In the patients the frequency of chromosomal aberrations was significantly higher than that of the control at the level of p > 0.001. These increased cellular chromosome aberrations may lead to phenotypic male reproductive disorder and may predispose to congenital abnormalities resulting in

increased infertility. The sperm count was less than 4 - 20 million per ml in the oligospermic patients and for controls more than 20 million/ml were observed.

The Y chromosome microdeletion frequency analysis

In Tiepolo and Zuffardi (1976) observed for the first time the involvement of Yq deletions in male infertility when they were analyzing cells from idiopathic infertile males. Since then, many structural abnormalities in the Y chromosome have been observed, including microdeletions detectable only by molecular methods. In a foreign population in England Pryor et al. (1997) showed Y chromosome microdeletions in azoospermia are 16%. Mole-cular studies have shown that microdeletions at Yq11 may represent the etiological factor in as many as 10 - 15% of cases with idiopathic azoospermia or severe oligozoospermia (Reijo et al., 1995; Vogt et al., 1996).

The microdeletion events appear in three critical regions of the long arm of the Y chromosome, initially considered non overlapping, called Azoospermia factor (AZFa, AZFb and AZFc) (Vogt et al., 1996). Around 71% of men with Y chromosomal microdeletions and severe oligozoospermia or idiopathic azoospermia were found to have AZFc deletions compared with 13% with AZFa and 31% with AZFb deletions (Ferlin et al., 1999; Reynolds and Cooke, 2005). Challenges for current research include the elucidation of the genomic mechanisms that generate such recurrent deletions and also the identification of the genes that cause infertility when deleted or damaged. Recombination between repetitive regions is believed to be the cause of the high incidence of de novo microdeletions in the Y chromosome long arm. For instance, Kuroda-Kawaguchi et al. (2001) demonstrated that 47 out of 48 men with AZFc deletions had the same proximal and distal breakpoints in 229 kb direct repeats flanking AZFc. Furthermore, Repping et al. (2002) demonstrated that recombination between repetitive regions in Yq can explain the majority of AZFb and AZFb + AZFc deletions.

Recently, it was shown that homologous recombination events are highly recurrent in the MSY region, especially at the AZFc locus (Repping et al., 2003; Skaletsky et al., 2003; Machev et al., 2004). The spermatogenic impairment caused by AZFb and AZFc deletions can be actually caused by genes mapped at these regions as proposed by several studies (Brown et al., 1998; Mahadevaiah et al., 1998). Since the first deletion mapping studies until the most recent and detailed physi-cal map of the human Y chromosome several genes rela-ted to spermatogenesis were discovered (Skaletsky et al., 2003).

The patients selected for this study from a new geographical ethnic population showed in 34 cases of azoospermia the microdeletion events in four critical regions of the long arm of the Y chromosome, namely azoospermia factor (AZFa, AZFb, AZFc and AZFd). 56% of azoospermia

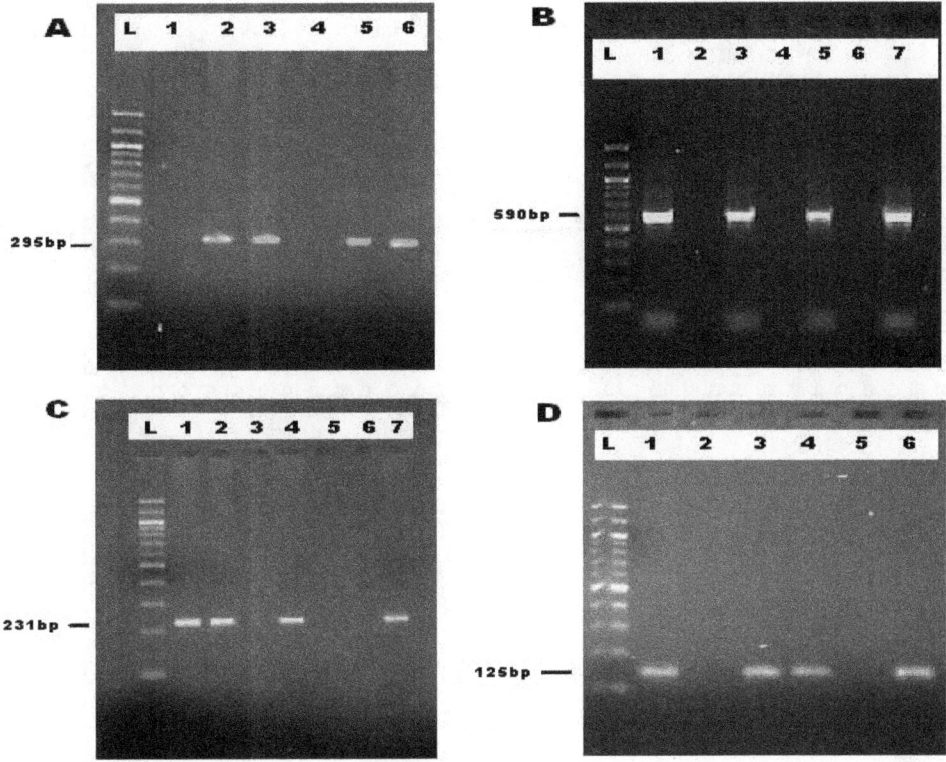

Figure 1. PCR analysis of Y chromosome microdeletions in azoospermic and oligospermic cases. (A) agarose gel electrophoresis analysis shows the microdeletion of AZFa region SY84 (295bp) STS marker of Y chromosome in azoospermia case (AZO17-1). L -100 bp DNA ladder, 1 - negative control, 2 - positive control; 3 - AZO16-1= azoospermia, 4 - AZO17-1= azoospermia (deletion), 5 - AZO19-1= azoospermia, 6 - AZO20-1= azoospermia. (B) agarose gel electrophoresis analysis shows the microdeletion of AZFb region SY164 (590bp) STS marker of Y chromosome in oligospermia cases (OLI33-1& 46-1). L - 100 bp DNA ladder, 1 - positive control, 2 - negative control, 3 - OLI32-1= oligospermia, 4 - OLI33-1= oligospermia (deletion), 5 - OLI34-1= oligospermia, 6 - OLI461= oligospermia (deletion), 7 - OLI47-1= oligospermia. (C) agarose gel electrophoresis analysis shows the microdeletion of AZFc region SY158 (231bp) STS marker of Y chromosome in oligospermia cases (OLI33-1& 35-1). L - 100 bp DNA ladder, 2 - OLI36-1= oligospermia, 3 - negative control, 4 - OLI32-1= oligospermia, 5 - OLI33-1= oligospermia (deletion), 6 - OLI35-1= oligospermia (deletion), 7 - OLI36-1= oligospermia. (D) agarose gel electrophoresis analysis shows the microdeletion of AZFd region SY145 (125bp) STS marker of Y chromosome in azoospermia case (AZO29-1). L -100 bp DNA ladder, 1 - positive control, 2 - negative control, 3 - AZO27-1= azoospermia, 4 - AZO28-1= azoospermia, 5 - AZO29-1= azoospermia (deletion), 6 - AZO31-1= azoospermia.

men with Y chromosomal microdeletions to have AZFc deletions as compared with 27% with AZFa, 4% with AZFb and 13% with AZFd deletions (Figures 1a, d and 2a and Table 1).

In oligospermia the microdeletion events in four critical regions (AZFa, AZFb, AZFc and AZFd) showed 48% of oligopermia men with AZFc deletions as compared with 33% with AZFa, 7% with AZFb and 12% with AZFd deletions (Figures 1b and c, 2b and Table 2). These results showed higher frequency when compared with European ethnic population.

Thus the overall frequency of Y chromosome micro-deletion detection in the infertile South Indian population in the present study was found to be (24.71%). The frequency of Y chromosome microdeletions was higher in azoospermic males (29.41%) compared with the cases of oligospermia (21.81%). In our report the frequencies of microdeletions were higher than that reported by Stuppia et al. (1996) and Yao et al. (2001) as 21 and 18.7% respectively.

AZFc region is one of the most important candidate genes involved in infertility, microdeletions in this region is known to cause sterility via meiotic arrest or absence of germ cells (Yao et al., 2001). In this study we observed 56% Y chromosome microdeletions in AZFc region in azoospermic cases and 48% in oligospermia cases. This

Table 1. The details of the karyotype, lifestyle and molecular analysis in azoospermia cases.

S. NO	Lab code	AGE	SMO	ALO	CHE	Karyotype	AZFa		AZFb			AZFc					AZFd	
							SY82	SY84	SY164	SY158	SY160	SY240	SY254	SY255	SY277	CDY	SY145	SY152
1	AZO1-1	28	–	–	–	46,XY	P	P	P	P	P	P	P	P	P	P	P	P
2	AZO2-1	29	–	–	–	46,XY	P	P	P	P	P	P	P	P	P	P	P	P
3	AZO3-1	29	–	–	–	46,XY	P	P	P	P	P	P	D	P	P	P	P	P
4	AZO4-1	30	–	–	–	46,XY	P	P	P	P	P	P	P	P	P	P	P	P
5	AZO5-1	33	–	–	–	46,XY	P	P	P	P	P	P	P	P	P	P	P	P
6	AZO6-1	31	–	–	–	46,XY	P	P	P	P	P	P	P	P	P	P	P	P
7	AZO7-1	32	–	–	–	46,XY	P	P	P	P	D	P	P	P	P	P	P	P
8	AZO8-1	31	–	–	–	46,XY	P	P	P	P	P	P	P	P	P	P	P	P
9	AZO9-1	29	–	–	–	46,XY	P	P	P	P	P	P	P	P	P	P	P	P
10	AZO10-1	28	–	–	–	46,XY	P	P	P	P	P	P	P	P	P	P	P	P
11	AZO11-1	31	†	–	–	46,XY	P	P	P	P	P	P	P	P	P	P	P	P
12	AZO12-1	34	†	–	–	46,XY	P	P	P	P	P	P	P	P	P	P	P	P
13	AZO13-1	35	†	–	–	46,XY	P	D	P	P	P	P	P	D	P	D	P	P
14	AZO14-1	32	†	†	–	46,XY	P	P	P	P	P	P	P	P	P	P	P	P
5	AZO15-1	36	†	–	–	46,XY	P	P	P	P	P	P	P	P	P	P	P	P
16	AZO16-1	26	†	–	–	46,XY	D	P	P	P	P	P	P	P	P	P	P	P
17	AZO17-1	28	†	–	–	46,XY	D	P	P	P	P	P	P	P	P	P	P	P
18	AZO18-1	31	†	–	–	46,XY	D	P	P	P	P	P	P	P	P	P	P	P
19	AZO19-1	36	†	–	–	46,XY	P	P	P	P	P	P	P	P	P	P	P	P
20	AZO20-1	32	–	–	†	46,XY	P	P	P	P	P	P	P	P	P	P	P	P
21	AZO21-1	37	†	–	–	46,XY	P	P	P	P	P	P	P	P	P	P	P	P
22	AZO22-1	38	†	–	†	46,XY	P	P	P	P	P	P	P	P	P	P	P	P
23	AZO23-1	32	–	–	†	46,XY	P	P	P	P	P	P	P	P	P	P	P	P
24	AZO24-1	24	†	–	–	46,XY	D	D	D	D	D	D	D	D	P	P	P	P
25	AZO25-1	29	–	–	–	46,XY	D	P	P	P	P	D	D	D	P	P	P	P
26	AZO26-1	27	†	–	–	46,XY	P	P	P	P	P	P	P	P	P	P	P	P
27	AZO27-1	31	†	–	–	46,XY	P	P	P	P	P	P	P	P	P	P	P	P
28	AZO28-1	32	†	†	–	46,XY	P	P	P	P	P	P	P	P	P	P	P	P
29	AZO29-1	35	–	–	†	46,XY	D	D	D	D	D	D	D	D	P	D	D	D
30	AZO30-1	36	†	†	–	46,XY	D	D	D	D	D	D	D	D	P	D	D	D
31	AZO31-1	34	†	–	–	46,XY	P	D	P	P	P	P	P	P	P	P	P	P
32	AZO32-1	37	†	–	–	46,XY	P	P	P	P	P	P	P	P	P	P	P	P
33	AZO33-1	38	–	–	†	46,XY	P	D	P	P	P	P	P	D	P	P	P	D
34	AZO34-1	29	†	–	–	46,XY	P	P	P	P	P	P	P	P	P	P	P	P

P, Present; **†**, Exposed; **SMO**, Smoking; **CHE**, Chemical exposure; **D**, Deleted; **–**, unexposed; **ALO**, Alcoholics.

is in agreement with the earlier studies showing that the incidence of deletions in the AZFc region was higher when compared with the AZFa and AZFb regions (Martinez et al., 2003; Peterlin et al., 2002). The size of the deletion on the Y chromosome in our study did not show any significant correlation with the amount of sperm production in oligospermia.

The effect of smoking, alcohol drinking and chemical exposure on the frequency of fertility

In recent years, there has been increasing

Table 2. The details of the sperm count, lifestyle and molecular analysis in oligospermia cases.

S.NO	Lab code	Age	SMO	ALO	CHE	Sperm count per ml	AZFa		AZFb				AZFc				AZFd	
							SY82	SY84	SY164	SY158	SY160	SY240	SY254	SY255	SY277	CDY	SY145	SY152
1	OLI 1–1	32	−	−	−	20	P	P	P	P	P	P	P	P	P	P	P	P
2	OLI 2–1	29	−	−	−	18	P	P	P	P	P	P	P	P	P	P	P	P
3	OLI 3–1	36	−	−	−	> 19	P	P	P	P	P	D	P	P	P	P	P	P
4	OLI 4–1	30	−	−	−	15	P	P	P	P	P	P	P	P	P	P	P	P
5	OLI 5–1	28	−	−	−	16	P	P	P	P	P	P	P	P	P	P	P	P
6	OLI 6–1	27	−	−	−	> 14	P	P	P	P	P	P	P	P	P	P	P	P
7	OLI 7–1	28	−	−	−	19	P	P	P	P	P	P	P	P	P	D	P	P
8	OLI 8–1	31	−	−	−	> 15	P	P	P	P	P	P	P	P	P	P	P	P
9	OLI 9–1	30	−	−	−	20	P	P	P	P	P	P	P	P	P	P	P	P
10	OLI 10–1	35	−	−	−	> 17	P	P	P	P	P	P	P	P	P	P	P	P
11	OLI 11–1	29	−	−	−	16	P	P	P	P	P	P	P	P	P	P	P	P
12	OLI 12–1	32	−	−	−	18	P	P	P	P	P	P	P	P	P	P	P	P
13	OLI 13–1	26	−	−	−	> 14	P	P	P	P	P	P	P	P	P	P	P	P
14	OLI 14–1	28	−	−	−	16	P	P	P	P	P	P	P	P	P	P	P	P
15	OLI 15–1	30	−	−	−	> 19	P	P	P	P	P	P	P	P	P	P	P	P
16	OLI 16–1	24	†	−	−	12	P	P	P	P	P	P	P	P	P	P	P	P
17	OLI 17–1	26	†	−	−	> 16	P	P	P	P	P	P	P	P	P	P	P	P
18	OLI 18–1	31	†	−	−	10	P	P	P	P	P	P	P	P	P	P	P	P
19	OLI 19–1	33	†	−	−	> 7	D	P	P	P	D	P	D	P	P	P	P	P
20	OLI 20–1	17	†	−	−	> 5	P	P	P	P	P	P	P	P	P	P	P	P
21	OLI 21–1	36	†	†	−	11	P	D	P	P	D	D	P	P	P	P	P	D
22	OLI 22–1	35	−	−	†	13	D	D	P	D	P	D	P	D	P	P	D	P
23	OLI 23–1	16	†	−	−	> 9	P	P	P	P	P	P	P	P	P	P	P	P
24	OLI 24–1	21	†	−	−	> 6	P	P	P	P	P	P	P	P	P	P	P	P
25	OLI 25–1	26	†	−	−	12	P	P	P	P	P	P	P	P	P	P	P	P
26	OLI 26–1	19	†	−	−	13	P	P	P	P	P	P	P	P	P	P	P	P
27	OLI 27–1	19	†	−	−	15	P	P	P	P	P	P	P	P	P	P	P	P
28	OLI 28–1	32	†	−	−	4	P	D	P	P	P	P	P	D	P	P	D	P
29	OLI 29–1	26	−	−	†	> 8	P	P	P	P	P	P	P	P	P	P	P	P
30	OLI 30–1	28	†	†	−	16	P	P	P	P	P	P	P	P	P	P	P	P
31	OLI 31–1	31	−	−	†	> 10	P	P	P	P	P	P	P	P	P	P	P	P
32	OLI 32–1	36	†	−	−	13	D	D	P	D	P	P	D	P	P	P	P	P
33	OLI 33–1	34	−	−	†	> 8	D	D	D	D	P	D	P	D	P	P	P	P
34	OLI 34–1	31	−	−	†	> 4	P	P	P	D	P	P	P	P	P	P	P	P
35	OLI 35–1	31	†	†	−	10	D	D	P	D	P	P	P	D	P	P	P	P

Table 2. Contd.

36	OLI 36-1	35	†	–	>7	P	P	P	P	P	P	P	P	P	P	P	P	P	P
37	OLI 37-1	29	–	–	12	P	P	P	P	P	P	P	P	P	P	P	P	P	P
38	OLI 38-1	35	–	†	16	D	P	P	D	P	P	P	P	P	P	P	P	P	P
39	OLI 39-1	37	–	†	11	P	P	P	P	P	P	P	P	P	P	P	P	P	P
40	OLI 40-1	36	†	†	19	D	P	D	P	P	P	P	P	P	P	P	P	P	D
41	OLI 41-1	32	†	–	15	P	P	P	P	P	P	P	P	P	P	P	P	P	P
42	OLI 42-1	29	†	†	>14	P	P	P	P	P	P	P	P	P	P	P	P	P	P
43	OLI 43-1	28	†	†	>6	P	P	P	P	P	P	P	P	P	P	P	P	P	P
44	OLI 44-1	30	†	†	>9	P	P	P	P	P	P	P	P	P	P	P	P	P	P
45	OLI 45-1	29	†	–	17	P	P	P	P	P	P	P	P	P	P	P	P	P	P
46	OLI 46-1	37	–	†	>6	P	D	P	D	P	P	P	P	P	P	P	P	P	P
47	OLI 47-1	31	†	–	16	P	P	P	P	P	P	P	P	P	P	P	P	P	P
48	OLI 48-1	28	–	†	>11	P	P	P	P	P	P	P	P	P	P	P	P	P	P
49	OLI 49-1	29	†	–	14	P	P	P	P	P	P	P	P	P	P	P	P	P	P
50	OLI 50-1	30	†	–	17	P	P	P	P	P	P	P	P	P	P	P	P	P	P
51	OLI 51-1	34	†	–	>4	P	P	P	P	P	P	P	P	P	P	P	P	P	P
52	OLI 52-1	29	†	–	>10	P	P	P	P	P	P	P	P	P	P	P	P	P	P
53	OLI 53-1	18	†	–	14	P	P	P	P	P	P	P	P	P	P	P	P	P	P
54	OLI 54-1	20	†	–	16	P	P	P	P	P	P	P	P	P	P	P	P	P	P
55	OLI 55-1	31	†	†	>8	P	D	D	D	D	P	D	P	P	P	P	P	D	P

P, Present; †, Exposed; SMO, smoking; E, Chemical exposure; D, Deleted; –, unexposed; ALO, Alcoholics.

concern about the possible deleterious effects of environmental factors on sperm quality. Amongst others, cigarette smoke is one factor which can theoretically influence male fertility in several ways. Suggested mechanisms include mutagenic effects of aromatic hydrocarbons (Kier et al., 1974), toxic effects of heavy metals such as cadmium (Ostergaard, 1977), reduced availability of haemoglobin due to carbon monoxide (Kaufman et al., 1983), accumulation of radio-active particles in the testes (Ravenkolt, 1982) and toxic effects of nicotine (Mattison, 1982). Sig-

nificantly elevated leukocytes have been reported in the peripheral blood of smokers (Parry et al., 1997). Leukocytes are the major source of reactive oxygen species (ROS) in the ejaculate (Sharma and Agarwal, 1996). Elevated leukocytes may impair fertility by formation of ROS (Ochsendorf, 1999). ROS are harmful to sperm DNA (Shen et al., 1999) and membrane phospholipids (Kim and Parthasarathy, 1998) because of oxidation. The effects of excessive oxidation on sperm function have been suggested as detrimental. The role of ROS, however and

whether ROS concentrations were elevated in the semen of smokers, has not been studied yet. Unfortunately, the results of several studies in this area have been contradictory. In some studies, no relation could be demonstrated between tobacco consumption and sperm quality whilst in other reports, an association has been described between smoking and low sperm count a relatively higher proportion of abnormal spermatozoa and reduced sperm motility. In one study, enhanced sperm movement has been associated with smoking, at least within the first hour of

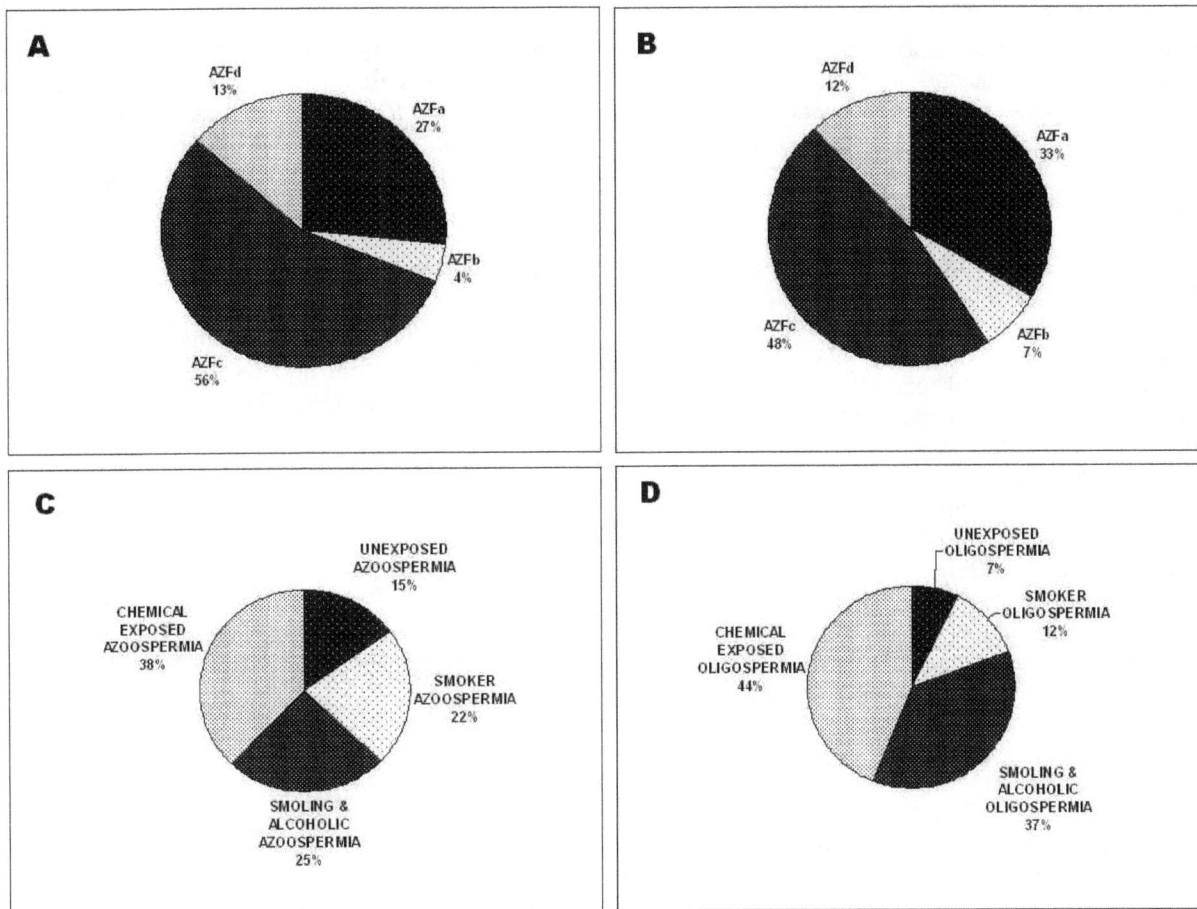

Figure 2. Pie diagram showing the percentage of Y chromosome deletion observed in AZF regions of azoospermia and oligospermia. (A) In azoospermia the microdeletion events in four critical regions of the long arm of the Y chromosome, namely azoospermia factor (AZFa, AZFb, AZFc and AZFd). 56% of azoospermia men with Y chromosomal microdeletions to have AZFc deletions as compared with 27% with AZFa, 4% with AZFb and 13% with AZFd deletions. (B) In oligospermia the microdeletion events in four critical regions (AZFa, AZFb, AZFc and AZFd) showed 48% of oligopermia men with AZFc deletions as compared with 33% with AZFa, 7% with AZFb and 12% with AZFd deletions. (C) In this new ethnic population the percentage of microdeletion observed in unexposed azoospermia had 15%, smoker azoospermia 22%, smoker and alcoholic azoospermia 25% and chemical exposed azoospermia had 38%. (D) The percentage of microdeletion observed in unexposed oligospermia had 7%, smoker oligospermia 12%, smoker and alcoholic oligospermia 37% and chemical exposed oligospermia had 44%.

ejaculation (Saaranen et al., 1987). Earlier studies have generally found an association between female smoking and pro-longed time to pregnancy (Howe et al., 1985; Alderete et al., 1995), but no association for male smoking (Baird and Wilcox, 1985; Suonio et al., 1990; Bolumar et al., 1996). Studies of caffeine use among women have found both decreased fecundability (Joesoef et al., 1990; Hatch et al., 1993) and no association (Joesoef et al., 1990; Florack et al., 1994), while studies examining male caffeine use have not found an association with fecundability (Florack et al., 1994). No associations between alcohol use and fecundability have been found (Weinberg et al., 1989; Florack et al., 1994). Limitations of some of these studies include failure to examine dose-response gradients and lack of control of spouse's behaviors. In this new ethnic

Dravidian population the percentage of microdeletion observed in unexposed azoospermia had 15%, azoospermia smokers 22%, azoospermia smokers and alcoholics 25% and chemical exposed azoospermia had 38% (Figure 2C). The percentage of microdeletion observed in unexposed oligospermia had 7%, oligo-spermia smokers 12%, oligospermia smokers and alcoholics 37% and chemical exposed oligospermia had 44% (Figure 2D). The consumption of alcohol shows little effect on fecundability (Olsen et al., 1983). Using the endpoint of TTP (time to pregnancy, length of time to achieve pregnancy), Sallmen et al. (1998) found limited support for the hypothesis that paternal exposure to organic solvents might be associated with decreased fertility. Farmers and agricultural workers are exposed to a variety of potentially harmful chemicals. No association was previously found between exposure to chemicals and infertility (Gerber et al., 1988). By contrast, occupational exposure to pesticides in fruit growers in The Netherlands (De Cock et al., 1994) was found. In an infertility-consulting population, environmental exposure, particularly to pesticides and solvents, is associated with dramatic changes in seminal characteristics (Oliva et al., 2001).

It was known that a comparison of the baseline characteristics of infertile and fertile men revealed that infertile men were more likely than fertile men to be Caucasian population, employed in blue-collar jobs, and less educated. Cases and controls were not evenly distributed among the clinical sites, and significant overall differences remained even when sites were categorized by geographic location. Among infertile subjects, socioeconomic characteristics were similar for persons with malefactor infertility and persons with unexplained infertility (Gracia et al., 2005).

Further it has been reported that in a certain ethnic men with a particular haplotype (II) have a lower sperm concentration compared with haplotype (III) and (IV) and the frequency of haplotype (II) is more common in azoospermia men compared with normal men (Kuroki et al., 1999).Thus it seems that genetic contribution towards male fertility on the account of a decreased sperm concentration might be significant in some ethinic group. Encouraged by those observations, we planned to look into the azoospermia and oligospermia patients from a new geographical and ethnic Dravidian population, Chennai, South east India. The assumption was proved to be correct that the frequency of infertility is high in the new Dravidian population of Chennai, South east India.

Conclusion

The high frequency of Y microdeletions suggests that the Y chromosome is susceptible to spontaneous loss of genetic material. Aberrant recombination events occur between areas of homologous or similar sequence repeats between X and Y chromosome or within Y chromosome itself by unbalanced sister chromatid exchanges (Yen et al., 1990). The instability of the Y chromosome may be related to a high frequency of repetitive elements clustered along the length of the chromosome (Krausz and McElreavey, 1999). Although it is very clear that microdeletions in the AZF region are responsible for spermatogenic failure, further studies are worthwhile to delineate the exact function of the genes present in AZF region and their role in spermatogenesis and fertility. However, etiologies of a large number of azoospermic men are still unknown. Analyzing the remaining azoospermic men with additional Y chromosome STS, X chromosome, and autosomal markers would help in identifying the etiology of the remaining azoospermic individuals (Brandell et al., 1998; Wang et al., 2001). In light of this study, we believe that the etiology of male

infertility may differ between ethnic populations. Therefore, researchers need to keep this in mind and define the strategies for analyzing infertile samples. This data will be useful for infertility clinics for genetic counseling by advising them to choose a female child in case of Y chromosome deletion and to adopt appropriate methods for assisted reproduction.

ACKNOWLEDGMENTS

The authors are indebted to patients and extremely grateful to Dr. Dhanapal and Dr. Sudhakar, Stanley Medical College and Hospitals, Chennai, India for providing us with the samples. The authors would also like to thank the management of VIT University for providing all the facilities to carry out this work.

REFERENCES

Affara NA, Mitchell MJ (2000). The role of human and mouse Y chromosome genes in male infertility. J. Endocrinol. Invest., 23: 630-645.

Affara NA, Wong J, Blanco P (1999). An exon map of the AZFc male infertility region of the human Y chromosome. Mamm. Genome., 10: 57-61.

Alderete E, Eskenazi B, Sholtz R (1995). Effect of cigarette smoking and coffee drinking on time to conception. Epidemiology, 6: 403-408.

Baird DD, Wilcox AJ (1985). Cigarette smoking associated with delayed conception. JAMA., 253: 2979-2983.

Benoff S, Jacob A, Hurley IR (2000). Male infertility and environmental exposure to lead and cadmium. Hum. Reprod., 6: 107-121.

Bolumar F, Olsen J, Boldsen J (1996). Smoking reduces fecundity a European multicenter study on infertility and subfecundity. Am. J. Epidemiol., 143: 578-87.

Bonde JP, Giwercman A, Ernst E (1996). Identifying environmental risk to male reproductive function by occupational sperm studies: logistics and design options. Occup. Environ. Med., 53: 511-519.

Brandell RA, Mielnik A, Liotta D, Ye Z, Veeck LL, Palermo GD, Schlegel PN (1998). AZFb deletion predicts the absence of spermatozoa with testicular sperm extraction: preliminary report of a prognostic genetic test. Hum. Reprod., 13: 2812-2815.

Brown GM, Furlong RA, Sargent CA, Erickson RP, Longepied G, Mitchell M, Jones MH, Hargreave TB, Cooke HJ, Affara NA (1998). Characterization of the coding sequence and fine mapping of the human DFFRY gene and comparative expression analysis and mapping to the Sxr^b interval of the mouse Y chromosome of the Dffry gene. Hum. Mol. Genet., 7: 97-107.

Chandley AC (1998). Chromosome anomalies and Y chromosome microdeletions as causal factors in male infertility; Hum. Reprod., 13: 45-50.

Cohn BA, Overstreet JW, Fogel RJ, Brazil CK, Baird DD, Cirillo PM (2002). Epidemiologic studies of human semen quality: considerations for study design. Am. J. Epidemiol., 155: 664-671.

De Cock J, Westveer K, Heederik D, Te Velde E, Van Kooij R (1994). Time to pregnancy and occupational exposure to pesticides in fruit growers in the Netherlands. Occup. Environ. Med., 51: 693-699.

Diemer T, Desjardins C (1999). Developmental and genetic disorders in spermatogenesis. Hum. Reprod., 5: 120-140.

Egozcue S, Blanco J, Vendrell J M, Garcia F, Veiga A, Aran B, Barri P N, Vidal F, Egozcue J (2000). Human male infertility: chromosome anomalies meiotic disorders, abnormal spermatozoa and recurrent abortion. Hum. Reprod., 6: 93-105.

Ferlin A, Moro E, Garolla A, Foresta C (1999). Human male infertility and Y chromosome deletions: role of the AZF candidate genes DAZ, RBM and DFFRY. Hum. Reprod., 14: 1710-716.

Florack EM, Zielhous GA, Rolland R (1994). Cigarette smoking, alcohol consumption and caffeine intake and fecundability. Prev. Med. 23: 175-180.

Gerber WL, De la Pena VE, Mobley WC (1988). Infertility, chemical exposure, and farming in Iowa: absence of an association. Urology, 31: 46-50.

Hatch EE, Bracken MB (1993). Association of delayed conception with caffeine consumption. Am. J. Epidemiol., 138: 1082-92.

Howe G, Westhoff C, Vessey M, Yeates D (1985). Effects of age, cigarette smoking, and other factors on fertility: findings in a large prospecuve study. Br. Med. J., 290: 1697-1700.

Hungerford DA (1965). Leucocytes cultured from small inocula of whole blood and the preparation of metaphase chromosomes by treatment with hypotonic KCl. Stain. Technol., 40: 333-338.

Joesoef MR, Beral V, Rolfs RT, Aral SO, Cramer DW (1990). Are caffeinated beverages risk factors for delayed conception?. Lancet., 335: 136-37.

Kaufman DW, Helmrich SP, Rosenberg L, Meittinen OS, Shapiro S (1983). Nicotine and carbon monoxide content of cigarette smoke and the risk of myocardial infarction in young men. N. Engl. J. Med., 30: 409-414.

Kier LD, Yamasaki E, Ames BN (1974). Detection of mutagenic activity in cigarette smoke condensates. Proc. natn. Acad. Sci USA., 71: 4159-4163.

Kim JG, Parthasarathy S (1998). Oxidation and the spermatozoa. Semin. Reprod. Endocrinol., 16: 235-239.

Krausz C, McElreavey K (1999). Y chromosome and male infertility. Front. Biosci., 4: 1-8.

Kuroda-Kawaguchi T, Skaletsky H, Brown LG, Minx PJ, Cordum HS, Waterston RH, Wilson RK, Silber S, Oates R, Rozen S, Page DC (2001). The AZFc region of the Y chromosome features massive palindromes and uniform recurrent deletions in infertile men. Nat. Genet., 29: 279-286.

Kuroki Y, Iwamoto T, Lee J (1999). Spermatogenic ability is different among males in different Y chromosome lineage. Hum. Genet., 44: 289-292.

Lahdetie J (1995). Occupation and exposure related studies on human sperm. J. Occup. Environ. Med., 37: 922-930.

Lotery AJ, Namperumalsamy P, Jacobson SG, Weleber RG, Fishman GA, Musarella MA, Hoyt CS, Héon E, Levin A, Jan J, Lam B, Carr RE, Franklin A, Radha S, Andorf JL, Sheffield VC, Stone EM (2000). Mutation Analysis of 3 Genes in Patients with Leber Congenital Amaurosis. Arch. Ophthalmol., 118: 538-543.

Machev N, Saut N, Longepied G, Terriou P, Navarro A, Levy N, Guichaoua M, Metzler-Guillemain C, Collignon P, Frances AM, Belougne J, Clemente E, Chiaroni J, Chevillard C, Durand C, Ducourneau A, Pech N, McElreavey K, Mattei MG, Mitchell MJ (2004). Sequence family variant loss from the AZFc interval of the human Y chromosome, but not gene copy loss, is strongly associated with male infertility. J. Med. Genet., 41: 814-825.

Mahadevaiah SK, Odorisio T, Elliott DJ, Rattigan A, Szot M, Laval SH, Washburn LL, McCarrey JR, Cattanach BM, Lovell-Badge R, Burgoyne PS (1998). Mouse homologues of the human AZF candidate gene RBM are expressed in spermatogonia and spermatids, and map to a Y chromosome deletion interval associated with a high incidence of sperm abnormalities. Hum. Mol. Genet., 7: 715-727.

Martinez MC, Bernabe MJ, Gomez E, Ballesteros A, Landeras J, Glover G, Gil-Salom M, Remohi J, Pellicer A (2000). Screening for AZF deletion in a large series of severely impaired spermatogenesis patients. J. Androl., 21: 651-655.

Mattison DR (1982). The effects of smoking on fertility from gametogenesis to implantation. Environ. Res., 28: 410-433.

Mehta RH, Anand Kumar TC (1997). Declining semen quality in Bangaloreans: a preliminary report. Curr. Sci., 72: 621-622.

Mosher WD, Pratt WF (1991). Fecundity and infertility in the United States: incidence and trends. Fertil. Steril., 56: 192-193.

Ochsen-dorf FR (1999). Infection in the male genital tract and reactive oxygen species. Hum. Reprod. Update, 5: 399-420.

Oldereid NB, Rui H, Clausen OPF, Purvis K (1989). Cigarette smoking and human sperm quality assessed by laser-Doppler spectroscopy and DNA flow cytometry J. Reprod. Fert., 86: 731-736.

Oliva A, Spira A, Multigner L (2001). Contribution of environmental

factors to the risk of male infertility. Hum. Reprod., 16: 1768-1776.

Olsen J, Rachootin P, Schiodt AV, Damsbo N (1983). Tobacco use, alcohol consumption and infertility. Int.J. Epidemiol., 12: 179-84.

Ostergaard K (1977). The concentration of cadmium in renal tissue from smokers and nonsmokers. Acta Med. Scand., 202: 193-195.

Parry H, Cohen S, Schlarb JE, Tyrrell DA, Fisher A, Russell MA, Jarvis MJ (1997). Smoking, alcohol consumption, and leukocyte counts. Am. J. Clin. Pathol., 107: 64-67.

Peterlin B, Kunej T, Sinkovec J, Gligorievska N, Zorn B (2002). Screening for Y chromosome microdeletions in 226 Slovenian subfertile men. Hum. Reprod., 17: 17-24.

Pryor JL, Kent-First M, Muallem A, Van Bergen AH, Nolten WE, Meisner L, Roberts KP (1997). Microdeletions in the Y chromosome of infertile men. New. Engl. J. Med., 336: 534-539.

Ravenkolt RT (1982). Radioactivity in cigarette smoke. New. Engl. J. Med., 307: 312.

Reijo R, Lee TY, Salo P, Alagappan R, Brown LG, Rosenberg M, Rozen S, Jaffe T, Straus D, Hovatta O, de la Chapelle A, Silber S, Page DC (1995). Diverse spermatogenic defects in humans caused by Y chromosome deletions encompassing a novel RNA-binding protein gene. Nat. Genet., 10: 383-393.

Repping S, Skaletsky H, Brown L, van Daalen SK, Korver CM, Pyntikova T, Kuroda-Kawaguchi T, de Vries JW, Oates RD, Silber S, van der Veen F, Page DC, Rozen S (2003). Polymorphism for a 1.6-Mb deletion of the human Y chromosome persists through balance between recurrent mutation and haploid selection. Nat. Genet., 35: 247-51.

Repping S, Skaletsky H, Lange J, Silber S, Van Der Veen F, Oates RD, Page DC, Rozen S (2002). Recombination between palindromes P5 and P1 on the human Y chromosome causes massive deletions and spermatogenic failure. Am. J. Hum. Genet., 71: 906-922.

Reynolds N, Cooke HJ (2005). Role of the DAZ genes in male fertility. Reprod. Biomed. Online., 10: 72-80.

Saaranen M, Suonio S, Kauhanen O, Saarikoski S (1987). Cigarette smoking and semen quality in men of reproductive age. Andrologia. 19: 670-676.

Sallmen M, Lindbohm ML, Anttila A, Kyyronen P, Taskinen H, Nykyri E, Hemminki K (1998). Time to pregnancy among the wives of men exposed to organic solvents. Occup. Environ. Med. 55: 24-30.

Seshagiri PB (2001). Molecular insights into the causes of male infertility. J. Biosci., 26: 429-435.

Shaffer LG, Tommerup N (2005). An International System for Human Cytogenetic Nomenclature (ISCN 2005): Recommendations of the International Standing Committee on Human Cytogenetic Nomenclature. Basel, Switzerland: S. Karger Publishers.

Sharma RK, Agarwal A (1996). Role of reactive oxygen species in male infertility. Urology, 48: 835-850.

Sharpe RM (2000). Lifestyle and environmental contribution to male infertility. Br. Med. Bull., 56: 630-642.

Sheiner EK, Sheiner E, Hammel RD, Potashnik G, Carel R (2003). Effect of occupational exposures on male fertility: literature review. Ind. Health., 41: 55-62.

Shen HM, Chia SE, Ong CN (1999). Evaluation of oxidative DNA damage in human sperm and its association with male infertility. J. Androl., 20: 718-723.

Skakkebaek NE, Giwercman A, de Kretser D (1994). Pathogenesis and management of male infertility. Lancet., 343: 1473-1478.

Skaletsky H, Kuroda-Kawaguchi T, Minx PJ, Cordum HS, Hillier L, Brown LG, Repping S, Pyntikova T, Ali J, Bieri T, Chinwalla A, Delehaunty A, Delehaunty K, Du H, Fewell G, Fulton L, Fulton R, Graves T, Hou SF, Latrielle P, Leonard S, Mardis E, Maupin R, McPherson J, Miner T, Nash W, Nguyen C, Ozersky P, Pepin K, Rock S, Rohlfing T, Scott K, Schultz B, Strong C, Tin-Wollam A, Yang SP, Waterston RH, Wilson RK, Rozen S, Page DC (2003). The male-specific region of the human Y chromosome is a mosaic of discrete sequence classes. Nature, 423: 825-837.

Stuppia L, Mastroprimino G, Calabrese G, Peila R, Tenaglia R, Palka G (1996). Microdeletions in interval 6 of the Y chromosome detected by STS-PCR in 6 of 33 patients with idiopathic oligo or azoospermia. Cytogenet. Cell Genet., 72: 155-158.

Suonio S, Saarikoski S, Kauhanen O, Metsäpelto A, Terho J, Vohlonen

I (1990). Smoking does affect fecundity. Eur. J. Obstet. Gynecol. Reprod. Biol., 34: 89-95.

Templeton AA, Penney GC (1982). The incidence, characteristics, and prognosis of patients whose fertility is unexplained. Fertil. Steril., 32: 175-180.

Tiepolo L, Zuffardi O (1976). Localization of factors controlling spermatogenesis in the nonfluorescentportion of the human Y chromosome long arm. Hum. Genet., 34: 119-124.

Vogt PH, Edelmann A, Kirsch S, Henegariu O, Hirschmann P, Kiesewetter F, Köhn FM, Schill WB, Farah S, Ramos C, Hartmann M, Hartschuh W, Meschede D, Behre HM, Castel A, Nieschlag E, Weidner W, Gröne HJ, Jung A, Engel W, Haidl G (1996). Human Y chromosome azoospermia factors (AZF) mapped to different subregions in Yq11. Hum. Mol. Genet., 5: 933-943.

Vogt PH, Edelmann A, Kirsch S, Henegariu O, Hirschmann P, Kiesewetter F, Kohn FM (1996). Human Y chromosome azoospermia factors AZF mapped to different subregions in Yq11. Hum. Mol. Genet., 5: 933-943.

Wang PJ, Mc Carrey JR, Yang F, Page DC (2001). An abundance of X-linked genes expressed in spermatogonia. Nat. Genet., 27: 422-426.

Weinberg CR, Wilcox AJ, Baird DD (1989). Reduced fecundability in women with prenatal exposure to cigarette smoking. Am. J. Epidemiol., 29: 1072-1078.

Yao G, Chen G, Pan T (2001). Study of microdeletions in the Y chromosome of infertile men with idiopathic oligo-or azoospermia. J. Assist. Reprod. Genet., 18: 612-616.

Yen PH, Li XM, Tsai SP, Johnson C, Mohandas T, Shapiro LJ (1990). Frequent deletions of the human X chromosome distal short arm result from recombination between low copy repetitive elements. Cell. 61: 603-610.

19

Low genetic diversity of *Hypophthalmus marginatus* from the Tocantins River based on cytochrome b sequence data

Emil José Hernández-Ruz[1], Evonnildo Costa Gonçalves[2], Artur Silva[3] and Maria Paula Cruz Schneider[3]

[1]Laboratório de Zoologia, Faculdade de Ciências Biológicas, Universidade Federal do Pará/UFPA, Campus Universitário de Altamira, Rua Coronel José Porfírio, 2515 CEP 68.372-040 - Altamira - PA, Brazil.
[2]Laboratório de Tecnologia Biomolecular, Universidade Federal do Pará, Instituto de Ciências Biológicas, Rua Augusto Corrêa, 01, Guamá, Belém, Pará, Brasil, 66075-900.
[3]Laboratório de Polimorfismo de DNA, Universidade Federal do Pará, Instituto de Ciências Biológicas, Rua Augusto Corrêa, 01, Guamá, Belém, Pará, Brasil, 66075-900.

In this study, we used a portion of mitochondrial cytochrome *b* to evaluate the current genetic structure of the *Hypophthalmus marginatus* Valenciennes 1840, from the Tocantins River in the eastern Amazonia. Genetic diversity was measured in two downstream (Abaetetuba and Cametá), and two upstream (Tucuruí and Itupiranga) stocks under influence of the Tucuruí Hydroelectric Dam. Additionally, one stock from the Araguaia River system was included in the analysis. Our findings provide evidence for the existence of reduced levels of genetic diversity in *H. marginatus* of the Tocantis basin. The pattern of distribution of haplotypes observed in the present study appears to reflect the migratory characteristics of *H. marginatus* (short distance migrator), suggesting gene flow mainly in the upstream to downstream. Alternatively, presence of rare sequences only in Cametá and Abaetetuba may result from recent mutations, since they differ from connecting haplotypes by only one nucleotide substitution.

Key words: Hydroelectric dam impact, *Hypophthalmus marginatus*, population genetic structure.

INTRODUCTION

The use of neutral markers has revealed historical and present-day barriers to gene flow in widespread marine species that formerly were believed to be homogeneous (Rocha et al., 2005). Santos et al. (2006) using the mitochondrial cytochrome b gene found deep genetic divergence without morphological change in the Neotropical species *Macrodon ancylodon* (Bloch and Schneider,

1801). Migratory freshwater fishes are Vulnerable to a variety of anthropogenic impacts, including harvesting, pollution and habitat disturbance.

Dams can be especially problematic, given that they not only convert riverine habitats into lacustrine environments, but also isolate populations from their spawning or feeding grounds (Barthem et al., 1991; Wei et al., 1997;

Ruban, 1997), thus directly affecting survival and reproduction of migratory freshwater fish by changing thermal and hydro-dynamic conditions in their habitat (Agostinho et al., 1992).

Six hydroelectric projects have now been completed in the Brazilian Amazon, yet long-term monitoring of fish populations is available for only Tucuruí Dam in the Tocantins River (Carvalho and Mérona, 1986; Mérona et al., 1987; Ribeiro et al., 1995; Cetra and Petrere, 2001; Santos et al., 2004), where alterations following impound-ment included reduction in fish diversity from the reservoir along with excessive population increase of predators such as *Cichla* Schneider 1801 and *Serrasalmus* Lacepède 1803 (Leite and Bittencourt, 1991). Furthermore, a drastic reduction in fish production has been observed downstream of the dam, probably due to poor water quality that runs through the turbines and the blocking of fish migration (Carvalho and Mérona, 1986; Fearnside, 1999).

Odinetz-Collart (1993) reported that in Cametá, harvest of freshwater shrimp has dropped from 179 tons in 1981 to 62 tons in 1988; while fish landings declined from 4,726 tons in 1985 to 831 tons in 1987 (the dam was built between 1984 and 1985). Catches in the reservoir had increased to pre-flooding levels by the early 1990s (Ribeiro et al., 1995); although, migratory species like *Hypophthalmus marginatus* were still not abundant such as nowadays. Therefore, due to great economic impor-tance to local fisheries, the current genetic structure of *H. marginatus* stocks in low to medium Tocantins River was investigated here. In addition to understand the impact of the Tucuruí Dam on this variability, the results will contri-bute to the eventual development of population mana-gement strategies.

The Tocantins is considered to be a plateau river, flowing for most of its length within an enclosed valley, draining an area of 343,000 km^2, which includes several large tributaries on its east margin. Over the past three decades, this basin has suffered consi-derable anthropo-genic pressure, including widespread deforestation, mi-ning and the construction of the Tucuruí Dam, one of the world's largest hydroelectric dams (Cetra and Petrere, 2001; Santos et al., 2004). It was built between 1984 and 1985, and flooded a reservoir of 2,840 km^2, most of which was covered in primary *terra firme* forest.

H. marginatus were sampled from 4 different points of the Tocantins River: Itupiranga (05° 06' 51.5" S, 49° 21' 34.9" W), Tucuruí (04° 18' 43.06" S, 49° 19' 58.1" W), Cametá (02° 03' 27.5" S, 49° 20' 31.9" W) and Abaetetuba (01° 40' 42.6" S, 49° 00' 16.6" W). Addi-tionally, samples from one point of Araguaia River (Conceição do Araguaia, 07° 58' 10, 8" S, 49° 11' 0, 6" W) were included (Figure 1). Our objective was to evaluate the levels of genetic diversity and differentiation among five populations of *H. marginatus* after cons-truction of the Tucuruí hydroelectric Dam.

MATERIALS AND METHODS

A total of 127 samples (17 to 31 per site) obtained from muscle or liver tissue were placed in absolute ethanol and frozen at -20°C until genomic DNA extraction, which was performed using Sambrook et al. (1989) standard protocol. The specimens were persevered in 4% formaldehyde and deposited in the Ichthyological collection of The Museu Paraense Emílio Goeldí (MPEG 13375, 17486, 17499 and 17578). A fragment of the mitochondrial cytochrome *b* gene was amplified through polymerase chain reaction (PCR) using primers GLUDG (CGAAGCT-TGACTTGAARAACCAYCGTTG) (Willis et al., 2007) and Cb3 (GGCAAATAGGAARTATCATTC) (Palumbi, 1996). The PCR amplifications were performed in 25 µL reaction volumes which contained 5 to 10 ng of genomic DNA, 50 mM KCl, 2 mM MgCl$_2$, 10 mM Tris-HCl, 20 µM of each dNTP, 5 pmol of each primer and a unit of *Taq* DNA polymerase (Invitrogen).

Amplification proceeded with a first denaturation step at 94°C for 7 min, and then 34 cycles of denaturation at 94°C for 1 min, annealing at 55°C for 1 min and extension at 72°C for 1 min (followed by a final extension at 72°C for 5 min). The amplified products were sequenced in both directions using the BigDye® Terminator v3.1 Cycle Sequencing Kit, according to the manufacturer's instructions (Applied Biosystems). Both forward and reverse primers were used for sequencing reactions in order to confirm nucleotide sequences. BioEdit software (Hall, 2007) was used to align sequences by eye.

Polymorphism levels within stocks were evaluated through nucleotide (π) and haplotype (h) diversity, estimated in Arlequin 3.5 (Excoffier and Lischer, 2010), using the "open unphase data file" option. In order to test the null hypothesis of no genetic differentiation among stocks, we used Analysis of Molecular Variance (AMOVA), available in Arlequin 3.5, which uses genotype frequencies and the number of mutations between different haplotypes to test the significance of the components of the variance associated with three hierarchical levels of genetic population (stock) structuring: (a) intra-population; (b) between populations of the same geographic group; and (c) between populations of different groups (Excoffier et al., 1992).

Arlequin 3.5 was also used to calculate (Fu, 1997) Fs, which significant values may indicate that sequences are evolving in a non-neutral way (lack mutation-drift equilibrium), or that populations were previously subdivided and/or have experienced past demographic fluctuations. We tested its significance by comparing the Fs values with a distribution generated from 1000 random samples under the hypotheses of selective neutrality and population equilibrium. The population demography (for example, bottlenecks or expansions) of all samples and each population were examined using two different approaches. First, the demographic history was investigated by comparing mismatch distributions in each geographic sample with those expected in stationary and expanding populations using DnaSP Version 4.10.7 (Rozas et al., 2003).

The shape of the mismatch distribution of pairwise differences is usually ragged or multimodal for populations at stationary demographic equilibrium, but is typically smooth or unimodal for populations that have passed through a recent expansion or bottleneck (Rogers and Harpending, 1992). The classic Mantel test (Mantel, 1967) is designed to evaluate the independence of the elements of two matrices (Raufaste and Rousset, 2001), taking into account the fact that the coefficients are not independent. In our case, it was used to evaluate possible isolation among the four populations, and was calculated using version 1.52 of the IBD software (Bohonak, 2002), generated from 1000 randomizations. The pattern of spatial variation of the populations was analyzed using the estimate of the correlation between the matrix of genetic

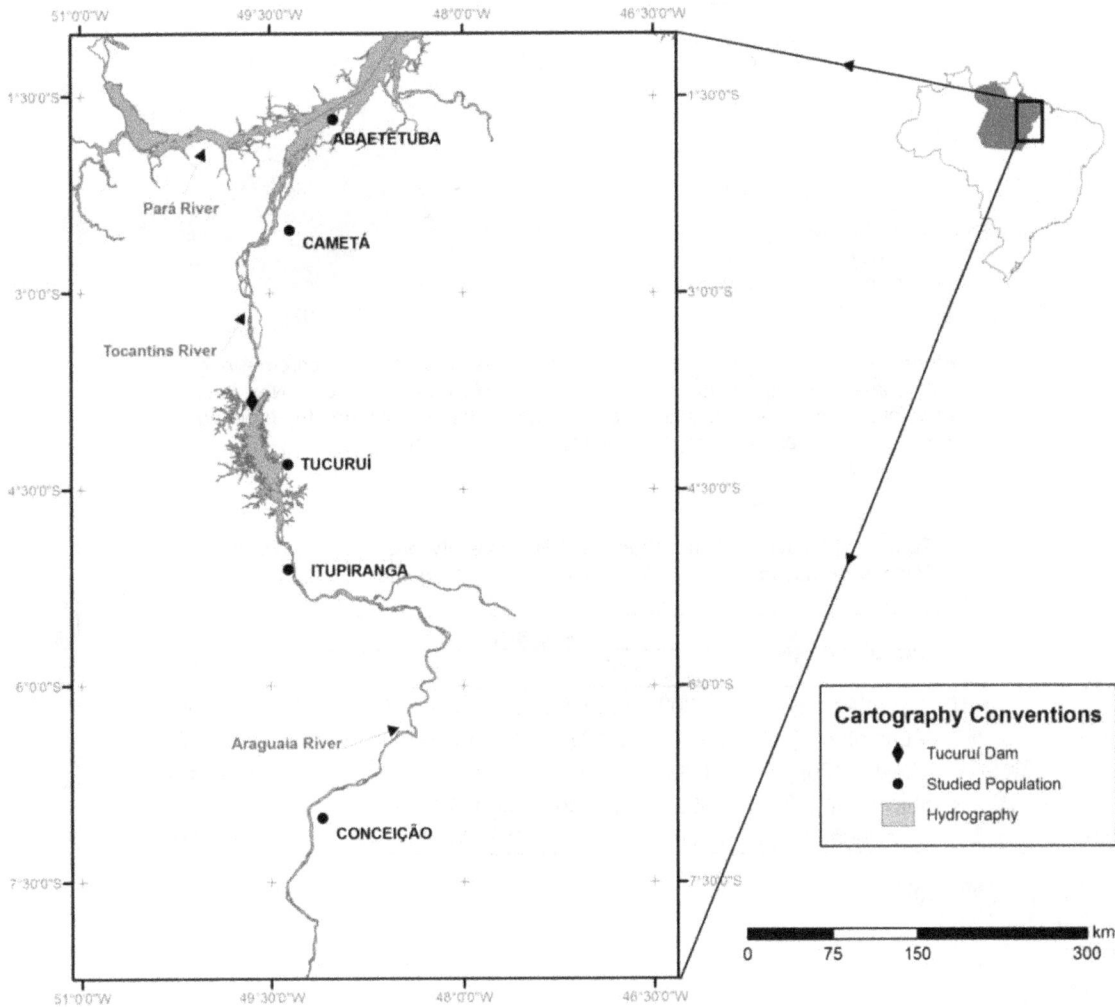

Figure 1. Distribution the five stocks of *Hypophthalmus marginatus* sampled on the Tocantins and Araguaia Rivers in Eastern Amazonia.

distances, based on *Fst*, and that of pairwise geographical distances. The significance of this correlation was tested by Mantel's *Z* (Manly, 1997), based on 1,000 randomizations. To provide a graphical representation of the haplotype relationships, a median-joining haplotype network was constructed using the Network software (Bandelt et al., 1999).

RESULTS

Sequences of a 745 bp portion of the cytochrome *b* gene were obtained in the 127 specimens, revealing nine distinct haplotypes and eight polymorphic sites, six of which presented two variants (informative to parsimony) and two were singletons (Figure 2). The most frequent haplotypes (Hap1 and Hap2) were observed in all stocks. Haplotype 3 was found in three stocks in both down- and up-stream sites, the haplotypes 4 to 6 were found in Cametá and Abaetetuba, the other haplotypes were found only in Abaetetuba.

Both haplotype (*h*) and nucleotide (*π*) diversities were greater in Abaetetuba and Cametá (Tables 1 and 3), which indicates that the genetically least diverse stocks are those upstream to the Tucuruí Dam. Fu's *Fs* values were negatives for Cametá, Abaetetuba and Tucuruí, yet they differed significantly from zero only in Cametá and Abaetetuba, suggesting a population expansion (Table 1). These data are sup-ported by monomodal mismatch distribution (Figure 4).

The Mantel test found no significant correlation between genetic variability and geographic distance (*Z* = 1.49, r = 0.37, P = 0.16), and the median-joining haplo-type net-work (Figure 3) presented no distinct clades. However, significant separation of the upstream (Conceição, Itupiranga and Tucuruí) and downstream (Cametá and Abaetetuba) stocks was supported by AMOVA (Table 2).

```
         22224455          Absolute frequency
         12471802
         38007055 Itu(30) Tuc(25) Cam(30) Aba(25) Con(17)

Hap 1(86) AGAAAGTT    19      18      20      13      15
Hap 2(22) .A......    11      06      02      01      02
Hap 3(06) ..G.....    -       01      04      01      -
Hap 4(04) .....A..    -       -       02      02      -
Hap 5(02) ..G....C    -       -       01      01      -
Hap 6(04) G.......    -       -       01      03      -
Hap 7(01) ..G...CC    -       -       -       01      -
Hap 8(01) ..G.G...    -       -       -       01      -
Hap 9(02) ...G.A..    -       -       -       02      -
```

Figure 2. Polymlorphic sites in 745 bases pairs of the cytochrome *b* gene of *Hypophthalmus marginatus* from Tocantins and Araguaia Rivers. Non informative sites for parsimony analysis are shaded in grey. Hap, Haplotype; Itu, Itupiranga; Tuc, Tucuruí; Cam, Cametá; Aba, Abaetetuba; Con, Conceição.

Table 1. Measures of mitochondrial DNA diversity based on cytochrome b of *Hypophthalmus marginatus* from Tocantins and Araguaia Rivers.

Population (N)	Diversity		Fu's *Fs*
	π	h	(P-value)
Cametá (30)	0.0009+/-0.0008	0.5448 +/- 0.1013	-2.876 (0.009)
Abaetetuba (25)	0.0018 +/-0.0013	0.7233 +/- 0.0918	-4.460 (0.002)
Tucuruí (25)	0.0006+/-0.0006	0.4400 +/- 0.0950	-0.217 (0.347)
Itupiranga (30)	0.0007 +/-0.0006	0.4805 +/- 0.052	1.561 (0.718)
Conceição (17)	0.0003 +/-0.0004	0.2206 +/- 0.121	0.035 (0.243)

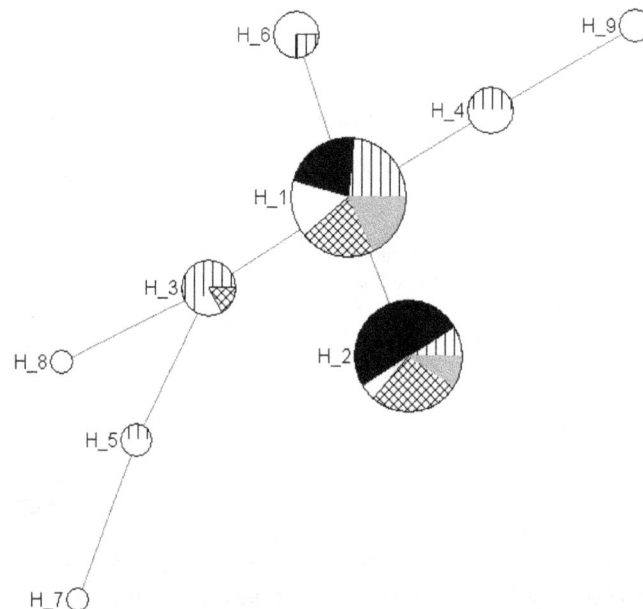

Figure 3. Median-joining network of mtDNA haplotypes derived from the polymorphic nucleotides of *Hypophthalmus marginatus* from the Tocantins and Araguaia Rivers. Circle size corresponds to overall haplotype frequency. Circles in grey = Conceição, diagonal cross = Tucuruí, black = Itupiranga, vertical = Cametá and white = Abaetetuba.

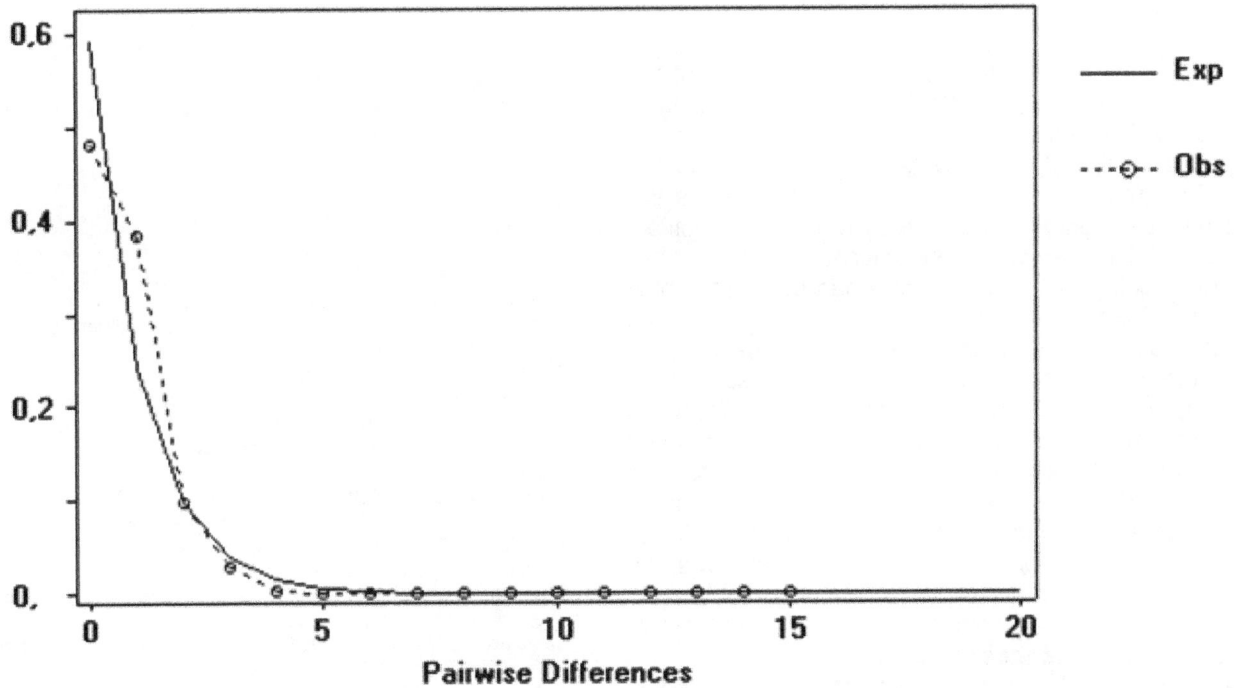

Figure 4. Pairwise mismatch distributions in 745 bases pairs of the cytochrome b gene of *Hypophthalmus marginatus* from Tocantins and Araguaia Rivers

Table 2. Analysis of molecular variance (AMOVA) for groupings of populations estimated using θ-statistics based on cytochrome *b* sequences of *Hypophthalmus marginatus* from Tocantins and Araguaia Rivers. Va = Among groups, Vb = Among populations within groups, Vc = Within populations.

Group	θst	θsc	θct	Percentage of variation		
				Va	Vb	Vc
{Cam, Aba}	0.002	-	-	-	0.19	99.81
{ Tuc, Itu, Con}	0.030	-	-	-	2.97	97.03
{ Tuc, Itu} {Con}	0.056	-0.006	0.062	6.17	-0.54	94.37
{Cam, Aba} {Tuc, Itu}	0.059*	-0.003	0.062	6.15	-0.27	94.12
{Cam, Aba, Tuc, Itu, Con}	0.047*	-	-	-	4.71	95.29
{Cam, Aba} {Tuc, Itu, Con}	0.066*	0.015*	0.052	5.23	1.38	93.39
{Cam, Aba} {Tuc, Itu} {Con}	0.061*	0.000*	0.061	6.07	-0.01	93.93

*P < 0.05.

Table 3. Pairwise comparisons between stocks, with *Fst* values calculated accordind Weir & Cockerham (1984).

Population	Conceição	Abaetetuba	Itupiranga	Cametá	Tucuruí
Conceição		0.28*	0.05*	0.04	0.10*
Abaetetuba			0.24*	0.28*	0.27*
Itupiranga				0.005	0.006
Cametá					0.008
Tucuruí					

*P < 0.05

DISCUSSION

One of the most noteworthy aspects of the results of this study was the reduced genetic diversity of *H. marginatus* populations from the Tocantins and Araguaia Rivers, in comparison with other freshwater fishes (Sivasundar et al., 2001; Turner et al., 2004), which may reflect a low mutation rate. Unfortunately, despite on Pimelodidae, nuclear loci data are scarce, but at least to mitochondria l protein-encoding loci, mutation rate has been estimated at 1.0 to 1.3% per Myrs (Bermingham et al., 1997), and nevertheless appears to be lower - 0.5% per Myrs - in some other taxa, such as prochilodontids (Sivasundar et al., 2001; Turner et al., 2004). According to Hardman and Lundberg (2006) and Torrico et al. (2009), which found low rates of molecular substitution for the cytochrome *b* equal to 0.38 to 0.53% per Myrs for the 'phracto-cephalines' (Pimelodidae) and 0.25% per Myrs within the related genus *Pseudoplatystoma* Bleeker 1862, respec-tively, lower rates of molecular evolution may charac-terize Neotropical Siluriforms, where *Hypophthalmus* Cuvier 1829 lineages were included.

The pattern of distribution of haplotypes observed in the present study appears to reflect the migratory charac-teristics of *H. marginatus*, thus the two more frequent haplotypes have been distributed to all collecting sites. The broad occurrence of haplotypes 1 and 2 represent the putative genetic pool of a common ancestor stock.

Additionally, it is worth noting that exclusive haplotypes - which probably influenced the trend of separation between the upstream and downstream stocks - were found only in downstream ones, which might be subject to gene flow with metapopulations composed of distinct breeding units from other aquatic systems such as the lower Amazon and the Marajó hydric system. Alternatively, considering the evidence of population expansion detected to Cametá and Abaetetuba, these rare sequences might result from recent mutations, since they differ from connecting haplotypes by only one nucleotide substitution. If low differentiation is prevalent, a larger sample of the species' geographic distribution range would be recommended. It would also be useful to assess the levels of genetic variability within and among populations from different basins for a better understan-ding of population dynamic of these species.

ACKNOWLEDGEMENTS

This study was supported by CAPES and CNPq. We are also grateful to IBAMA, to provide authorization 013-2007 for capture, collection and transport of biological material; to Centrais Elétricas do Norte do Brasil S. A. (ELETRONORTE S. A.) for providing logistic support at Tucuruí, and EMATER/PA and Comunidade Jaraqüera Grande for logistics in Cametá. Special thanks to Soraya Andrade, Silvanira Ribeiro Barbosa, Cassio Andrade and Stephen Ferrari for helping in Laboratory, logistic support and English translation. EJHR thanks Victor González for helping with the English translation of a preliminary version of this paper.

REFERENCES

Agostinho AA, Júlio HF Jr, Borghetti JR (1992). Considerações sobre os impactos dos represamentos na ictiofauna e medidas para sua atenuação - um estudo de caso: reservatório de Itaipu. Rev. UNIMAR. 14:89-107.

Bandelt HJ, Forster P, Röhl A (1999). Median-joining networks for inferring intraspecific phylogenies. Mol. Biol. Evol. 16:37-48.

Barthem R, Lambert M, Petrere M (1991). Life strategies of some long-distance migratory catfish in relation to hydroelectric/dams in the Amazon Basin. Biol. Conserv. 55:339-345.

Bermingham E, McCafferty SS, Martin AP (1997). Fish biogeography and molecular clocks: Perspectives from the Panamanian Isthmus. In:Kocher TD, Stepien CA (eds.). Molecular. Systematics of Fishes. San Diego: Academic Press. pp. 113-128.

Bohonak AJ (2002). IBD (Isolation By Distance): a program for analyses of isolation by distance. J. Hered. 93:153-154.

Carvalho JL, Mérona B (1986). Estudos sobre dois peixes migratórios do baixo Tocantins, antes do fechamento da barragem de Tucuruí. Amazoniana 9:595-607.

Cetra M, Petrere M (2001). Small-scale fisheries in the middle River Tocantins, Imperatriz (MA), Brazil. Fish. Manage. Ecol. 8:153-162.

Excoffier L, Lischer HEL (2010). Arlequin suite ver 3.5: a new series of programs to perform population genetics analyses under Linux and Windows. Mol. Ecol. Resour. 10:564-567.

Excoffier L, Smouse PE, Quattro JM (1992). Analysis of molecular variance inferred from metric distances among DNA haplotypes: application to human mitochondrial DNA restriction data. Genet. 131:479-491.

Fearnside PM (1999). Social impacts of Brazil's Tucuruí Dam. Environ. Manage. 24:485-495.

Fu YX (1997). Statistical tests of neutrality of mutations against population growth, hitchhiking, and background selection. Genet. 147:915-925.

Hall TA (2007). Bioedit v709: Biological sequence alignment editor analysis program for Windows 95/98/Nt. Nucl. Acids. Symp. Ser. 41:95-98.

Hardman M, Lundberg JG (2006). Molecular phylogeny and chronology of diversification for "phractocephaline" catfishes (Siluriformes: Pimelodidae) based on mitochondrial DNA and nuclear recombination activating gene sequences. Mol. Phylogenet. Evol. 40:410-418.

Leite RAN, Bittencourt MM (1991). Impacto de hidrelétricas sobre a ictiofauna amazônica: O exemplo de Tucuruí. In: Val AL, Figiuolo R and Feldberg E (eds.). Bases Científicas para Estratégias de Preservação e Desenvolvimento da Amazônia: Fatos e Perspectivas. Instituto Nacional de Pesquisas da Amazônia (INPA), Manaus. 1:85-100

Manly BFJ (1997). Randomization, Bootstrap and Monte Carlo Methods in Biology. 399 p. Chapman & Hall, London.

Mantel NA (1967). The detection of disease clustering and a generalized regression approach. Cancer Res. 27:209-220.

Mérona B, Carvalho JL, Bittencourt MM (1987). Les effets immédiats de la fermeture du barrage de Tucuruí (Brésil) sur l'ichtyofaune en aval. Rev. Hydrobiol. Trop. 20:73-84.

Odinetz-Collart O (1993). Ecologia e potencial pesqueiro de camarão-canela, *Macrobrachium amazonicum*, na Bacia Amazônica. In: Ferreira EJG, GM Santos, Leão ELM, Oliveira LA (eds.). Bases Científicas para Estratégias de Preservação e Desenvolvimento da Amazônia. Vol 2. Instituto Nacional de Pesquisas da Amazônia (INPA), Manaus. pp. 147-166.

Palumbi SR (1996). Nucleic acid II: the polymerase chain reaction. In: Hillis DM, Morits C, Mable BK (eds.). Molecular Systematics Sunderland, Sinauer. pp 205-247.

Raufaste N, Rousset F (2001). Are partial mantel tests adequate? Evolutn. 55:1703-1705.

Ribeiro MCLB, Petrere M, Juras AA (1995). Ecological integrity and fisheries ecology of the Araguaia-Tocatins River Basin, Brazil. Regul. Rivers Res. & Manage. 11:325-350.

Rocha LA, Robertson DR, Roman J, Bowen BW (2005). Ecological speciation in tropical reef fishes. Proc. R. Soc. Lond. B. Biol. Sci. 272:573-579.

Rogers AR, Harpending HC (1992). Population growth makes waves in the distribution of pairwise genetic differences Mol. Biol. Evol. 9:552-569.

Rozas J, Sánchez-DelBarrio JC, Messeguer X, Rozas R (2003). DNAsp, DNA polymorphism analyses by the coalescent and other methods. Bioinformatics 19:2496-2497.

Ruban GI (1997). Species structure, contemporary distribution and status of the Siberian sturgeon, Acipenser baerii. Eniviron. Biol. Fis. 48:221-230.

Sambrook J, Fritsch EF, Maniatis T (1989). Molecular cloning: A Laboratory Manual 2nd edn. Cold Spring Harbor Laboratory Press, Cold Spring Harbor, NY, USA.

Santos GM, Mérona B, Juras AA, Jégu M (2004). Peixes do baixo río Tocantins: 20 anos depois da Usina Hidrelétrica de Tucuruí Brasília. ELETRONORTE. p. 216.

Santos S, Hrbek T, Farias IP, Schneider H, Sampaio I (2006). Population genetic structuring of the king weakfish, Macrodon ancylodon (Sciaenidae), in Atlantic coastal waters of South America: deep genetic divergence without morphological change. Mol. Ecol. 15:4361-4373.

Sivasundar A, Bermingham E, Ortí G (2001). Population structure and biogeography of migratory freshwater fishes (Prochilodus: Characiformes) in major South American rivers. Mol. Ecol. 10:407-418.

Torrico JP, Hubert N, Desmarais E, Duponchelle F, Nuñez F, Rodriguez J, Montoya-Burgos J, García Dávila C, Carvajal-Vallejos FM, Grajales AA, Bonhomme F, Renno JF (2009). Molecular phylogeny of the genus Pseudoplatystoma (Bleeker 1862): Biogeographic and evolutionary implications. Mol. Phyl. Evol. 51:588-594.

Turner TF, Mcphee MV, Campbell P, Winemiller KO (2004). Phylogeography and intraspecific genetic variation of prochilodontid fishes endemic to rivers of northern South America. J. Fish Biol. 64:186-201.

Wei Q, Ke F, Zhang J, Zhuang P, Luo J, Zhou R, Yang W (1997). Biology, fisheries, and conservation of sturgeons and paddlefish in China. Environ. Biol. Fish. 48:241-255.

Willis SC, Nunes MS, Montaña CG, Farias IP, Lovejoy NR (2007). Systematics, biogeography, and evolution of the Neotropical peacock basses Cichla (Perciformes: Cichlidae). Mol. Phyl. Evol. 44:291-307.

20

Mapping *Rf3* locus in rice by SSR and CAPS markers

M. Alavi[1*], A. Ahmadikhah[1], B. Kamkar[2] and M. Kalateh[3]

[1]Department of Plant Breeding and Biotechnology, Gorgan University of Agricultural Sciences and Natural Resources, Gorgan, Iran.
[2]Department of Agronomy, Gorgan University of Agricultural Sciences and Natural Resources, Gorgan, Iran.
[3]Agricultural Selection Station, Gorgan, Iran.

Cytoplasmic male sterility (CMS) is a common phenomenon that has been extensively used for production of hybrid seeds in various crops. *Rf* genes are needed for restoring fertility to CMS lines. Searching for and molecular tagging of restorer genes is of high importance where phenotyping is very time consuming and requires the determination of spikelet sterility in testcross progeny. In this study we attempted to map a fertility restorer gene using SSR and CAPS markers in rice line IR36 in a F_2 population developed from the cross Neda-A×IR36. The genetic linkage analysis indicated that thee SSR markers (RM1, RM3233, RM3873) and one CAPS marker (RG140/*Eco*RI) on the short arm of chromosome 1 were linked to *Rf3*. *Rf3* flanked by tow SSR markers RM1 and RM3873 at distances of 5.6 and 14 cM, respectively. The use of identified markers give promise for their application in molecular marker assisted selection (MAS).

Key words: CMS, fertility restoration, SSR marker, Rf3 gene.

INTRODUCTION

Rice is one of the most important agricultural products in the world earning substantial foreign exchange and is a staple food crop in densely populated Asia. Rice has been one of the most important plants locating at the forefront of plant genomics because of its small genome size and relatively low amount of repetitive DNA, its diploid nature and its ease of manipulation in tissue culture. In the 1990s, many advances occurred in the application of molecular markers in rice (see reviews in Mackill and Ni, 2001; Temnykh et al., 2001). Cytoplasmic Male Sterility (CMS) in plants caused by lesion or rearrangement of the mitochondrial genome is unable to produce functional pollens (Jing et al., 2001). Nuclear genes are required to restore pollen fertility to CMS lines. Therefore, the CMS systems are widely used for hybrid seed production. In hybrid seed production using three-line system, the combination of a CMS line, a maintainer line and a restorer line carrying the restorer gene (*Rf*) to restore fertility is indispensable for the development of hybrid varieties (Virmani et al., 2003).

Cytoplasmic male sterility (CMS) and nucleus controlling fertility restoration are widespread plant reproductive features that provide useful tools to exploit heterosis in crops. Rice wild abortive (WA) type cytoplasmic male sterility (CMS) is commercially used for production of hybrid seeds in Asia. Within rice there are several types of CMS/Rf system, among them important is wild abortive (WA), BaoTai (BT) and Honglian (HL) which popularly are applied in commercial hybird rice seed production (Virmani and Shinjyo, 1998). In rice, hybrid rice varieties developed based on wild abortive (WA) type CMS accounted for approximately 90% of hybrid rice in china. So, inheritance of fertility restoration in the WA type CMS has been extensively investigated. Most of the investigators tended to agree that the restoration of WA type CMS is controlled by two nuclear genes and their chromosomal locations have resolved (Yao et al., 1997; Zhang et al., 1997; Komori et al., 2003; Ahmadikhah et al., 2006).

WA-type CMS is due to the cytoplasm derived from wild rice, called WA acting in a sporophytic manner and is widely used for hybrid seed production in the indica subspecies. Tow major fertility-restorer genes, *Rf3* and *Rf4*, are required for the production of viable pollen in WA-type CMS and the genes have been mapped to chromosomes 1 and 10 respectively (Yao et al., 1997; Zhang et al., 1997; Ahmadikhah et al., 2006; Ahmadikhah and Alavi, 2009). Searching for restorer genes is a good

*Corresponding author. E-mail: alavimehrian@gmail.com.

approach where phenotyping is very time-consuming and requires the determination of spikelet sterility in testcross progeny.

Molecular markers are particularly useful for accelerating the process of introducing a gene or Quantitative Trail Loci (QTL) into an elite cultivar or breeding line via backcrossing. When introducing a gene or a trait, markers linked to the gene can be used to select plants possessing the desired trait, and markers throughout the genome can be used to select plants that are genetically similar to the recurrent parent (Young and Tanksley, 1989; Hospital et al., 1992; Semgan et al., 2006). A significant advance in the practical utilization of molecular marker was the development of SSR markers, also referred to microsatellite markers (McCouch et al., 2002). These markers are highly polymorphic and easy to detect. The high polymorphism means that these markers can be used in germplasms that is closely related (Ni et al., 2002; Yang et al., 1994). Recently, a fairly dense SSR map of rice has been published (McCouch et al., 2002). Mapping agronomically important genes can provide useful information for plant breeders. In this line, our objective was to map an important *Rf* locus in rice using molecular markers.

MATERIALS AND METHODS

Mapping population and fertility scoring

Mapping population was derived from a cross between Neda-A (CMS) and IR36 and consisted from 6 extremely sterile and 85 extremely fertile individuals Neda-A is an elite male sterile line of WA-type. IR36 is a strong restorer line for WA-CMS (Ahmadikhah et al., 2007). For the analysis of pollen fertility, panicle from the main tiller of each plant was selected and several spikelets were randomly selected from different positions in the panicle. The anthers from each spikelet were squashed in a drop of 1% Iodine Potassium Iodide (I-KI) solution on a glass slide separately and observed under a light microscope. F2 plants were classified into different groups based on proportion of stained-round pollen grains as completely sterile (< 5% fertility), partially sterile (5.1 - 50% fertility), partially fertile (50.1 - 80%) and fertile (> 80%), as proposed by Ahmadikhah et al. (2006). Pollen fertility was investigated at flowering time. The seed setting rates of bagged panicles were evaluated at maturity. Sterile plants contained less than 5% stainable pollen and produced no fertile seed. All the others were treated as the fertile individuals.

Genomic DNA extraction

Young leaves were collected from the parental lines and respective 209 individual F2 plants from cross Neda-AxIR36 and subsequently screened for pollen fertility. Total genomic DNA was isolated from the leaves according to CTAB method (Saghai-Maroof et al., 1984) with some modifications (Ahmadikhah, 2009). Leaves were ground in liquid nitrogen using mortar and pestle to a very fine powder. It was then transferred to pre-warmed extraction buffer and incubated at 65°C for 45 min and vortexed for 60 s, then added an equal amount of chloroform: isoamyl alcohol (24:1), mixed well by gentle inversion and centrifuged. The supernatant was transferred to a fresh tube and DNA was precipitated by adding 0.7 volume of cold isopropanol to precipitate DNA. After centrifugation, the pellet was dried and dissolved in water. DNA was quantified on the agarose gel, diluted and used in PCR.

Bulked segregant analysis (BSA)

Tow bulk one sterile and one fertile were formed by mixing the DNA from 10 fully fertile (pollen fertility > 90%) and 10 fully sterile (pollen fertility 0%) from corresponding F2 plants was used to constitute fertile and sterile bulks, respectively, for BSA. Polymerase chain reaction (PCR) was performed using DNAs from sterile line Neda-A and restorer line IR36 corresponding sterile and fertile bulk and F2 individuals The parental lines IR36 and Neda-A along with the bulks were simultaneously screened with rice SSR and CAPS markers to find informative. In addition 3 SSR markers RM1, RM3233, RM2873 and one CAPS marker derived from RFLP clones RG140 reported to be linked with the fertility restorer genes in other source were also used in BSA

PCR condition

Polymerase chain reaction (PCR) was performed in 15 µl volumes containing 0.75 µM/l of each primer, 7.5 µl master mix (200 µM/l dNTPs, 50 mM/l KCl, 10 mM/l Tris HCl, 1.5 mM/l MgCl2, and 1 unit of Taq DNA polymerase (Cinnagen)) 5 µl H2O and 1 µl DNA. The PCR profile was 94°C for 5 min (denaturation), followed by 35 cycles of 94°C for 1 min, 50, 55 or 60°C (depend on the melting temperature of the primer pairs) for 1 min, 72°C for 2 min and finally 72°C for 7 min in the final extension. The products from PCR reaction were resolved by electrophoresis in 2/5% agarose gel containing 0.5 µg/ml ethidum bromide.

Linkage map construction

JoinMap software (Stam, 1995) was used to calculate the marker distances and to assign the linked markers to linkage groups on a personal computer. Map distances were based on the Kosambi function (Kosambi, 1944). Linkage groups were assigned to corresponding chromosomes based on SSR markers mapped by McCouch et al. (2002). For single-marker analysis, the recombination frequency between a positive marker and an *Rf* locus was calculated using maximum likelihood estimator (Allard, 1956), assuming that all the extremely sterile and fertile individuals were homozygous at the targeted *Rf* locus.

RESULTS AND DISCUSSION

Segregation of the F$_2$ population for fertility restoration

Mean pollen fertility for F$_1$ hybrids of Neda-A x IR36 was calculated as high as 86.3%, indicating that paternal parent IR36 is a strong restorer line. However, mean pollen fertility of F$_2$ was 73.6% and ranged between 0% and 99%. Phenotypic distribution in F$_2$ is shown in Figure 1 that shows pollen fertility skewed toward parent IR36. Since one major peak is seen in Figure 1, in this step of analysis we can conclude that one powerful locus exists in IR36 and confers fertility restoration to CMS.

The F$_2$ population was segregated into 202 fertile and 7 sterile plants that well fitted with 15:1 ratio (X^2_{fact} = 3.0;

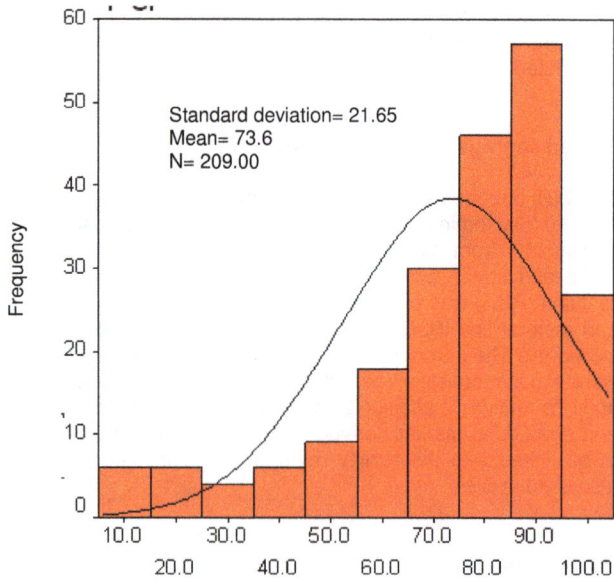

Figure 1. Frequency distribution of pollen fertility (Fr on vertical axis) in F_2 population of Neda-AxIR36.

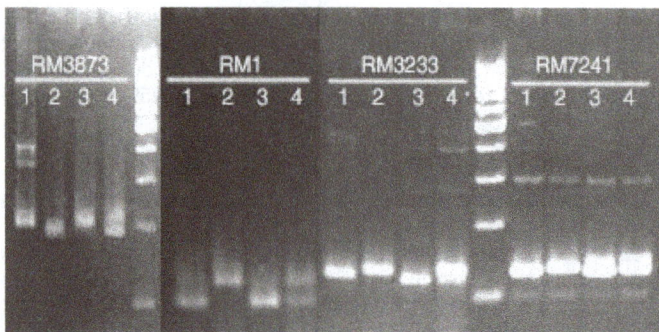

Figure 2. BSA analysis for confirmation of linkage of markers to *Rf3* locus.

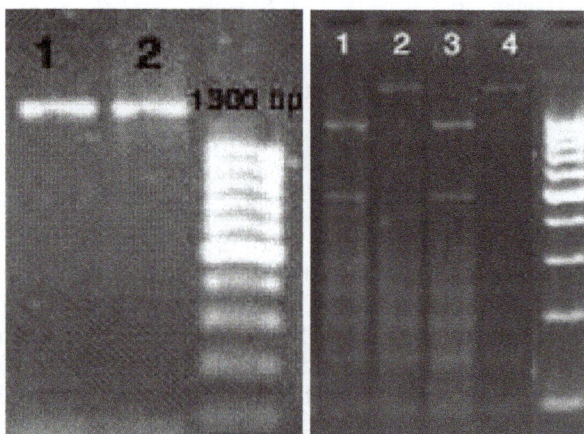

Figure 3. The banding pattern of RG140-based primer pair (left), and BSA analysis of RG140/*Eco*RI after digestion with restriction enzyme *Eco*RI (right); 1 = Neda-A, 2 = IR36, 3 = sterile bulk and 4 = fertile bulk

X^2_{05} = 3.84) for two dominant genes interacting in duplicate fashion. Therefore, in this step, we can conclude that IR36 carries two *Rf* genes. However, whether they act independently or dependently is yet unknown until test it with theoretical ratio 9:3:3:1 for two independent genes. By testing dependence/independence of these two genes using chi square statistic (X^2_{fact} = 72.74; $X^2_{05/df = 3}$ = 7.82), indicated that these genes were not independent and acted dependently; that is, these two genes locate on a linked segment of chromosome. Thus, according to first conclusion and considering second one, we can conclude that one complex locus consisted of two linked genes in IR36 restores fertility to CMS line.

Bulked segregant analysis (bsa)

SSR markers

For detection of the fertility-restoring genes in IR36, 26 pairs of microsatellite primers (Table 1) were selected to screen polymorphism between Neda-A and IR36. Among SSR primers, seven primer pairs (RM1, RM3783, RM 3233, RM7241 on chromosome 1 and RM171, RM6100, RM228 on the chromosome 10) showed polymorphism between Neda-A and IR36 (Table 2). To assess the possibility that a genomic region contains a locus for fertility restoration, tow bulks (Bf and Bs, for fertile bulk and sterile bulk, respectively), each consisting of 12 extremely fertile and 6 sterile plants was assayed with each pair of the polymorphic markers between population mapping parents.

To confirm the linkage of the mentioned primers to fertility restoration, they were tested in PCR with DNA from two parents and respective bulks (Bf and Bs). Results of BSA analysis revealed that microsatellite primers RM1, RM3233, RM3783 on the short arm of chromosome 1 not only showed polymorphic bands between Neda-A and IR36, but also were linked to a fertility-restorer locus *Rf3*.

Development of a caps marker

In addition to SSR markers mentioned above, we used one RFLP-based primer pair (named RG140/*Eco*RI), for detection of polymorphism among parental lines. After amplification of the DNA from Neda-A and IR36 in PCR with RG140-based primers, we did not detect any polymorphism between parental lines and in PCR was produced a monomorphic band (Figure 2). Therefore, to detect probable polymorphism in the PCR product we digested it with restriction enzyme *Eco*RI. This digestion produced polymorphism between two parents; so that the PCR product from CMS line Neda-A was cleaved into two fragments and that of restorer line IR36 stayed intact (Figure 3). Then, BSA analysis was conducted with RG140/*Eco*RI that revealed this CAPPS marker was linked to fertility restoration locus *Rf3*.

Table 1. The sequence information of primer pairs used in this study.

Name	'Sequence (5' to 3')	Chr.
RM7241	CAGTCGCACTAACTGAACAACACC CACGGACAGATCAGTTTCTTTCG	1
RM1003	GATTCTTCCTCCCCTTCGTG TTCCTGTCAGAACAGGGAGC	1
RM272	AATTGGTAGAGAGGGGAGAG ACATGCCATTAGAGTCAGGC ACAAGGCCGCGAGAGGATTCCG	1
RM134	GCTCTCCGGTGGCTCCGATTGG	7
RM1335	GCATGCATGAATATGATGG AGATCGAACAAGAAGAGTGG	7
RM226	GAAGCTAAGGTCTGGGAGAAACC AATGGCCTTAACCAAGTAGGATGG	1
RM171	AACGCGAGGACACGTACTTAC ACGAGATACGTACGCCTTTG	10
RM3019	AAAGGTGTTGTAGGAGTCGAGGTTGG ACGCATTCGCCTTTGACATGC	1
RM7180	GTGTTTATAGGGGTGCCACG TGTTGGTGGTGCAGGTAAAG	1
RM6737	GCACGTAAATGATAGGCACCATTGC CACAAGGTGGTGTGGGCTAGACG	10
RM6100	TTCCCTGCAAGATTCTAGCTACACC TGTTCGTCGACCAAGAACTCAGG	10
RM1146	TCTCCCTATTCCCGTGTAAATCG CCCGATGATCGATTGTACCTAGC	10
RM228	TCTAACTCTGGCCATTAGTCCTTGG AAGTAGACGAGGACGACGACAGG	10
RM5841	TGAGAGTTACCGTCCATCTAGC GAGTACAGTGAGTGCCCTACG	10
RM10355	GGACCCATATGCTTCATGTCACC CTCACGTTCTCCTTCACCAAAGC	1
RM10346	GCTTGATCTGCCCTTGTTTCTTGG AACTCGAGCGGCCTTCTCAGC	1

Table 1. Contd.

Name	'Sequence (5' to 3')	Chr.
M302	TCATGTCATCTACCATCACAC ATGGAGAAGATGGAATAC	1
RM543	CTGCTGCAGACTCTACTGCG AAATATTACCCATCCCCCCC	1
RM319	ATCAAGGTACCTAGACCACCAC TCCTGGTGCAGCTATGTCTG	1
RM522	ACCAGAGAAGCCCTCCTAGC GTTCTGTGGTGGTCACGTTG	1
RM311	TGGTAGTATAGGTACTAAACAT TCCTATACACATACAAACATAC	1
SSR061	GTCTAATTTTCCTCCCTCCT TGATGTTGGCTTTGTTATTG	1
RM151	GGCTGCTCATCAGCTGCATGCG TCGGCAGTGGTAGAGTTTGATCTGC	1
RM3233	GAAATTCGAAATGGAGGGAGAGC GGTGAGTAAACAGTGGTGGTGAGC	1
RM3873	GCTATAGACGCCTCCTCCTTATCC AAAGCTAGCTAGGACCGACATGC	1
RM1	GCGAAAACACAATGCAAAAA GCGTTGGTTGGACCTGAC	1
RG140	GTACATAGTAGCACCTGCTC TCCCTAGTTTGTGCTACTC	1

Table 2. SSR markers which showed polymorphism between parental lines.

Location (chromosome)	Size (base pair)	Name
1	100-113	RM1
1	120-130	RM3233
1	185-200	RM3873
1	120-130	RM7241
10	320-340	RM171
10	160-170	RM6100
10	110-120	RM228

Mapping SSR and CAPS markers

After confirmation of linkage of the SSR primer pairs (RM1, RM3233 and RM3873 and CAPS marker Rg140/*Eco*RI)

Table 3. Result of single marker analysis.

	RM1	RG140	RM3873	RM3233
Rf3	6.49	12.53	15.83	16.96
	(29.72)	(14.49)	(12.24)	(10.8)
RM1	-	10.43	18.48	15.21
		(15.54)	(9/59)	(11.39)
RG140		-	23.31	18.13
			(6.55)	(9.14)
RM3873			-	32.13
				(2/31)

Note: Distances are expressed in cM and LOD scores are shown in parenthesis.

to *Rf3*, we selected 91 F$_2$ plants (including 6 extremely homozygous sterile individuals and 85 extremely homozygous fertile ones) for calculating linkage distance in the mapping population. In single marker analysis RM1 in distance ~6.5 cM, RG140 in distance of ~12.5 cM, RM3873 in distance of ~15.8 cM and RM3233 in distance of ~17 cM were mapped with *Rf3* locus (Table 3).

However, when all the marker loci together with *Rf3* locus were selected for mapping purpose in JoinMap3.0, SSR markers RM1, RM3233 and RG140 located on one side and RM3783 alone on the other side of *Rf3* (Figure 4).

As seen in Figure 3 the *Rf3* was flanked by two SSR markers RM1 and RM3783, at distances of 5.6 and 14 cM, respectively.

In this study, molecular markers linked to fertility restoration in rice WA-CMS system were successfully identified and mapped on the short arm of chromosome 1 by the use of SSR markers. Some workers located Rf3 on the short arm of chromosome 1, too (Yao et al., 1997; Zhang et al., 1997; He et al., 2002). For example, He et al. (2002) found a SSR marker RM1 linked o Rf3 gene in the short arm of chromosome 1. In our work also we showed that RM1 was linked to Rf3 gene. We also developed a CAPS marker namely RG140/EcoRI on the basis of RFLP sequences in this region and showed that this marker was linked to Rf3. Zhang et al. (1997) mapped Rf3 between RFLP markers RG140 and RG532. Ahmadikhah and Alavi (2009) also reported that CAPS marker RG140/EcoRI was linked to Rf3 on the short arm of chromosome 1. The identification of closely linked markers would be valuable for use in MAS strategy and finally help in map-based cloning of the fertility restorer gene in near future (Ahmadikhah and Karlov, 2006). For example, expected efficiency of RM1 and RM3873 in MAS when they are used alone is 89.1 and 74.0%, respectively. However, when they are used together, their expected efficiency in MAS will be 99.2%.

ACKNOWLEDGMENT

This work was financially supported by Gorgan University of Agricultural Sciences and Natural resources.

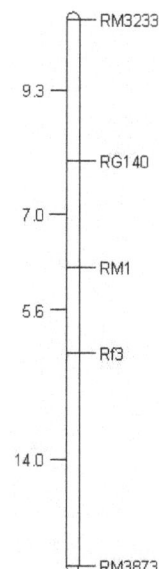

Figure 4. Molecular mapping of *Rf3* locus on the short arm of chromosome 1 of Indica rice. Marker loci and *Rf3* are shown on the right and distances (in cM) on the left.

REFERENCES

Ahmadikhah A (2009). A rapid mini-prep DNA extraction method in rice (*Oryza sativa*). Afr. J Biotechnol. 8(2): 234-238.

Ahmadikhah A, Alvi M (2009). A cold-inducible modifier QTL affecting fertility restoration of WA CMS in rice. Int. J. Genet. Mol. Biol., 1(5): 089-093.

Ahmadikhah A, Karlov GI (2006). Molecular mapping of the fertility – restoration gene Rf4 for WA-cytoplasmic male sterility in rice. Plant Breed., 125(4), 67-71.

Ahmadikhah A, Karlov GI, Nematzadeh GH, Ghasemi Bezdi K (2007). Inheritance of the fertility restoration and genotyping of rice lines at the restoring fertility (Rf) loci using molecular markers. Int. J. Plant Prod., 1(1): 13-21.

Allard RW (1956). Formulas and tables to facilitate the calculation of recombination values in heredity. Hilgardia, 24: 235- 278.

He GH, Wang WM, Liu GQ, Hou L, Xiao YH (2002). Mapping of two fertility-restoring gene for WA cytoplasmic male sterility in minghui63 using SSR markers. Acta Genet. Sinica, (2999): 798-802.

Hospital F, Chevalet C, Mulsant P (1992). Using markers in gene introgression breeding programs. Genetics 132, 1199-1210.

Jing R, Li X, Yi P, Zhu Y (2001). Mapping fertility-restoring genes of rice WA cytoplasmic male sterility using SSLP markers. Bot. Bull. Acad. Sin., 42, 167-171.

Komori T, Yamamoto T, Takemori N, Kashihara M, Matsushima H et al (2003). Fine genetic mapping of the nuclear gene, Rf-1, that restores the BT-type cytoplasmic male sterility in rice (Oryza sativa L.) by PCR-based markers. Euphytica, 129: 241-247.

Kosambi DD (1944): The estimation of map distance from recombination values. Ann. Eugen, 12, 172-175.

Mackill DJ, Ni J (2001). Molecular mapping and marker assisted selection for major-gene traits in rice. In: Khush GS, Brar DS, Hardy B, eds. Rice genetics IV. Los Baños (Philippines), Int. Rice Res. Inst., 137-151.

McCouch SR, Teytelman L, Xu Y, Lobos KB, Clare K, Walton M, Fu B, Maghirang R, Li Z, Zing Y, Zhang Q, Kono I, Yano M, Fjellstrom R, DeClerck G, Schneider D, Cartinhour S, Ware D, Stein L (2002). Development and mapping of 2240 new SSR markers for rice (Oryza sativa L.). DNA Res., 9: 199-207.

Ni J, Colowit PM, Mackill DJ (2002). Evaluation of genetic diversity in rice subspecies using microsatellite markers. Crop Sci., 42: 601-607.

Semgan K, Bjornstad A, Ndjiondjp MN (2006). Progress and prospects of marker assisted backcrossing as a tool in crop breeding programs. Afr. J. Biotechnol., 5(25): 2588-2603.

Stam (1995). JoinMap 2.0 deals with all types of plant mapping populations. Plant Genome III Abstracts, World Wide Web site: www.intl-pag.org.

Temnykh S, DeClerck G, Lukashova A, Lipovich L, Cartinhour S, McCouch SR (2001). Computational and experimental analysis of microsatellites in rice (Oryza sativa L.). Frequency, length variation, transposon associations, and genetic marker potential. Genome Res., 11: 1441-1452

Virmani SS, Shinjyo C (1998). Current status of analysis and symbols for male-sterile cytoplasms and fertility-restoring genes. Rice Genet. Newslett., 5: 9-15.

Virmani SS, Sun ZX, Mou TM, Jauhar AA, Mao CX (2003). Two-line Hybrid rice breeding manual. Los Baños (Philippines), International Rice Research Institute.

Yao FY, Xu CG, Yu SB, Li XJ, Gao YJ, Li XH, Zhang QF (1997). Mapping and genetic analysis of two fertility restorer loci in the wild-abortive cytoplasmic male sterility system of rice (Oryza sativa L.). Euphytica, 98: 183-187.

Zhang Q, Bharaj TS, Virmani SS, Huang H (1997). Mapping of the Rf-3 nuclear fertility restoring gene for WA cytoplasmic male sterility in rice using RAPD and RFLP markers. Theor. Appl. Genet., 94: 27-33.

Yang GP, Saghai-Maroof MA, Xu CG, Zhang Q, Biyashev RM (1994). Comparative analysis of microsatellite DNA polymorphism in landraces and cultivars of rice. Mol. Gen. Genet., 245: 187-194.

Young ND, Tanksley SD (1989). Restriction fragment length polymorphism maps and the concept of graphical genotypes. Theor. Appl. Genet., 77: 95-101.

21

T-DNA direct repeat, vector backbone and gene trap counter selection by a new vector (pNU435) for high throughput functional genomics

R. S. Bhat[1]*, S. S. Biradar[1], R. V. Patil[2], V. U. Patil[1] and M. S. Kuruvinashetti[1]

[1]Department of Biotechnology, University of Agricultural Sciences, Dharwad 580 005, India.
[2]Division of Horticulture, of Agricultural Sciences, Dharwad 580 005, India.

The nature of T-DNA/*Ds* insertion decides the utility of launch pad lines for i*Ac*/*Ds* based insertional mutagenesis. Direct or inverted T-DNA/*Ds* repeats and insertions with vector backbone lead to poor recovery of flanking sequences; whereas T-DNA/*Ds* insertion leading to gene trap would limit the use of such launch pad lines. A new *Ds* tagging/trapping vector, pNU435 containing two copies of intron-interrupted *barnase* was used in tomato to counter select such T-DAN/*Ds* insertions. T$_1$ and T$_2$ plants generated in this study were devoid of direct repeats, vector backbone and gene traps as evidenced by the lack of barnase expression. Further evidence based on the LB flanking sequence recovered through TAIL-PCR did not show any vector backbone or direct repeats. But two out of six plants showed inverted repeats. Genome search with LB flanking sequence indicated that the insertions were not in genic region, and hence not led to gene traps. pNU435 with features for counter selecting undesirable T-DNA/*Ds* insertions can be employed for high throughput functional genomics.

Key words: pNU435 vector, T-DNA/*Ds* insertion, *barnase*, tomato, flanking sequence, functional genomics.

INTRODUCTION

Functional genomics aims at assigning function to genes and their regulatory elements of a genome by various approaches (Hirochika et al., 2004; Krishnan et al., 2009) such as comparative genomics, transcriptome analysis, mutagenesis, gene silencing and FOX (full length cDNA over-Expression) hunting (Nakamura et al., 2007), etc. Mutagenesis is a direct way of discovering novel genes and regulatory elements involved in various biological processes. High throughput functional genomics using maize *Ac*/*Ds* system has been widely employed for insertional tagging in various crops (Bancroft et al., 1992; Enoki et al., 1999). Insertional inactivation with *Ds* requires large scale production of launch pad lines harbouring T-DNA/*Ds*. Further, mobilization of *Ds* is brought about by crossing with i*Ac* lines. Launch pad line with a single copy, clean T-DNA/*Ds* insertion [without vector backbone (VB) and repeats] and insertion not being

a gene trap is considered as most useful in insertional inactivation. Earlier efforts have observed that about 30-60% of launch pad lines contain T-DNA repeats (direct and inverted) and vector backbones, and such plants are prone to post-integration rearrangements and gross deletions (Jeon et al., 2000; Upadhyaya et al., 2002; Kim et al., 2003; Eamens et al., 2004; Sallaud et al., 2004). Also rescuing the sequence flanking T-DNA is very difficult or in some cases impossible in such lines.

Since T-DNA has preferential insertion in gene-rich regions when compared with repetitive DNA, significantly large number of launch pad lines can be expected to be gene traps (Jeon et al., 2000; Upadhyaya et al., 2002; Kim et al., 2003; Eamens et al., 2004; Sallaud et al., 2004). In rice, ~23% of launch pad lines were gene traps (Eamens et al., 2004). T-DNA/*Ds* gene traps might result in untagged mutations in the stable mutant derived from such launch pad lines. Though the copy number is not under control, recovering clean integration of T-DNA/*Ds* without any gene traps could be manipulated by improvising the vectors used for developing launch pad

---

*Corresponding author. E-mail: bhatramesh12@rediffmail.com

Ds5'	SA	eyfp - nosT	CaMV 35SP – bar -ocsT	ColE Ori and bla	nosT - uidA	SA	Ds3'

RB	bn (1) -nosT	Hph (i)-tmlT	Ubi P (i)	LB	(I) bn (i)-nosT

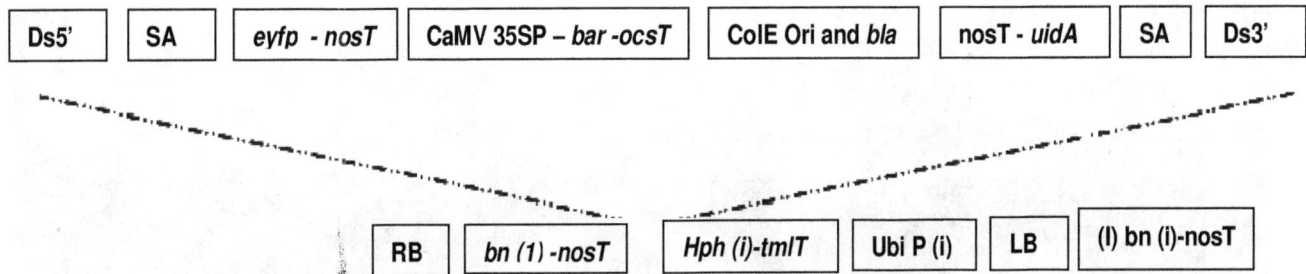

Figure 1. T-DNA/Ds of pNU435. 1. A promoterless intron interrupted barnase-nosT cassette placed next to RB to serve as T-DNA/Ds direct repeat (RB–LB–RB and LB) counter selector and T-DNA/Ds gene trap counter selector; 2: maize ubiquitin promoter-first exon-modified intron (with LB repeat sequences incorporated)-intron interrupted barnase [bn (i)]-nosT, to serve as vector backbone counter selector.

lines. Bidirectional gene trap constructs (Eamens, et al. 2004) utilizing an intron-interrupted barnase gene as a VB counter selector outside the T-DNA region of pEU334AN or pEU334BN could reduce the VB containing T-DNA/Ds lines (Hanson et al., 1999). However, ~27% lines still contained direct or inverted repeats of T-DNA. Therefore, Upadhyaya et al. (2006) constructed a Ds vector, pNU435 (Gen Bank Acc. No. DQ225750) by incorporating features that can counter select lines with T-DNA/Ds direct repeat, insertion with VB, and insertion in genic region (gene trap).

However, till date pNU435 has not been tested and validated for its aforesaid activities in any of the systems. In the present study, an effort was made to check T-DNA/Ds direct repeat, vector backbone and gene trap counter selection activity of Ds vector, pNU435 in tomato. pNU435 (kindly donated by Dr. Narayana Upadhyaya, CSIRO Plant Industry, Canberra, Australia) contained two copies of intron-interrupted barnase; one in the VB immediately after ubiquitin promoter-LB sequence and the other immediately after the RB sequence (Figure 1) to counter-select transformants with either VB or direct repeat of T-DNA/Ds. barnase placed next to RB would also counter select regenerants with T-DNA/Ds integration in genic region (gene trap). Seeds of Pusa Ruby cultivar of tomato were sown in vitro on half strength MS media (Murashige and Skoog, 1962). Seven-day old cotyledonary leaves were used for co-cultivation by following the standard protocol of Agrobacterium-mediated transformation (McCormick, 1991). Co-cultivated cotyledonary leaves were transferred to regeneration medium (MS with 3% sucrose, 2 mg/l zeatin, 0.1 mg l^{-1} IAA, 200 µg ml^{-1} cephotaxime). Shoots were transferred to MS basal medium supplemented with 0.05 mg/l of IBA for rooting. Transgenic plants were confirmed by gus-specific PCR (RB59_GUS_F 5' TCACCGAAGTTCATGCCAGTCC 3' and RB59_GUS_R2 5' ACGCTCACACCGATACCATCAG 3') (Figure 2).

Barnase is a bacterial protein with 110 amino acids possesses ribonuclease activity (Hartley, 1988). It is lethal to

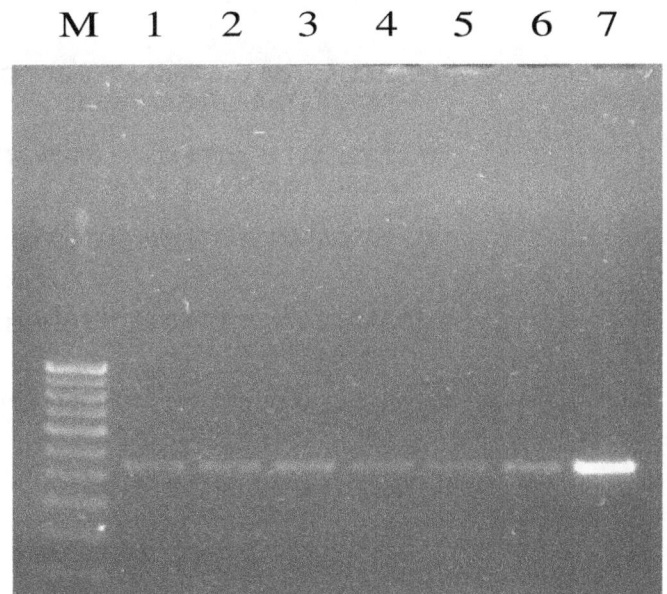

Figure 2. Agarose gel electrophoresis of products amplified by gus-specific PCR. Lane M, 100 bp DNA ladder; lanes 1-6, promoter trap lines (PT1, PT2, PT3, PT4, PT5 and PT6); lane 7, Pnu 435 (positive control).

the cell when expressed without its inhibitor barstar. T-DNA/Ds integration in genic region leading to either promoter or gene trap would express barnase resulting in cell death. Also in any T-DNA insertion with direct repeat (RB-LB-RB and LB) or with vector backbone (LB-VB), ubiquitin promoter would drive barnase gene. But in this study, transgenic plants were normal (without cell/tissue death) and did not show barnase expression upon RT-PCR (data not shown) indicating that T-DNA/Ds insertion was neither a gene trap, nor associated with direct repeat or vector backbone. Further confirmation that T-DNA/Ds insertion was not in geneic region came from right border flanking sequence tag (FST) of T-DNA/Ds as recovered by TAIL-PCR (Liu et al., 1995) in randomly selected two plants (PT4 and PT5). These FSTs showed homology to

Figure 3. T$_2$ plants resistant to Basta. A, Parent (Pusa Ruby); B, Progenies of PT4.

A

1 CAACGACTGA CTGTAGTATT AGGGGATTAG AGTGTCACGT TCCGACACAA
TAAGAATAAA GAGAATGAAT

71 CTGGAATTAT GTTAATATAC TCAATTTAAA GAACCTATTT CCCAAGTGAG TATGGTGTGG
AGGCTTGAGT

141 CCTCATAGGT GTGCTCGGTG TTGACGCCTA TCCTGAAAAA

B

1 ACGAGCTGAC ATGTAGTATT AGGGGATTAG AGTGTCACGT TCCGACACAA TAAGACTAAA
GAGAATGAAT

71 CTGGAATTAT GTTAATATAC TCAATTTAAA GAACCTATTT CCCAAATGAG TATGGTGTGG
AAGCTTGTGA

141 CCCCTCATAG GTGTGCTCAA AGTTGACCCC CTATCCTGCA

Figure 4. Left border flanking sequence tag obtained from PT4 (A) and PT5 (B).

RB, indicating the possibility of inverted repeats (LB-RB-RB and LB). This was further tested among the progenies (T$_2$) of PT4 and PT5. T$_2$ plants were first confirmed for the presence of T-DNA/*Ds* by spraying BASTA. Majority of the plants could survive the selection (Figure 3). Left border FSTs (Figure 4) recovered using TAIL-PCR in T$_2$ progenies showed neither repeat nor vector backbone. PT4 and PT5 were therefore confirmed to contain LB-RB-RB and LB inverted repeats without any vector backbone.

Regeneration of such lines with inverted repeats is possible since pNU435 can counter select only direct repeats, but not inverted repeats. BLAST search of these FSTs against tomato genome showed that T-DNA/*Ds* insertion was in retrotransposons like Tork-1 and Jinling-2, but not in any genic region. Therefore, PT4 and PT5, like other plants tested in this study, were confirmed not to be gene traps. Launch pad line with already one of its genes insertionally-tagged/trapped with T-DNA/*Ds*, is not generally employed for generating *Ds* tagged mutants for functional genomics (Sallaud et al., 2004). Reason being,

such mutants might carry empty T-DNA (without *Ds*) and *Ds* tags in two different genes leading to untagged mutations. To the best of our knowledge, pNU435 is the only T-DNA/*Ds* vector available to counter select T-DNA/*Ds* direct repeat, vector backbone and gene trap. Majority of the cells upon transformation with pNU435 may not regenerate due to complex insertions, where barnase would express and kill the cell. This was reflected in this study by the relatively low regeneration frequency compared to that obtained with other constructs (data not shown). Thus pNU435 with features for counter selecting undesirable T-DNA insertions at cell level would save time and resources in checking the events for generating launch pad lines for high throughput functional genomics.

ACKNOWLEDGEMENT

The authors wish to thank Dr. N. M. Upadhyaya of CSIRO

T-DNA direct repeat, vector backbone and gene trap counter selection by a new vector (pNU435) for high throughput...

153

Plant Industry, Canberra for kindly providing pNU435, and also for the critical guidance.

REFERENCES

An S, Park S, Jeong DH, Lee DY, Kang HG, Yu JH, Hur J, Kim SR, Kim YH, Lee M, Han S, Kim SJ, Yang J, Kim E, Wi SJ, Chung HS, Hong JP, Choe V, Lee HK, Choi JH, Nam J, Kim SR, Park PB, Park KY, Kim WT, Choe S, Lee CB, An G (2003). Generation and analysis of end sequence database for T-DNA tagging lines in rice. Plant Physiol., 133: 2040-2047.

Bancroft I, Bhatt AM, Sjodin C, Scofield S, Jones JD, Dean C (1992). Development of an efficient two-element transposon tagging system in Arabidopsis thaliana. Mol. Gen. Genet., 233: 449-461.

Chen S, Jin W, Wang M, Zhang F, Zhou J, Jia Q, Wu Y, Liu F, Wu P (2003). Distribution and characterization of over 1000 T-DNA tags in rice genome. Plant J., 36: 105-113.

Eamens AL, Blanchard CL, Dennis ES, Upadhyaya NM (2004). A bidirectional gene trap construct suitable for T-DNA and Ds-mediated insertional mutagenesis in rice (Oryza sativa L.). Plant Biotechnol. J., 2: 367-380.

Enoki H, Izawa T, Kawahara M, Komatsu M, Koh S, Kyozuka J, Shimamoto K (1999) Ac as a tool for the functional genomics of rice. Plant J, 19: 605-613

Hanson B, Engler D, Moy Y, Newman B, Ralston E, Gutterson N (1999). A simple method to enrich an Agrobacterium-transformed population for plants containing only T-DNA sequences. Plant J., 19: 727-734

Hartley RW (1988) Barnase and barstar. Expression of its cloned inhibitor permits expression of a cloned ribonuclease. J. Mol. Biol., 202: 913-915.

Hirochika H, Guiderdoni E, An G, Hsing YI, Eun MY, Han CD, Upadhyaya N, Ramachandran S, Zhang Q, Pereira A, Sundaresan V, Leung H (2004). Rice mutant resources for gene discovery. Plant Mol. Biol., 54: 325-334.

Hsing YI, Chern CG, Fan MJ, Lu PC, Chen KT, Lo SF, Sun PK, Ho SL, Lee KW, Wang YC (2007). A rice gene activation/knockout mutant resource for high throughput functional genomics. Plant Mol. Biol., 63: 351-364.

Jeon JS, Lee S, Jung KH, Jun SH, Jeong DH, Lee J, Kim C, Jang S, Lee S, Yang K, Nam J, An K, Han MJ, Sung RJ, Choi HS, Yu JH, Choi JH, Cho SY, Cha SS, Kim SI, An G (2000). T-DNA insertional mutagenesis for functional genomics in rice. Plant J., 22: 561-570.

Jeong DH, An S, Park S, Kang HG, Park GG, Kim SR, Sim J, Kim YO, Kim MK, Kim SR, Kim J, Shin M, Jung M, An G (2006). Generation of a flanking sequence-tag database for activation-tagging lines in japonica rice. Plant J., 45: 123-132.

Kim SR, Lee J, Jun SH, Park S, Kang HG, Kwon S, An G (2003). Transgene structures in T-DNA-inserted rice plants. Plant Mol. Biol., 52: 761-773.

Krishnan A, Guiderdoni E, An G, Hsing YC, Han C, Lee MC, Yu SM, Upadhyaya N, Ramachandran S, Zhang Q (2009). Mutant Resources in Rice for Functional Genomics of the Grasses. Plant Physiol., 149: 165.

Liu YG, Mitsukawa N, Oosumi T, Whittier RF (1995). Efficient isolation and mapping of Arabidopsis thaliana T-DNA insert junctions by thermal asymmetric interlaced PCR. Plant J., 8: 457-463.

McCormick S (1991). Transformation of tomato with Agrobacterium tumefaciens. Plant Tiss. Cult. Manual., 6: 1-9.

Murashige T, Skoog F (1962). A revised medium for rapid growth and bioassay with tobacco tissue cultures. Physiol. Plant., 15: 473-497.

Nakamura H, Hakata M, Amano K, Miyao A, Toki N, Kajikawa M, Pang J, Higashi N, Ando S, Toki S, Fujita M, Enju A, Seki M, Nakazawa M, Ichikawa T, Shinozaki K, Matsui M, Nagamura Y, Hirochika H, Ichikawa H (2007). A genome-wide gain-of-function analysis of rice genes using the FOX-hunting system. Plant Mol. Biol., 65: 357-371.

Sallaud C, Gay C, Larmande P, Bes M, Piffanelli P, Piegu B, Droc G, Regad F, Bourgeois E, Meynard D, Perin C, Sabau X, Ghesquiere A, Glaszmann JC, Delseny M, Guiderdoni E (2004). High throughput T-DNA insertion mutagenesis in rice: a first step towards in silico reverse genetics. Plant J., 39: 450-464.

Upadhyaya NM, Zhou XR, Zhu QH, Ramm K, Wu L, Eamens AL, Sivakumar R, Kato T, Yun DW, Santhoshkumar C, Narayanan KK, Peacock JW, Dennis ES (2002). An iAc/Ds gene and enhancer trapping system for insertional mutagenesis in rice. Funct. Plant Biol., 29: 547-559.

Upadhyaya NM, Zhu QH, Zhou XR, Eamens AL, Hoque MS, Ramm K, Shivakkumar R, Smith KF, Pan ST, Li S, Peng K, Kim SJ, Dennis ES (2006). Dissociation (Ds) constructs, mapped Ds launch pads and a transiently-expressed transposase system suitable for localized insertional mutagenesis in rice. Theor. Appl. Genet., 112: 1326-1341.

22

Estimation of heritability and genetic advance of yield traits in wheat (*Triticum aestivum* L.) under drought condition

Manal H. Eid

Botany Department, Faculty of Agriculture, Suez Canal University, 41522, Ismailia, Egypt.
Email: eid_manl@hotmail.com.

Four varieties/lines of wheat and their crosses namely, *Sakha8 , Sids1, line 1 Line3 and Line1x Sakha8, Line3x Sakha8, Line1x Sids1, Line3x Sids1* were evaluated to estimate heritability and genetic advance for yield traits. The experimental material was planted under irrigation as well as drought stress conditions. The mean average for plant height, spike length, number of spikes per plant, number of grains per spike, 50% heading date and 1000 grain-weight revealed highly significant differences among genotypes and crosses under both sowing conditions. Low, medium and high heritability was found in different yield traits under study. High heritability accompanied by high genetic advance was observed for spike length and 1000 grain-weight. Low heritability coupled with low genetic advance was for plant height and number of grains per spike. However, the heritability was generally found to be lower under drought stress conditions. Greater magnitude of heritability coupled with higher genetic advance in some traits under study provided that these parameters were under the control of additive genetic effects. This indicates that selection should lead to fast genetic improvement of the material. Moreover, the genetic correlations in study were high for most of the traits, suggesting a strong inherent association among these traits at the genetic level. These traits therefore deserve better attention in future breeding programs for evolving better wheat in stress environments.

Key words: Wheat, heritability, genetic advance, yield traits.

INTRODUCTION

Wheat is the largest grain crop in the world. It provides food to 36% of the global population, and contributes 20% of food calories. With progressive global climatic change and increasing shortage of water resources and worsening eco-environment, wheat production is influenced greatly (Singh and Chaudhary, 2006). The Increasing yield potential has indisputable importance in solving world hunger issue. The grain yield of wheat is determined by three yield components: productive spikes per unit area, kernel per spike and kernel weight. The product of the first two components gives the total kernel numbers per unit area (Collaku, 1994). Its increase is one of the major factors that have contributed to wheat yield improvement, which mainly resulted from increase in kernels per spike or from increase in kernel weight. Yield and its component traits are controlled by polygene, whose expression is greatly affected by environments

(Ahmed et al., 2007). Thus, determination of the number, locations, and effects of this polygene is desired for obtaining optimal genotypes in breeding practice.

However, it is interesting but not surprising that genetic increase in wheat yields in dry areas has not been as great as in more favorable environments or where irrigation is available. A likely reason for this is that those dry environments are characterized by unpredictable and highly variable seasonal rainfall and hence highly variable yields. Since yield has a complex trait and is strongly influenced by the environment, sever losses can be caused by drought, a stress common in most arid and semi arid areas. Accordingly, drought tolerance is one of the main components of yield stability and its improvement is a major challenge to geneticists and breeders.

Therefore, sufficient genetic information regarding the yield traits of wheat under drought is essential and important

important to get progress in plant breeding program. Moreover, understanding of the genetic control of these economic traits through quantitative traits locus analysis allows the identification of discrete chromosome segments controlling complex traits (Frova et al., 1999). Besides, other genetic analysis could determine whether the correlations between the traits are genetics or phonetics, and how its onward transmission from parents to progeny, like heritability takes place. Heritability, a measure of the phenotypic variance attributable to genetic causes, has predictive function of breeding crops (Songsri et al., 2008). It provides an estimate of genetic advance a breeder can expect from selection applied to a population under certain environment. The higher the heritability estimates, the simpler are the selection procedures (Khan et al., 2008). It has been emphasized that heritability alone is not enough to make sufficient improvement through selection generally in advance generations unless accompanied by substantial amount of genetic advance. The utility of heritability therefore increases when it is used to calculate genetic advance, which indicates the degree of gain in a character obtained under a particular selection pressure. Thus, genetic advance is yet another important selection parameter that aids breeder in a selection program (Shukla et al., 2004).

Keeping in view, the genetic studies on wheat were undertaken to estimate heritability and genetic advance of yield traits under drought conditions.

MATERIALS AND METHODS

Field experiments were conducted at the Experimental Farm, Faculty of Agriculture Suez Canal University, Ismailia, Egypt during 2006/2007 and 2007/2008 seasons. The experimental material comprised four varieties/lines of wheat namely, Sakha8, Sids1, line 1 and Line3.

During the crop season, 2006/2007, the following cross combinations were made to raise F1 namely: Line1x Sakha8, Line3x Sakha8, Line1x Sids1, and Line3x Sids3. The experimental design used randomized complete block design (RCBD) with three replicates. The experimental plot consisted of 6 rows, 3 m long and 5 cm apart in which grains were drilled by hand. The normal recommended agricultural practices of wheat production were applied at the proper time.

During the next crop season (2007/2008), two experiments, one under irrigation condition and the other under drought stress condition, were done on October 2007. Irrigation water was supplied by sprinklers to provide two water regimes during plant growth. Control regime was watered every 7-10 day throughout the growing season till harvest time. Drought was created in this rain-free environment by withholding irrigation after 30 days from sowing and giving supplementary irrigations every three weeks during 90 days post-sowing. Water application was monitored via a water meter and the control treatment (well- watered) received 420 mm, while the drought experiment (severe stress) received 140 mm.

However, in crop season (2007/2008), ten guarded plants were randomly selected from each plot, genotype and cross for agronomic traits analysis. Date of heading was recorded as the date when 50% of the shoots had reached this stage. At physiological maturity, plant height was measured from the soil surface to the top of the spike on the main shoot. Spike length of the main spike of the selected plant excluding awn was measured

in centimeters: number of spikes per plant. Number of grains per spike was counted from the spikes used for measuring spike length. The total number of grains recorded was divided by the number of spikes, and the average was computed. For 1000 grain-weight, a single sample of 1000 grains was counted in grams from the yield of the selected plants.

Analyses of variance (ANOVA) were carried out on grain yield and its components, using the software Genstat version 6.1 (Lawes Agricultural Trust, Roth Amsted Experimental Station).

Narrow sense heritability (h^2) was estimated from parent offspring regression according to the method of Anderson et al. (1991).

$$Y = a + b_{xi} + e_i,$$

Where e_i = error and b = the regression coefficient.

$$b = \frac{\sigma_{xy}}{\sigma^2_x}$$

Where σ_{xy} = covariance of parent-offspring and σ^2_x = total variance of parental measurements.

Genetic correlation between six traits was determined according to the method of Kwon and Torrie (1964):

$$rg = \frac{Cov_{gij}}{(\sigma^2_{gi} \times \sigma^2_{gj})}$$

Cov_{gij} and $\sigma2_{gi}$ are the estimates of covariance and variance for traits i and j.

The genetic advance was calculated according to Allard's (1964), and was estimated from the following formula:

$$GA = i\, h^2 Vp$$

Where i = 1.76 (10 % selection intensity), Vp: phenotypic variances, h^2: heritability.

RESULTS AND DISCUSSION

Mean performance

The analysis of variance showed that the mean squares for genotypes were significant for all traits studied. This indicates the existence of a high degree of genetic variability in the material to be exploited in breeding program, and that also reflected the broad ranges observed for each trait (Table 1).

Plant height in wheat has been observed to be affected by drought stress to a considerable extent (Table 2). Line 1 was superior with respect to plant height (69.4 cm), followed by cross Line3xSakha8 (66.6 cm) and Line 3xSids1 (65.9 cm), whereas, Sids1 and Sakha 8 recorded the lowest plant height under drought condition. However, the varieties that adapted to the water stress environments are generally short in stature, as compared to the ones which adapted to optimal moisture conditions (Foulkes et al., 2004). The height of the culms, size of the leaves, the distance between the veins and the stomata openings are all affected when they are developing under

Table 1. Significance of mean squares due to different sources of variation for evaluating 4 wheat genotypes and crosses.

Source	df	Plant height	Spike-length	No. spike	No. grains/spike	50% heading date	1000-grain weight
Main plots Blocks	2	37.5*	12.04**	0.27ns	2.04ns	2.43 ns	78.2*
Water- reg	1	47.7**	1.2ns	13.3*	227.6**	1136.1**	1822.1**
genotypes	15	396.9**	2.4**	0.72*	64.2**	13.06**	19.7**
genotypes * Water reg	15	3.3**	0.85*	0.66*	31.7**	27.9**	9.8**

*and ** denote significance at 0.05 and .01 % level of probability

Table 2. Mean values of various traits of wheat genotypes and crosses under irrigation and drought conditions.

Genotypes/ crosses	Plant height		Spike-length		No. spike		No. grains/spike		50% heading date		1000-grain weight	
	con	stress	con	stress	con	stress	con	stress	con	stress	con	stress
Line1	72.4a	69.4b	9.7ab	9.3b	2bcdef	2bcdef	17.3abc	17.3bc	95ab	89.5d	25.6ab	23.1de
Line3	59e	58.7e	8c	8c	2.8a	2cde	17.2abcd	16.5cdef	94.3bc	94.5abc	25.5abc	23.3de
Sakha8	45.1f	45.0f	8.1c	7.9c	2.2bcde	1.5f	17bcde	16ef	94.2bc	90.8d	24.2cd	19.6f
Sids1	45.3f	45.1f	9.7ab	9.4ab	1.5ef	1.5ef	14.3g	13.4g	95ab	90.5d	24.3abc	20.4f
Line1xSakha8	67.7c	58.5e	8.3c	9.6ab	2.2bcde	2.2bcde	17.3abc	17.3abc	96.7a	94bc	26.2a	20.5f
Line3xSakha8	72a	66.6d	9.8a	8.3c	2.5abc	1.5f	16.2ef	16.5def	94.8ab	93.7bc	26a	21.5ef
Line1xSids1	70.3b	59e	9.7ab	9.5ab	2.3bcd	2def	17.7ab	14.5g	94.7ab	91.5cd	25.9a	23.1de
Line3xSids1	66.6d	65.9d	9.6ab	9.5ab	2.7ab	1.5ef	17.8a	16f	95ab	94bc	25.9	23de
CV	1.4		4.3		21.3		4.8		1.9		4.8	
LSD 0.05%	2.2		1.0		1.2		1.9		3.5		2.3	

Means not sharing a common letter in a column differ significantly at 0.05% level of probability.

limited water supplies. When wheat plants are deprived of water at vegetative and flowering stages, shorter plants are obtained as a result of low moisture absorption, lower soil absorption, lower soil nutrient uptake, reduced cell size and reduced photosynthesis (Ahmed et al., 2007).

Drought caused reduction in spike length and number of spikes per plants. It was observed *Sakha8* was the lowest genotype for spike length (7.9 cm) and number of spike per plant (1.5) under drought conditions.

The number of grains per spike ranged between 13.4 and 17.3 under drought stress condition (Table 2). It is evident from the results that line1 and cross *Line1x Sakha8* produced maximum number of grains per spike (17.3). *Sids1* produced statistically lower number of grains per spike (13.4) than others under drought stress. However, the decreased number of grains per spike in wheat under stress may be attributed to accelerated apex development and reduced number of leaves at the main stem. At the stage of early emergence, water stress affects the reproductive development, grains initiation and hence final number of grains (Riaz and Chwodhry, 2003).

However, compared to well- watered genotypes, drought caused reductions in days to 50% heading. Therefore, genotypes which flowered and matured earlier may have been favored by partial escape from drought and have an ability to complete their life before get ting dehydrated by high summer temperatures 1000 –grain

weight is vital yield component and with similar weight under drought is more or less stable characters of wheat cultivars. Under drought, this trait may be affected to a greater extent and genotypes showing high 1000-grains weight under irrigated conditions may not be able to produce grains with similar weight (Table 2). This is possible due to the shortage of moisture which forces the grains to complete their formation in relatively lesser time. Based on mean performance, comparison of the four crosses under drought condition (Table 2), the highest value with respect to 1000-grain weight (23.1 g) was obtained from the cross *Line1xSids1*, followed by the cross *Line3xSids1* (23 g). It was noted that these two crosses involving genotypes line 1 and line 3 produce higher 1000 grain weights under drought condition as well. The good mean performance in the two crosses reflected that effective selection for this character is possible in appropriate cross combinations and new genotype may evolve, possessing higher 1000 grain weight along with resistance against drought.

Larger estimates of variation coefficients presented in Table 2 number of spike and number of grain and 1000-grain weight and spike length. The magnitude of variation coefficients might indicate that four genotypes and their progenies had exploitable genetic variability for yield characters under investigation. However, these results partially coincided with earlier findings of Riaz and Chwodhry (2003); this is perhaps due to differences in

Table 3. Genetic correlation among the studied traits under irrigation and drought conditions.

Traits	Treatment	Spike length	No. spike	No. grains/spike	50% heading date	1000 grains-weight
Plant height	control	0.545437	0.194535	0.305937	-0.03853	0.423622
	stress	0.195252	0.425327	0.163353	0.29461	0.858464
Spike length	control		-0.10388	-0.02141	-0.08288	0.627319
	stress		0.071888	-0.24338	-0.14403	0.236863
No. spike	control			0.609886	0.213458	0.074922
	stress			0.065956	0.321663	0.529082
No. grains/spike	control				0.424029	0.12229
	stress				0.40878	0.251132
50% heading date	control					-0.14338
	stress					0.380185

breeding material or variation in environment or interaction.

GENETIC CORRELATION AMONG TRAITS UNDER IRRIGATION AND DROUGHT CONDITIONS

Since the correlations have to be made in the light of their genetic behavior, genotypic correlation values are used for further analysis. Genetic relation of traits may result from pleotropic effects of a gene, linkage of two genes, chromogema and regimental affiliation or due to the environmental influences (Sgro and Hoffmann, 2004).

The relationship of plant height was positive with spike length, number of spikes, and number of grains per spike and 1000-grain weight under control and drought conditions (Table 3).

Also, number of spike per plant and number of grains per spike had positive relationship with the other yield traits under study. These traits might not be independent in their action and are interlinked likely to bring simultaneous change for other characters. They can be effectively used as selection criteria for wheat yield under drought conditions.

Spike length had negative correlation with number of grains per spike and heading date, whereas it was positive with 1000-grain weight under control and drought conditions.

Moreover, spike length had negative correlation with number of spikes under control but positive under drought condition. Similar genetic behavior, with heading date, had negative correlation with 1000 grain weight under control, whereas positive under drought condition. There is evidence that change in conditions can influence genetic interactions among traits as well as genetic variance in traits themselves. Exposure to drought conditions may induce positive correlation among traits

because the expression of new gene will breakdown negative correlations (Sgro and Hoffmann, 2004).

The genetic correlations in general were high for most of the traits, suggesting a strong inherent association among these traits at the genetic level (Table 3).

HERITABILITY AND GENETIC ADVANCE

It has been emphasized that without genetic advance, the heritability values would not be of practical importance in selection based on phenotypic appearance. So, genetic advance should be considered along with heritability in coherent selection breeding program.

In the present investigation, high heritability values coupled with high genetic advance (Table 4) were recorded for spike length under irrigated and drought condition. This indicated the additive nature of genetic variation was transmitted from the parents to the progeny. Also, this trait can easily be fixed in the genotypes by selection in early generations. These results find support from the earlier studies reported (Riaz and Chwodhry, 2003; Hasssan, 2004). However, spike length is a character of considerable importance, as the larger spike is likely to produce more grains and eventually higher yields per plants. Genotypes retaining larger spikes under moisture stress are likely to be more productive under stress environment. Better heritability values recorded point to the possibility of improvement in this parameter (Ahmed et al., 2007; Songsri et al., 2008).

Moderate heritability was accompanied by high genetic advance for number of spike per plant under control, but it had a low value under drought condition.

High heritability accompanied by low genetic advance for days of 50% heading is indicative of non-additive gene actions' predominance which could be exploited through heterosis breeding. Further explanation by Sardana

Table 4. Estimation of heritability and genetic advance for the studied traits under irrigation and drought conditions.

Traits	Heritability	Genetic advance %	Heritability	Genetic advance %
	Control	Control	Stress	Stress
Plant height	0.29249	6.7	0.1631	3.72
Spike length	0.5961	32.8	0.548	15.16
No. spike/ plant	0.339	21.4	0.12	2.02
No. grains /spike	0.2692	1.07	0.25	1.77
50% heading date	0.6269	0.422	0.75	5.3
1000 grain-weight	0.7379	24.07	0.429	6.17

Sardana et al. (2007), suggested that high heritability may not necessarily lead to increased genetic gain, unless sufficient genetic variability existed in the germplasm.

Low heritability with low genetic advance values was found for plant height and number of grains, indicating slow progress through selection for thesis traits. The reason for the low heritability for these two components is a result of some variances constituting the environment variance.

It is obvious from Table 4 that the heritability estimate for 1000-grian weight was high, coupled with high genetic advance under control condition, whereas its values was low with genetic advance under drought condition. This differential response could be explained by the fact that this study that used genotypes could have different sets of alleles and possibly different loci are being expressed under different environmental conditions. Collaku (1994) reported in his study that low heritability is as a result of drought stress. Nonetheless, Rana et al. (1999) stated that the 1000 grain weight under arid conditions is significant for the phonotypical selection.

However, the comparison of heritability for all the traits was done under drought stress and irrigated conditions. This indicates that except for heading date, heritability increases with better input condition environment. This may be due to influence of environment on genotypes under drought stress condition. Similarly, it was resulted that heritability for yield traits in faba bean was higher in well-watered treatment than drought stress condition (Link et al., 1999; Toker, 2004).

Conclusion

In general, it is considered that if a character is governed by non-additive gene action, it may give heritability but low genetic advance, whereas if the character is governed by additive gene action, both heritability and genetic advance would be high. In the present study expected genetic advance values were based on narrow sense heritability, which incorporates additive portion of the total phenotypic variance. Thus, narrow sense heritability is more useful for measuring the relative

importance of additive portion of genetic variance that can be transmitted to the offspring. The traits under study which had high heritability and also showed high expected genetic advance could be substantially considered or making selections as these traits were mainly influenced by the major effects of additive gene action. With knowledge of inheritance traits it could introduce specific traits into more widely adapted genotypes and thus meet a goal for developing cultivars better adapted to dry land conditions. But the final test for any wheat variety or areas subjected to limited moisture supply will be found in whether it has ability to yield adequate returns under relatively dry conditions over a period of years.

In conclusion, the genetic parameters discussed here are functions of environmental variability, so estimates may differ in other environment. Based on the high heritability and high genetic advance shown by the different characters, especially, spike length, 1000 grain weight, it could conclude that the determinant genetic effects of the phenotypic expression of these characters are fundamentally of the additive type. For this reason, a high response should be achievable after several selection cycles.

The development of varieties adapted to the arid conditions depends on improvement of potential yield and yield evaluation in different environments. However, the inherent understating of the limits of improving potential yield suggested that long range solution of yield improvement cannot be sustained by improving yield potential alone. Many other environmental variables should be controlled and optimized, so as to minimize the relative effect of genotype and environment interaction/ in addition, the most promising genotypes and cross combinations, *Line1* and *Line3* and; *Line1xSids1* and *Line3xSids1*. These genotypes and traits under study therefore deserve better attention in future breeding programs for evolving better wheat for stress environments.

ACKNOWLEDGEMENT

The author is thankful to Prof. Dr. Abdel-Raheem Ahmed at the Faculty of Agriculture, Suez Canal University, Ismailia, Egypt for kindly providing seed samples in this study.

REFERENCES

Ahmed NCM, Khaliq IMM (2007). The inheritance of yield and yield components of five wheat hybrid populations under drought conditions. Indonesian J. Agric. Sci., 8(2): 53-59.

Allard R (1964). Principles of plant breeding. John Wiley and Sons. Inc.New York, London.

Anderson W, Holbrook C, Wynne J (1991). Heritability and selection or resistance to early and late leafspot in peanut. Crop Sci., 31: 588-593.

Collaku A (1994). Selection for yield and its components in a winter wheat population under different environmental conditions in Albania. Plant Breed., 112(1): 40-46.

Foulkes M, Verma R, Sylvester R, Weightman R, Snap J (2004). Traits For improved drought tolerance of winter wheat in the UK. Proceeding of the 4 th international crop science congresses. Brisbane, Australia.

Frova C, Krajewski P, Fonzo N, Villa M (1999). Genetic analysis of drought tolerance in maize by molecular markers. 1. Yield components. Theory Appl. Genet., 99:280-288.

Hassan G (2004). Diallel analysis of some important parameters in wheat (Triticum aestivum L.) under irrigated and rainfed conditions. PhD. Agricultural University, Peshawar, Pakistan.

Khan H, Rahman H, Ahmed H, Ali H (2008). Magnitude of heterosis and heritability in sunflower over environments. Pak. J. Bot., 1: 301-308.

Kwon S, Torrie J (1964). Heritability and inter-relationship among traits of two soybean population. Crop. Sci., 4: 196-198.

Link W, Abdelmula A, VonKittlitz E (1999). Genotypic variation for drought tolerance in Vicia faba L. Plant Breeding.188:477-483.

Rana V, Shaama S, Sethi I (1999). Comparative estimates o genetic variation in wheat under normal and drought stress conditions. J. Hill. Res., 12 (2): 92-94.

Riaz R, Chowdhry M (2003). Estimation of variation and heritability of some physio-morphic traits of wheat under drought conditions. Asian J. Plant Sci., 2(10): 748-755.

Sardana S, Mahjan R, Gautam N, Ram B (2007). Genetic variability in pea (Pisum sativum L.) germplasm for utilization. SABRAO J. Breed. Genet., 39 (10):31-41.

Sgro C, Hoffmann A (2004). Genetic correlations, tradeoffs and environmental variation. Heredity, 93: 241-248. .

Shukla S, Bhargava A, Chatterjee A, Singh S (2004). Estimates of genetic parameters to determine variability for foliage yield and its different quantitative and qualitative traits in vegetable amaranth (A. tricolor). J. Genet. Breed., 58: 169-176.

Singh G, Chaudhary H (2006). Selection parameters and yield enhancement of wheat (Triticum aestivum L)under different moisture stress condition. Asian J. Plant Sci., 5: 894-898.

Songsri P, Joglloy S, Kesmala T, Vorasoot N, Akkasaeng CPA, Holbrook C (2008). Heritability of drought resistance traits and correlation of drought resistance and agronomic traits in peanut. Crop Sci., 48:2245-2253.

Toker C (2004). Estimates of broad-sense heritability or seed yield and yield criteria in faba bean (Vicia faba L.). Hereditas. 20:222-225.

23

Partitioning and distribution of random amplified polymorphic DNA (RAPD) variation among eggplant *Solanum* L. in Southwest Nigeria

Sifau, Mutiu Oyekunle[1,3], Ogunkanmi, Liasu Adebayo[1], Adekoya, Khalid Olajide[1], Oboh, Bola Olufunmilayo[1] and Ogundipe, Oluwatoyin Temitayo[2]

[1]Department of Cell Biology and Genetics, University of Lagos, Lagos, Lagos State, Nigeria.
[2]Department of Botany, University of Lagos, Lagos, Lagos State, Nigeria.
[3]Molecular Biology Laboratory, Biotechnology Unit, National Centre for Genetic Resources and Biotechnology (NACGRAB), PMB 5382, Ibadan, Oyo State, Nigeria.

Solanum L., the largest genus of the Solanaceae family, vary morphologically, is diverse in number and is ecogeographically distributed. In Nigeria, previous studies had focused mainly on chromosome morphology, genome description and medicinal values, which are insufficient for genetic affinities. This study used four highly polymorphic random amplified polymorphic DNA primers to describe both the genetic relatedness and variability among 25 accessions of eggplant from Southwestern Nigeria. At a truncated line of 65%, five clusters and two ungrouped samples are distinguishable from the dendrogram. The data reveals that *Solanum dasyphyllum* Schum. & Thonn. is more closely related to *Solanum macrocarpon* L. than to *Solanum melongena* L. The relatedness between *Solanum incanum* L. and *Solanum melongena*, a probability of being progenitors from a common ancestral lineage was also shown. Occurrence of *Solanum scabrum* L. and *Solanum nigrum* L. in the same clusters different from *S. melongena*, is an indication of distant relatedness to *S. melongena* but close relatedness between them. High level of polymorphism was observed in this study going by the coefficient of variation which exhibited a good separation from a conserved region of the genome. This study, therefore, reveals a wide and diverse genetic base in Nigerian eggplant *Solanum*.

Key words: Eggplant, genome, synonymy, polymorphism, phylogenetic.

INTRODUCTION

Solanum L., a complex and large genus of the family Solanaceae has an unresolved proper delineation of the species. The genus contains roughly between 1,500 and 2,000 species (Bohs, 2001). They are morphologically varied, numerically diversed and vastly ecogeographically distributed. Several species of vegetable *Solanum* important for human diet and health are referred to as eggplant (Daunay et al., 2001). Examples include *Solanum melongena, Solanum aethiopicum, Solanum macrocarpon, Solanum quitoense* Lam., *Solanum sessiliflorum* Dunal and related species. The taxonomy of eggplant *Solanum* has remained challenging due to species' large size, overlapping ecogeographical distribution (Levin et al., 2005), morphological plasticity, similar genomes (Okoli, 1988) and existence of swamps of natural hybrids (Obute et al., 2006; Oyelana and Ugborogho, 2008). The inconsistencies and misconceptions generated by these factors have made past attempts at taxonomically resolving the complexities associated with the genus difficult. The taxonomic uncertainties still persist in this genus largely because previous studies to address the taxonomic problem of vegetable *Solanum* have focused mainly on morphology (Karihaloo and Rai, 1995; Kumar et al., 2013), crossability and F1 fertility (Baksh, 1979; Hassan and Lester, 1990a; Lester and Hassan, 1991; Furini and Wunder, 2004) and anatomy (Hassan and Lester, 1990b). Establishing genetic affinities on such parameters are insufficient, as *Solanum* makes successful crosses with putative progenitors as well as distantly related species.

The advent of molecular biology has revolutionized the field of plant systematics and has been used successfully in phylogenetic relationships at all taxonomic levels (Bohs, 2005) as well as in DNA fingerprinting of plant genomes (Cervera et al., 1998) and in genetic diversity studies (Isshiki et al., 2008; Fory et al., 2010). The use of molecular techniques in genetic diversity studies is supported by the finding that evolutionary forces such as natural selection and genetic drift produce divergent phylogenetic branching which can be recognized because the molecular sequences, on which they are based, share a common ancestor (Singh et al., 2006). Random amplified polymorphic DNA (RAPD), when compared with other molecular markers, is more effective in this regard as it is simple, rapid, requires only a small quantity of DNA and it is well adapted for nonradioactive DNA fingerprinting of genotypes (Cao et al., 1999). It is also able to generate numerous polymorphisms (Williams et al., 1990). Karihaloo et al. (1995) focused directly on nuclear genomic diversity of *Solanum* by undertaking RAPD analysis. Karihaloo and Gottlieb (1995) also reported that greater DNA polymorphism exists in weedy *Solanum insanum* than in advanced cultivars of eggplants. RAPD data were used in several other studies such as Miller and Spooner (1999), Stedje and Bukenya-Ziraba (2003) and Singh et al. (2006) to clarify phylogenetic relationships. Other molecular markers have also been previously used to study the variability as well as relatedness among eggplant *Solanum* species. For instance, Nunome et al. (2003a) and Ge et al. (2013) both employed microsatellite markers or simple sequence repeat (SSR) markers, Behera et al. (2006) used STMS markers, Fory et al. (2010) worked on Colombian collection of *Solanum* using amplified fragment length polymorphism (AFLP), and more recently, Ali et al. (2013) studied the diversity among samples of Chinese *Solanum* by comparing results of RAPD and SSR markers.

In Nigeria, not many works have been done on the nature of genetic diversity and characterization of vegetable *Solanum*, especially using molecular methods. Many vegetable *Solanum* species that occur in Nigeria are sources of food and of medicinal importance (Gbile and Adesina, 1988). Taxonomic studies on the vegetable *Solanum* species in Nigeria have been based on chromosome morphology (Oyelana and Ugborogho, 2008), genome description (Okoli, 1988), medicinal and food values (Gbile and Adesina, 1988). These have not resolved the problems of synonymy and taxa misidentification common to the genus. As a result, this study attempts to resolve to a larger extent the taxonomic difficulties associated with vegetable *Solanum* especially among the species found in Southwestern Nigeria using RAPD molecular marker.

MATERIALS AND METHODS

Sample collection and identification

Fresh leaves (young and matured), fruits and seeds of eggplant *Solanum* samples of different species were collected from different locations in Southwestern Nigeria (Longitude 3° 20'E - 5° 10'E and Latitude 6° 15'N - 9° 00'N) especially in areas known for eggplant diversity. Each sample was labelled accordingly. The fresh leaves were prepared for molecular analysis while mature leaves were prepared for herbarium. A total of 25 samples were collected and analyzed in this study. Their authenticated names and places of collection are shown in Table 1. The breakdown showed that the collections consists of 10 different *Solanum* species made of 2 samples of *Solanum dasyphyllum*, 2 of *Solanum nigrum*, 3 *Solanum macrocarpon*, 2 *Solanum torvum*, 1 *Solanum erianthum*, 3 *Solanum melongena*, 7 *Solanum gilo*, 2 *Solanum scabrum*, 2 *Solanum aethiopicum* and 1 of *Solanum incanum*. Figure 1 shows some of the samples with variations in shapes and colours.

Voucher specimens were prepared from the samples following the method of Ogundipe et al. (2009) and sent to Forestry Herbarium Ibadan (FHI) where they were authenticated by taxonomists. These specimens were then deposited at both the University of Lagos Herbarium (LUH) and Forestry Herbarium Ibadan (FHI) for reference purposes.

Total genomic DNA extraction

Total genomic DNA extraction was carried out on young fresh leaves of each sample (Dellaporta et al., 1983). This was followed by additional purification in a silica-column inserted into vacuum manifold connected to a vacuum pump using QIAquick purification kit (Promega). Verification of the quality of the purified DNA samples was achieved by electrophoresis on a 1% Agarose gel.

Polymerase chain reaction (PCR)

Twenty seven (27) Operon primers (Operon Technologies Inc., USA) were screened based on higher GC content (between 60 - 70%) and their previous workability. Only four (4) that are highly polymorphic and gave reproducible bands were selected and used in the analysis of all the 25 genotypes. Total reaction volume for PCR was 10 µl containing 1.0 µl of 10x TAE buffer, 2 µl of 10 mg/µl sample DNA, 1.0 µl MgCl$_2$, 0.8 µl mixture of 10 mM dNTP, 20 (5% Tween), 20 polyoxyethylene sorbitan monolaurate with 20 ethylene oxide units, 4.6 µl of distilled water, and 5 U Taq DNA polymerase (1 U final conc.). Amplification was accomplished on the Techne TC- 412 thermal cycler (Model FTC41H2D, Barloworld Scientific Ltd, Staffordshire, UK), using the following temperature profile: Initial strand separation step of 3 min at 94°C followed by 40 cycles each consisting of a denaturing step of 20 s at 94°C, annealing step of 40 s at 35°C and an extension step of 1 min at 72°C. The last cycle was followed by 5 min extension at 72°C to allow complete extension of the PCR products with a final hold at 4°C till electrophoresis. The reaction was repeated two times for each

Table 1. Eggplant *Solanum* samples and places of collection.

Sample I.D no.	Identification Name	Place of collection	State of collection
OG02	*Solanum dasyphyllum*	Wasinmi	Ogun
OG03	*S. nigrum*	Wasinmi	Ogun
OG04	*S. dasyphyllum*	Joga orile	Ogun
OG05	*S. nigrum*	Joga orile	Ogun
OG06	*S. macrocarpon* (White fruit)	Abulemaria	Ogun
OG07	*S. macrocarpon* (Green fruit)	Abulemaria	Ogun
OG08	*S. torvum*	Wasimi-Imasai	Ogun
OG09	*S. erianthum*	Wasimi-Imasai	Ogun
OG10	*S. melongena* (Green fruit)	Wasinmi-Imasai	Ogun
OY11	*S. gilo* Raddi (White fruit)	Igboho	Oyo
OY12	*S. gilo* Raddi (White fruit)	Igboho	Oyo
OY13	*S. gilo* Raddi (White fruit)	Igboho	Oyo
OY14	*S. gilo* Raddi (White fruit)	Igboho	Oyo
OY15	*S. incanum* L. (Green small fruit)	Igboho	Oyo
OY16	*S. scabrum*	Igboho	Oyo
OY17	*S. aethiopicum*	Igboho	Oyo
OY18	*S. scabrum*	Igboho	Oyo
OY19	*S. melongena* (White fruit)	Igbope	Oyo
OY20	*S. aethiopicum*	Igboho	Oyo
OS21	*S. torvum*	Iwo	Osun
OG22	*S. melongena* (Green fruit)	J3 Camp, Ijebu Ode	Ogun
LA23	*S. gilo* Raddi (Green egg-shaped fruit)	Bariga, Lagos	Lagos
LA24	*S. gilo* Raddi (Green round fruit)	Agbowa-Ikosi	Lagos
LA25	*S. gilo* Raddi (Green round fruit with greenish purple stem)	Agbowa-Ikosi	Lagos
LA26	*S. macrocarpon* (Green fruit)	Agbowa-Ikosi	Lagos

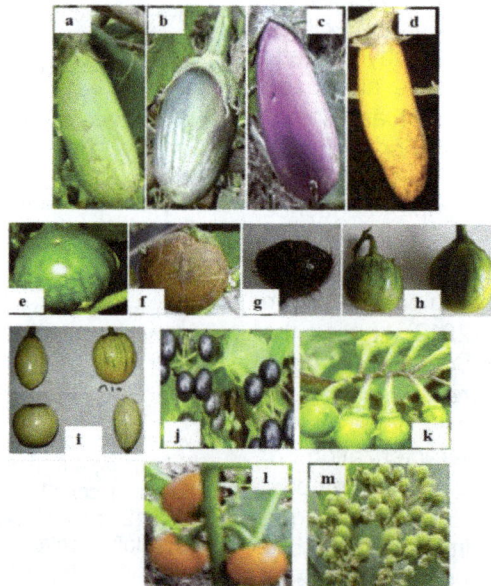

Figure 1. Variability in fruit colour and shape of some eggplant samples studied. Legend: (a - d) *S. melongena*; (e and f) *S. macrocarpon*; (g) *S. dasyphyllum*; (h and i) *S. gilo*; (j) *S. scabrum*; (k) *S. incanum*; (l) *S. aethiopicum*; (m) *S. erianthum*.

Table 2. Operon primers selected with their nucleotide sequence and their characteristic number of bands at amplification in samples analyzed.

Primer used	Primer sequence (5´- 3´)	Total number of bands	Number of polymorphic bands	Percentage polymorphic bands (%)
V-19	(5´- GGGTGTGCAG -3´)	12	10	83.3
B-18	(5´- CCACAGCAGT -3´)	12	10	83.3
OPU-13	(5'- GGCTGGTTCC -3´)	13	9	69.2
OPU-15	(5'- ACGGGCCAGT -3´)	15	13	86.7
Total		52	42	80.8
Average		13	10.5	

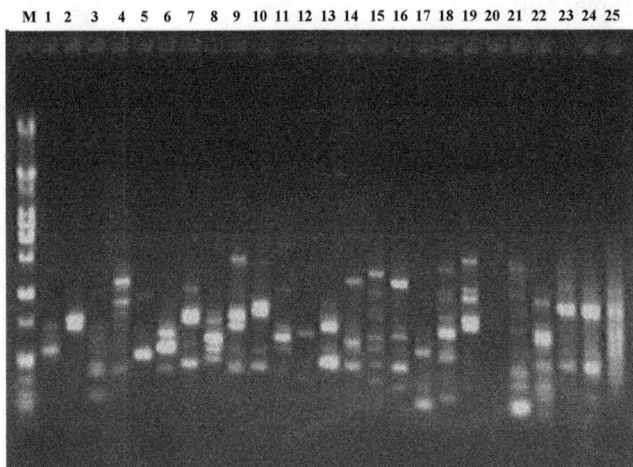

Figure 2. RAPD profiles generated by primer B-18 for *Solanum* samples studied. Legend: M represents the 100 bp DNA Ladder which serves as the reference point; 1 to 25 corresponds to bands produced by the amplified DNA from the 25 samples.

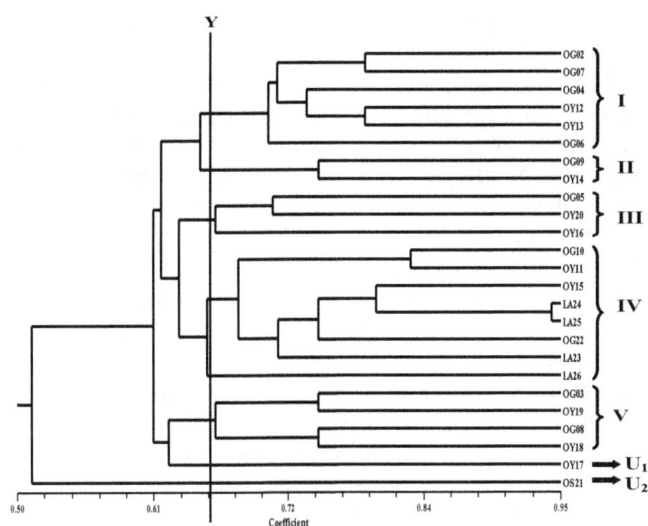

Figure 3. A UPGMA dendrogram showing genetic relationship among accessions of eggplants studied. Legend: Y represents truncated line at a co-efficient of similarity 0.65; I to V represent the five clusters that were distinguishable from the dendrogram while U_1 and U_2 represent ungrouped samples at that co-efficient of similarity.

selected primer to make the result more reliable. 5 µl of each of PCR product (amplicon) were mixed with 3 µl of 10X loading dye (0.25% bromophenol blue, 0.25% xylene cyanol and 40% sucrose, w/v) and spun briefly in a micro centrifuge before loading on a1.5% agarose gel which has been previously stained with safe view. This was run for 1 h 30 min at 110 V/cm. Thereafter, the gel was viewed (with the aid of eye protector) and photographed in the Gel Documentation and Analysis Systems (UVdoc, GA-9000/9010 Version 12).

Data analysis

For each sample, only distinct, well-resolved and unambiguous bands were scored. Faint bands were discarded. The amplified fragments were scored as 1 (present) and 0 (absent) to generate binary matrices. From this matrix, similarity matrices were computed using Sequential Hierarchical and Nested (SAHN) clustering option of the NTSYS-pc 2.02j software package (Rohlf, 1996). The software generated a dendrogram, which grouped the test lines using Unweighted Pair Group Method with Mathematic Average (UPGMA) on the basis of genetic similarity and Jaccard's coefficient.

RESULTS

The RAPD analysis of the 25 samples revealed a total of fifty two (52) bands, amplified by four (4) different oligonucleotide primers namely OPU-13, OPU-15, B-18 and V-19 (Table 2). Forty two (42) of these bands were highly polymorphic with percentage polymorphism put at 80.8% (Table 2). The numbers of amplification products obtained were in the range 12-15. Primers V-19 and B-18 produced the minimum number of (12) bands each, OPU-13 produced 13 bands and primer OPU-15 produced the maximum number of (15) bands. Average of 13 bands was also obtained per primer as shown in Table 2. Figure 2 shows the RAPD profile produced by B-18 Operon primer for the 25 samples.

Jaccard's similarity coefficient matrix generated a dendrogram (Figure 3) based on polymorphism obtained with all the selected four primers using UPGMA clustering option of NTSYS-pc 2.02j software package (Rohlf, 1996). The scale of the dendrogram constructed from the data was between 0.50 and 0.95 with a mean value of 0.73 (Figure 3). At a truncated line of 65% (a similarity co-efficient of 0.65), five clusters (I - V) and two ungrouped samples (U_1 and U_2) are distinguishable from the dendrogram. Cluster IV is the largest consisting of 8 samples while Cluster II being the smallest is made up of 2 samples (Figure 3). All the samples of *S. dasyphyllum* occur in Cluster I together with 2 samples (out of 3) of *S. macrocarpon* and 2 of *S. gilo*. Cluster IV contains most samples of *S. gilo* together with 2 of *S. melongena* and 1 of *S. incanum* and *S. macrocarpon*, respectively. One sample each of *S. nigrum*, *S. aethiopicum* and *S. scabrum* grouped together in Cluster III; so also, Cluster V contains one sample each of *S. melongena*, *S. nigrum*, *S. torvum* and *S. scabrum*, respectively. The only sample of *S. erianthum* occurs with one *S. gilo* in Cluster II while the remaining samples of *S. aethiopicum* and *S. torvum* remained ungrouped U_1 and U_2 respectively. It is worthy of notice that just as the selected primers were able to detect inter-specific polymorphism, they equally did so intra-specifically. This accounted for the occurrence of samples of the same species in different clusters e.g. one sample each of *S. nigrum* and *S. scabrum* occurring in both clusters III and V.

DISCUSSION

Hammond (1979) stated that biosystematics and evolutionary studies have for long time, and for the most part, considered the morphological features of the mature organism. This observation is evident in earlier works on *Solanum* taxonomy such as that of Isshiki et al. (2008), Karihaloo and Rai (1995), Karihaloo and Gottlieb (1995) and Oyelana and Ugborogho (2008). Unfortunately, these and many other studies based on morphological features have not totally resolved the difficulties associated with *Solanum* taxonomy. Discontinuous markers such as random fragment length polymorphism (RFLP), RAPD, AFLP and Single Nucleotide Polymorphism (SNP) have been useful in providing a measure of genetic distances to establish both the taxonomy and phylogenetic relationships among *Solanum* taxa (Karihaloo et al., 1995; Rodriguez et al., 1999; Poczai et al., 2008; Polignano et al., 2009).

The dendrogram constructed based on RAPD data obtained from all the four primers used reflected the morphological variation observed on the samples of eggplant and related species during their collections. It is evident from the dendrogram that collections originating from various parts of the study area did not form well-defined distinct clusters. They were interspersed with each other, indicating no association between RAPD pattern and the area of collection of accessions. This however, contrasted with the finding of Ge et al. (2013) who used SSR markers to obtain clusters among Chinese eggplant accessions that resulted in clades corresponding to the geographic divisions.

The present data revealed that *S. dasyphyllum* is more closely related to *S. macrocarpon* than to *S. melongena*

as evident in cluster I. This observation is in agreement with the findings of Mace et al. (1999), and Isshiki et al. (2008). These workers used AFLP markers to determine the taxonomic position of *S. dasyphyllum* and *S. macrocarpon* both of series Macrocarpa outside section Melongena which comprises *S. melongena*. According to Mace et al. (1999), this close relationship between *S. macrocarpon* and *S. dasyphyllum* is also supported by earlier findings of Jaeger (1986) who considered *S. macrocarpon* to be a domesticated modification of the wild plants known as *S. dasyphyllum*. Mace et al. (1999) stated further that Jaeger (1986) then assigned the wild form of a subspecies status under *S. macrocarpon*, the earlier name.

The occurrence of most samples of *S. gilo* and two samples (out of three) of *S. melongena* in cluster IV is an indication of close relatedness and possibility of having a common ancestor. Occurrence of *S. incanum* together with *S. melongena* still in cluster IV also indicates relatedness and probably progenitors from a common ancestral lineage. This observation of closeness between *S. incanum* and *S. melongena* supports the earlier finding of Sakata and Lester (1994) that used chloroplast DNA, Karihaloo et al. (1995), Furini and Wunder (2004) and Singh et al. (2006). In fact, Karihaloo et al. (1995) had earlier observed that wild forms of *S. incanum* are regarded as belonging to the same species as *S. melongena*. Singh et al. (2006) also stated that at the species level the cultivable type of *S. melongena* is more closely related to *S. incanum* followed by *S. viarum* whereas *S. surattence* and *S. nigrum* showed a closer association among themselves in comparison with the cultivated *S. melongena*. *S. scabrum* and *S. nigrum* occur together in both Clusters III and V, an indication of similarity between the two. The implication of this is that they are only distantly related to *S. melongena* and are more closely related to each other.

The level of polymorphism observed in the present study was high going by the coefficient of variation. The correlation coefficient 0.95 for the highest similarity between genotypes and the least 0.50 exhibited a good separation from a conserved region of the genome. This is an indication that eggplant *Solanum* has a wide and diverse genetic base.

These results agreed with those obtained by previous workers on *Solanum* e.g. Furini and Wunder (2004), Singh et al. (2006) and Levin et al. (2006). However, these are not in agreement with some earlier workers; for instance, Karihaloo and Gottlieb, (1995) studied variation among the cultivated and weedy taxa of *S. melongena* by allozymes and RAPD analyses; also Ge et al. (2013) examined the genetic diversity and relationships among eggplant accessions collected from seven areas in China using SSR markers. These authors observed little or moderate amount of genetic polymer-phism among the genotypes studied; even Karihaloo and Gottlieb (1995) suggested the existence of a very small gene pool from which the cultivated forms of *S. melongena* arose.

However, RAPD has some disadvantages which may affect the reliability of these results. For example, it is non-reproducible; they are dominant thereby making it impossible to distinguish between homozygosity and heterozygosity, and also RAPD results can be difficult to interpret. To overcome these, Ali et al. (2013) for example, analyzed the diversity of Chinese eggplant using inter-simple sequence repeat (ISSR) and RAPD procedures. The results showed that ISSR markers were more effective than RAPD markers for detecting genetic diversity.

Notwithstanding, the overall results of the present study were satisfactory enough in terms of their statistical values and concordance with previously published data. However, the accuracy of the clustering result may be increased by increasing the data and sample numbers of eggplant accessions as well as employing other better markers such as SSR, AFLP, ISSR, etc, in the analysis.

Conclusion

The study provides species database of the vegetable, *Solanum* and related species in Southwestern Nigeria and by extension in the country as a whole with emphasis on variation patterns which is a major contribution to global biodiversity information system. From the study also, it is evident that RAPD and other discontinuous markers can be made use of as a means of genetic distances to establish *Solanum* taxonomy as well as phylogenetic relationships among taxa. Detection of genetic differences and discrimination of genetic relationship between *Solanum* species are for sustainable utilization and conservation of plant genetic resources.

ACKNOWLEDGEMENTS

The authors are grateful to the University of Lagos (UNILAG) Central Research Committee (CRC) for providing Grant (No: CRC 2005/03, UNILAG) for the work. We also appreciate Pa Daramola and Pa Odewo of Lagos University Herbarium (LUH) for their immense contribution in identification and determination of the plant samples used in this study.

REFERENCES

Ali Z, Xu ZL, Zhang DY, He XL, Bahadur, Yi JX (2013). Molecular Diversity Analysis of Eggplant (*Solanum melongena*) Genetic Resources. Genet. Mol. Res. 10(2):1141-1155.

Baksh S (1979). Cytogenetic Studies of the F1 Hybrid of *Solanum incanum* L. x *S. melongena* L. Variety "Giant of Benares" *Euphytica* 28:793-800.

Bohs L (2001). Revision of *Solanum* Section *Cyphomandropsis* (Solanaceae). Syst. Bot. Monogr. 61:1-85.

Bohs L (2005). Major Clades in Solanum Based on ndhF Sequence Data. In: Keating, R. C., Hollowell, V. C. and Croat, T. B. (Eds.). A festschrift for William G. D'Arcy: the legacy of a taxonomist. St. Louis Monogr. Syst. Bot. Missouri Bot. Gard. 104:27-49.

Cao W, Scoles G, Hucl P, Chibbar RN (1999). The use of RAPD analysis to classify *Triticum* accessions. Theor. Appl. Genet. 98:602-607.

Cervera MT, Cabezas JA, Sancha JC, Martínez de Toda F, Martínez-Zapater JM (1998). Application of AFLPs to the Characterization of grapevine *Vitis vinifera* L. genetic resources. A case study with Accessions from Rioja (Spain). Theor. Appl. Genet. 97:51-59.

Daunay MC, Lester RN, Gebhardt C, Hennart JW, Jahn M, Frary A, Doganlar S (2001). Genetic Resources of Eggplant (*Solanum melongena* L.) and Allied Species: a New Challenge for Molecular Geneticists and Eggplant Breeders. In: van den Berg RG, Barendse GW, van der Weerden GM, Mariani C (Eds.). *Solanaceae V*: Advances in Taxonomy and Utilization. Nijmegen University Press, Nijmegen, The Netherlands. p. 354.

Dellaporta SL, Wood J, Hicks JB (1983). A plant minipreparation: version II. Pl. Mol. Bio. Rep. 1:19-21.

Furini A, Wunder J (2004). Analysis of eggplant (*Solanum melongena*) related germplasm: morphological and AFLP data contribute to phylogenetic interpretations and germplasm utilization. Theor. Appl. Genet. 108:197-208.

Gbile ZO, Adesina SK (1988). Nigerian *Solanum* species of economic importance. Ann. Missouri Bot. Gard. 75(3):862-865.

Ge H, Liu Y, Jiang M, Zhang J, Han H, Cheng H (2013). Analysis of genetic diversity and structure of eggplant populations (*Solanum melongena* L.) in China using simple sequence repeat markers. Sci. Hort. 162:71-75.

Hammond DH (1979). Growth Regulator Interactions on Morphogenesis in *Solanum* Species. In: Hawkes JG, Lester RN, Skelding AD. (eds). The biology and taxonomy of the Solanaceae. Linnean Society Symposium Series (7), London Academic Press. pp. 357-369.

Hassan SMZ, Lester RN (1990a). Crossability relationships and *in vitro* germination of F1 hybrids between *Solanum melongena* L. X *S. panduriforme* E. Meyer (*S. incanum* L. *sensu ampl*). SABRAO J. 22:65-72

Hassan SMZ, Lester RN (1990b). Comparative Micromorphology of the Seed Surface of *Solanum melongena* L. (Eggplant) and Allied Species. Pertanika 13:1-8.

Isshiki S, Iwataa N, Khana MMR (2008). ISSR variations in eggplant (*Solanum melongena* L.) and related *Solanum* species. Sci. Hort. 117:186-190.

Karihaloo JL, Brauner S, Gottlieb LD (1995). Random amplified polymorphic DNA variation in the eggplant *Solanum melongena* L. Theor. Appl. Genet. 90:767-770.

Karihaloo JL, Gottlieb LD (1995). Allozyme variation in the eggplant *Solanum melongena* L. (Solanaceae). Theor. Appl. Genet. 90:578-583.

Karihaloo JL, Rai M (1995). Significance of Morphological Variability in *Solanum insanum* L. (*sensu lato*). Plant Genet Res Newslett. 103: 24-26.

Kumar RS, Arumugam T, Anandakumar CR (2013) Genetic Diversity in Eggplant (*Solanum melongena* L.). Plant Gene Trait 4(2):4-8.

Lester RN, Hassan SMZ (1991). Origin and Domestication of the Brinjal Eggplant *Solanum melongena*, from *S. incanum*, in Africa and Asia. In: Hawkes JG, Lester RN, Nee M, Estrada-R N (Eds.), Solanaceae III: Taxonomy, Chemistry, Evolution. The Royal Botanic Gardens, Kew, Richmond, UK. p. 500.

Levin RA, Myers NR, Bohs L (2006). Phylogenetic relationships among the "Spiny *Solanums*" (*Solanum* subgenus *Leptostemonum*, Solanaceae). Amer. J. Bot. 93(1):157-169.

Levin RA, Watson K, Bohs L (2005). A Four-gene Study of Evolutionary Relationships in *Solanum* Section Acanthophora. Amer. J. Bot. 92(4):603-612.

Mace ES, Lester RN, Gebhardt CG (1999). AFLP analysis of genetic relationships among the cultivated eggplant *Solanum melongena* L., and wild relatives (Solanaceae). Theor. Appl. Genet. 99:626-633.

Miller JT, Spooner DM (1999). Collapse of species boundaries in the wild potato *Solanum brevicaule* Complex (Solanaceae sect. Petota): Molecular Data. Pl. Syst. Evol. 214:103-130.

Obute GC, Benjamin CN, Okoli BE (2006). Cytogenetic studies on some Nigerian species of *Solanum* L. (Solanaceae). AJB 5(9):689-692.

Ogundipe OT, Ajayi GO, Adeyemi TO (2009). Phytoanatomical and antimicrobial studies on *Gomphrena celosioides* Mart. (Amaranthaceae). Hamdard Medicus 51(3):146-156.

Okoli BE (1988). Cytotaxonomic studies of five West African species of *Solanum* L. (Solanaceae). Feddes Repert. 99(5-6):183-187.

Oyelana OA, Ugborogho RE (2008). Phenotypic variation of F1 and F2 populations from three species of *Solanum* L. (Solanaceae). AJB 7(14):2359-2367.

Poczai P, Taller J, Szabo I (2008). Analysis of phylogenetic relationships in the genus *Solanum* (Solanaceae) as revealed by RAPD markers. Plant Syst. Evol. 275:59-67.

Polignano G, Uggenti P, Bisignano V, Gatta CD (2009). Genetic divergence analysis in eggplant (*Solanum melongena* L.) and allied Species. Genet. Res. Crop Evol. 9:1- 11.

Rodriguez JM, Berke T, Engle L, Nienhuis J (1999). Variation among and within *Capsicum* species revealed by RAPD markers. Theor. Appl. Genet. 99:147-156.

Rohlf FJ (1996). NTSYS-pc: Numerical taxonomy and multivariate System, Version 2.02j. Exeter Software. Setauket, New York.

Sakata Y, Lester RN (1994). Chloroplast DNA diversity in eggplant (*Solanum melongena*) and its related species *S. incanum* and *S. marginatum*. Euphytica 80:1-4.

Singh AK, Singh M, Singh AK, Singh R, Kumar S, Kalloo G (2006). Genetic diversity within the genus *Solanum* (Solanaceae) as revealed by RAPD markers. Curr. Sci. 90(5):711-716.

Stedje B, Bukenya-Ziraba R (2003). RAPD Variation in *Solanum anguivi* Lam. and *S. aethiopicum* L. (Solanaceae) in Uganda. Euphytica 131:293-297.

Williams JKF, Kubelik AR, Livak KG, Rafalski JA, Tingey SV (1990). DNA Polymorphisms amplified by arbitrary primers is useful as genetic markers. Nucl. Acids Res. 18:6531-6535.

24

Iron starvation induces expression of a putative xylanase gene in *Salmonella enterica* subsp. *enterica* serovar Enteritidis mini-transposon*5 lac*Z1 mutants

Kazhila C. Chinsembu* and André Faul

Department of Biological Sciences, Faculty of Science, University of Namibia, P/Bag 13301, Windhoek, Namibia.

In this study, *Salmonella enterica* subsp. *enterica* serovar Enteritidis mini-transposon*5 lac*Z1 mutant strains induced under iron starvation were analyzed. Inverse polymerase chain reaction (IPCR) was used to isolate iron starvation-induced DNA fragments upstream of mini-Tn*5* (containing a promoter-less *lac*Z gene) insertion in *S. enterica* subsp. *enterica* serovar Enteritidis mutants showing β-galactosidase activity during growth in Fe^{3+}-deprived media. Out of ten mutant strains analyzed, four (Ez188, Ez477, Ez1819 and Ez2508) were induced during growth in the presence of the iron chelator, 2,2'-dipyridyl. IPCR products of Ez188 and Ez2508 were reamplified by nested PCR. Sequence analysis of the Ez188 PCR product revealed that a putative xylanase gene was induced under iron starvation conditions.

Key words: *Salmonella* mutants, iron starvation, induced, genes, putative xylanase.

INTRODUCTION

Salmonella are gram-negative bacteria in the family *Enterobacteriaceae*. They are ubiquitous pathogens found in humans and their livestock, wild mammals, reptiles, birds and even insects (Falkow and Mekalanos, 1990). *Salmonella* species causes gastroenteritis, septicaemia and enteric fever (Falkow and Mekalanos, 1990). *Salmonella enterica* subsp. *enterica* serovar Enteritidis causes many episodes of bacterial food contamination all over the world. *Salmonella*, like most other microorganisms, require iron as an essential element in a variety of metabolic and informational cellular pathways (Miethke and Marahiel, 2007). Iron is needed for support of biological processes such as transport and storage of oxygen, reduction of ribonucleotides and dinitrogen, activation and decomposition of peroxides and electron transport (Bagg and Neilands, 1987). More than 100 enzymes acting in primary and secondary metabolism possess iron-containing cofactors such as iron-sulfur clusters or heme groups (Miethke and Marahiel, 2007). Although iron is the fourth most abundant element in the earth's crust, it is largely unavailable to microorganisms because under aerobic conditions at neutral pH, Fe (II) is oxidised to Fe (III) forming an insoluble hydroxide (Neilands, 1982).

The human body contains 3 - 5 g of iron but the element is bound intracellularly in haemoglobin, heme, ferritin and hemosiderin and extracellularly to transferrin and lactoferrin (Payne, 1998). The average amount of free iron in human serum is estimated at 10^{-24} M (Raymond et al., 2003); a concentration far less than the 0.06 µg/ml minimum required for survival of *Salmonella* species (Payne, 1988). When the intracellular iron concentration drops below the threshold of about 10^{-6} M (which is critical for microbial growth), siderophores are produced to scavenge for iron (Dertz and Raymond, 2003, 2004). Most pathogens possess an arsenal of iron acquisition systems which provide an advantage for microbial multiplication in different compartments with changing iron source composition and pH conditions (Rouault, 2006). The host uses several mechanisms to withhold iron including several iron-binding proteins such as ovalbumin, lactoferrin, ferritin and transferring (Rouault, 2006). The low iron concentration in the host makes *Salmonella* increase the expression of siderophores and virulence determinants (Litwin and Calderwood, 1993). Thus, the ability of *Salmonella* species to acquire iron in the host is

_____

*Corresponding author. E-mail: kchinsembu2008@yahoo.co.uk, kchinsembu@unam.na.

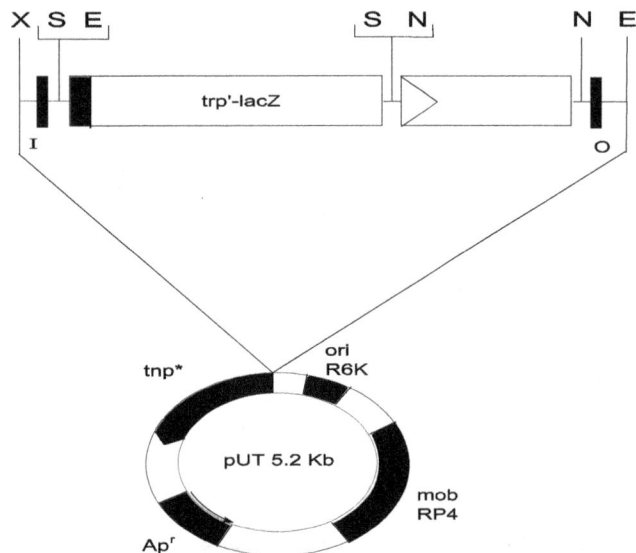

Figure 1. Organization of the promoter-probe minitransposon (mini-Tn5*lacZ*1) with its vector delivery plasmid (pUT/km). Details concerning the construction of pUT/km were described by Herrero et al. (1990). Adapted from de Lerenzo et al. (1990) and Herrero et al. (1990). Restriction sites: X, *Xba*I; S, *Sfi*I, E, *Eco*RI; N, *Not*I.

an essential adaptive component in pathogenesis (Litwin and Calderwood, 1993; Miethke and Marahiel, 2007).

Since the competition for iron is an important adaptive response that influences the outcome of the host-*Salmonella* interaction and pathogenesis, characterisation of iron starvation-induced genes would help elucidate the molecular basis underlying the induction of *Salmonella* genes during host infection. Iron starvation-induced proteins in *Salmonella* are also strong immunogens.

During the acute-phase reaction occurring upon inflammation, iron storage is enhanced by an interleukin-1- and tumor necrosis factor-dependent increase of macrophage ferritin pool (Jurado, 1997). During infection, the concentration of lactoferrin receptors on macrophages is also increased (Jurado, 1997). Given that the hypoferremic response and the development of siderophore binding proteins are an intriguing example of pathogen-directed host defense system (Jurado, 1997), isolation of iron starvation-induced promoters or their genes can yield practical benefits in the form of novel vaccines and treatments.

Salmonellae survive in phagocytic cells including macrophages. Thus iron starvation-induced promoters or genes may lead to the development of recombinant vaccines that produce antigens during host infection only; this can prevent problems associated with production of vaccines in culture (Chinsembu, 1996a, 1996b). There-fore, the aim of this study was to identify some of the *S. enterica* subsp. *enterica* serovar Enteritidis (*S. enteritidis*) genes induced under iron starvation conditions.

MATERIALS AND METHODS

Bacterial strains, media and growth conditions

S. enterica subsp. *enterica* serovar Enteritidis mini-transposon5 *lac*Z1 (mini-Tn5 containing the modified β-galactosidase reporter gene, *lac*Z1) mutants were used in this study. The mini-Tn5*lac*Z1 construct with its suicide vector delivery plasmid are shown in Figure 1. The mini-transposon contained a promoter-less *lac*Z gene. Construction of the vector was done according to Herrero et al. (1990) and that of *S. enterica* subsp. *enterica* serovar Enteritidis mini-Tn5*lac*Z1 mutants was done according to de Lorenzo et al. (1990). The eleven mutants were denoted as: Ez766, Ez033, Ez149, Ez188, Ez477, Ez1594, Ez1819, Ez1987, Ez2169, Ez2278 and Ez2508. The *lac*Z1 gene fusion in Ez766 was constitutively expressed; therefore Ez766 was used as a positive control. A *S. enterica* subsp. *enterica* serovar Enteritidis natural (wild type) isolate, 76 Sa88, was used as a negative control. Bacterial strains were streaked on Luria-Bertani (LB) medium (10 g/l bacto-tryptone, 5 g/l bacto-yeast extract, 10 g/l NaCl, pH adjusted to 7.4) containing 20 g/l bacto-agar. Overnight bacterial cultures were grown at 37 °C from single colonies in LB broth.

Screening assays

The strategy used to isolate iron starvation-induced DNA fragments from *S. enterica* subsp. *enterica* serovar Enteritidis mutants is shown in Figure 2. *S. enterica* subsp. *enterica* serovar Enteritidis mini-Tn5*lac*Z1 mutants including the negative (wild type) and positive (Ez766) controls were screened for the expression of β-galactosidase during growth on iron-deprived medium. Iron (III) was removed from LB-agar medium by adding 2,2'-dipyridyl to 2 mM final concentration. 100 µl of 250 mg/ml 5-bromo-4-chloro-3-indolyl-β-D-galactopyranoside (Xgal, 250 mg/ml in dimethyl formamide) were also added to the medium and mixed gently before pouring onto plates.

LB-gar and LB-agar/Xgal plates without 2,2'-dipyridyl were included as controls. Each mutant liquid culture (75 µl) was added to 75 µl of 80% glycerol in microtiter plate wells and transferred onto media using a sterilised replicator fork. Plates were allowed to dry for 5 min and incubated at 37 °C overnight. Iron starvation-induced β-galactosidase expression was scored on the basis of colour: deep blue colonies were strongly induced (+++), blue colonies were moderately induced (++), pale blue colonies were weakly induced (+) and white colonies were not induced (-). Screening tests were carried out in triplicate and 4 independent assays were done. Only mutants that produced deep blue colonies were analysed further.

*Taq*I digestion and Southern blotting

Total genomic DNA (5 µg) isolated from Ez188, Ez477, Ez1819, Ez2508 and *S. enterica* subsp. *enterica* serovar Enteritidis wild type (negative control) strains was digested with 20 units of *Taq*I restriction endonuclease at 65 °C for 2.5 h. Digested DNA (4 µg) was run on 0.8% agarose gel until all smaller fragments were (less than 800 bp) migrated into the electrophoresis buffer. Separated DNA fragments were blotted onto 0.45 µm pore size Hybond-N membranes (Amersham Biosciences). Membranes were prehybridised with 1 mg/ml of Herring sperm DNA and hybridised at 65 °C overnight with a probe of 25 ng of pUC18 DNA radio-labelled with [α^{32}P]dCTP (Amersham Biosciences). Hybridised filters were exposed to medical X-ray film (Fuji Photofilm, Japan) for 2 h at -70 °C with an amplifying screen.

Plasmid pUC18 contained *lac*Z sequences hence it was used as

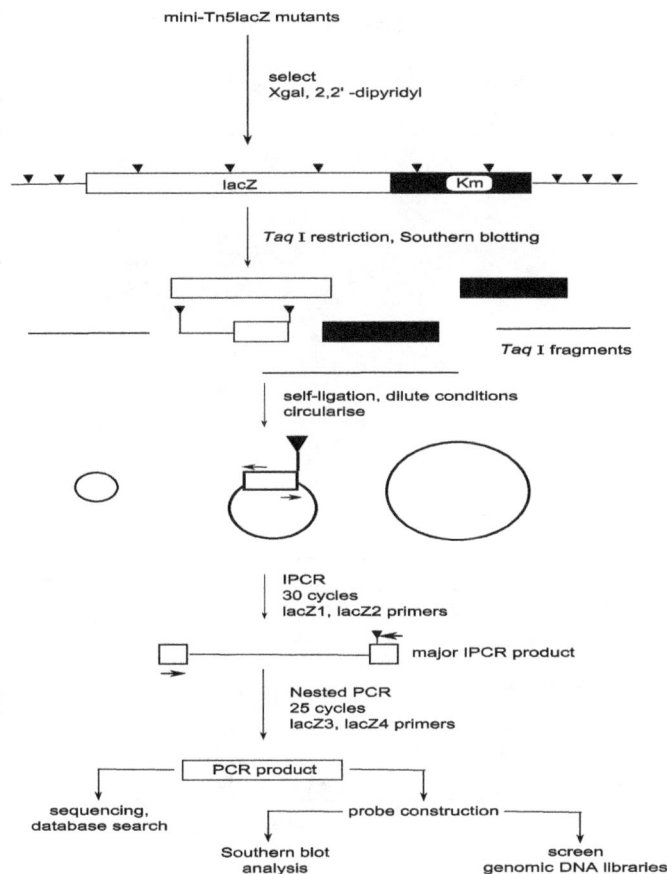

Figure 2. Flow chart showing the strategy used to analyse *S. enterica* subsp. *enterica* serovar Enteritidis mini-Tn*5lacZ*1 mutants. Mutant genomic DNA upstream of the lacZ region (arrowed box in circle) was amplified by IPCR and reamplified by nested PCR. The number and location of of TaqI restriction sites (black pointers) is hypothetical.

a probe to identify *Taq*I-digested fragments flanking the mini-transposon. *S. enterica* subsp. *enterica* serovar Enteritidis wild type did not contain sequences homologous to *lacZ*1 and *lacZ*2 primers and thereby allowed identification of *S. enterica* subsp. *enterica* serovar Enteritidis mini-Tn5 IPCR amplicons since such fragments were absent in the wild type control.

Self-ligation of *Taq*I-digested DNA

*Taq*I digested DNA (250 ng) of Ez188, EZ477, Ez1819, Ez2508, and *S. enterica* subsp. *enterica* serovar Enteritidis wild type (negative control) was self-ligated under dilute conditions in a volume of 50 µl (2 µl DNA [250 ng], 5 µl ligase buffer, 41 µl of sterile distilled water, and 5 µl [20 units] of T4 DNA ligase) at 12 - 14°C overnight.

Inverse polymerase chain reaction

Self-ligated DNA for Ez188, Ez477, Ez1819 and Ez2508 was amplified by inverse polymerase chain reaction (IPCR) using lacZ1 (5'-GGAATTCAAAGCGCCATTCGCCATTCAG-3') and lacZ2 (5'-

GGAAGCTTATGGCAGGGTGAAAGCAGG-3') primers. The IPCR mixture was as follows: 5 µl of 10× PCR buffer, 8 µl of deoxynucleotide triphosphates (NTPs, 1.25 mM), 1.5 µl of lacZ1 (20 mM), 1.5 µl of lacZ2 (20 mM), 24 µl of sterile distilled water, 10 µl of DNA template (50 ng), and 0.5 units of DNA polymerase. Amplification was done in the Perkin Elmer Gene Amp PCR system 9600 thermocycler in two stages: 94°C (denaturation), 53°C (annealing), and 72°C (extension) for three cycles of one minute each and a further 35 cycles of 30 s each. IPCR products (20 µl) were run on 0.8% agarose gel. Correct IPCR products flanking mini-Tn5 sequences were excised from gel and purified using the Jetsorb kit (Genomed).

Reamplification of IPCR products, sequencing, and database analysis

IPCR products (Ez188 and Ez2508) purified from gel were reamplified by nested PCR using lacZ3 (5'-CTAGACGTTTCCCAGTCACGAC-3') and lacZ4 (5'-GCGGATCCTTTCGGCGGTGAAATTATCG-3') primers. The PCR reaction mixture consisted of 2 µl of template DNA, 5 µl of 10 × PCR buffer, 8 µl of dNTPs (1.25 mM), 1.5 µl of lacZ3 (20 mM), 1.5 µl of lacZ4 (20 mM), 32 µl of sterile distilled water, and 0.5 units of DNA polymerase. Twenty five cycles (denaturation, 94°C / 20 s; annealing, 53°C / 20 s; extension, 72°C / 1 min) were performed using programmes 64 - 65 in the thermocycler described above. Amplicons were purified from gel, stored at -20°C and only Ez188 was later sequenced. Database search was done by using the BLASTN 2.2.23+ network service (Altschul et al., 1990). The search parameters were: program (blastn); word size (28); expect value (10); hitlist size (100); match/mismatch scores (1- 2); gapcosts (0, 0); low complexity filter (yes); filter string (L;m;), and genetic code (1).

RESULTS

Table 1 shows the induction of β-galactosidase activity in *S. enterica* subsp. *enterica* serovar Enteritidis mini-Tn*5lacZ*1 mutants cultured in iron-deprived medium. The results indicate that β-galactosidase expression in four *S. enterica* subsp. *enterica* serovar Enteritidis mutants (Ez188, Ez477, Ez1819 and Ez2508) was induced during growth in iron-deprived medium as demonstrated by the deep blue colonies. Ez033, Ez149 and Ez2278 produced pale blue colonies in LB-broth-Xgal. This showed that these mutants were weakly induced even in the presence of iron (III). Mutants that were weakly induced and non-induced in LB broth-dipyridyl-Xgal were not analysed further.

In order to identify *S. enterica* subsp. *enterica* serovar Enteritidis genomic DNA fragments containing the miniTn*5lacZ*1 construct, radio-labelled pUC18 DNA was hybridised to *Taq*I digested DNA in Southern analysis. Based on Southern blot results (Table 2), the 1.1 kb (Ez188), 1.0 kb (Ez2508), and 0.9 kb (Ez1819) fragments were selected as the correct IPCR products upstream of mini-Tn*5lacZ*1 (Figure 3). Ez477 DNA was difficult to amplify by IPCR. The 1.1 kb Ez188 and 1.0 kb Ez2508 IPCR amplicons were isolated from gel and reamplified by nested PCR (Figure 4). BLAST nucleotide results showed

Table 1. Iron starvation induced β-galactosidase expression in *S. enterica* subsp. *enterica* serovar Enteritidis mutants.

Bacterial strain	Iron starvation-induced β-galactosidase activity score		
	LB-dipyridyl-Xgal	LB-Xgal	LB
S. enterica subsp. *enterica* serovar Enteritidis wild type	-	-	-
Ez766	+++	+++	-
Ez033	+	+	-
Ez149	+	+	-
Ez188	+++	-	-
Ez477	++	-	-
Ez1594	+	-	-
Ez1819	+++	-	-
Ez1987	+	-	-
Ez2169	+	-	-
Ez2278	+	-	-
Ez2508	+++	-	-

+++ deep blue colonies; ++ moderately blue coloies; + pale blue colonies; - white colonies. Scores were taken after overnight incubation. Ez766 and *S. enterica* subsp. *enterica* serovar Enteritidis wild type were included as positive and negative controls, respectively.

Table 2. Sizes of expected IPCR fragments versus observed sizes in Southern and IPCR analyses.

Mutant strains	Molecular sizes of fragments in base pairs		
	Southern blot	Observed IPCR	Expected IPCR
Ez188	1,800	1,100	1,210
Ez477	2,500	*	1,910
Ez1819	1,600	900	1,010
Ez2508	1,700	1,000	1,110

Based on the length of fragments obtained in Southern analysis, sizes of expected IPCR products were calculated by subtracting 590 bp (unamplified 'core' region between lacZ1 and lacZ2 sequences) (Herrero et al., 1990; de Lorenzo et al., 1990) from corresponding fragments identified in Southern hybridisation. Observed and expected sizes of IPCR products were not exactly the same due to problems with resolution of fragments on gel. *Not amplified.

that Ez188 sequences had 96% similarity to a putative xylanase coding region of *S. enterica* subsp. *enterica* serovar Typhimurium strain LT2 (Figure 5).

DISCUSSION

The transposon Tn5 is a composite element in which inverted repeats of the mobile insertion element IS50 bracket a segment containing genes for resistance to antibiotics (De Bruijm and Lupski, 1984). It has been greatly utilised in the insertion mutagenesis of many gram-negative bacteria because of its advantages (Sasakawa and Yoshikawa, 1987; Simon et al., 1989). In order to simplify the generation of insertion mutants, a series of Tn5-derived mini-transposons were constructed by de Lorenzo et al. (1990). The mini-Tn5lacZ1 construct used for the generation of *S. enterica* subsp. *enterica* serovar Enteritidis mutants contained a promoterless β-galactosidase reporter gene and formed operon fusions

(Hererro et al., 1990). The enzyme β-galactosidase converted Xgal into a deep blue precipitate that was highly visible for scoring. Therefore, expression of β-galactosidase during growth in media containing the Fe^{3+} chelator (2,2'-dipyridyl) and Xgal suggested the existence of Fe^{3+} starvation-inducible promoter(s) upstream of the mini-transposon.

The strains Ez033, Ez149, Ez1819 and Ez2278 produced pale blue colonies on LB-Xgal without 2,2'-dipyridyl. These pale blue colonies were probably due to basal β-galactosidase induction. Ez188, Ez1819 and Ez2508 showed more β-galactosidase activity than Ez477. Increased activity in some mutants may have been due to increased transcription attributable to a stronger promoter. However, the same effect may mean that the intervening DNA region between the promoter and the lacZ gene may be shorter.

Since the 1990s, mutant fusions of lacZ and 2,2-dipyridyl have been widely employed in the analysis of iron regulated genes (Hassan and Sun, 1992; Postle,

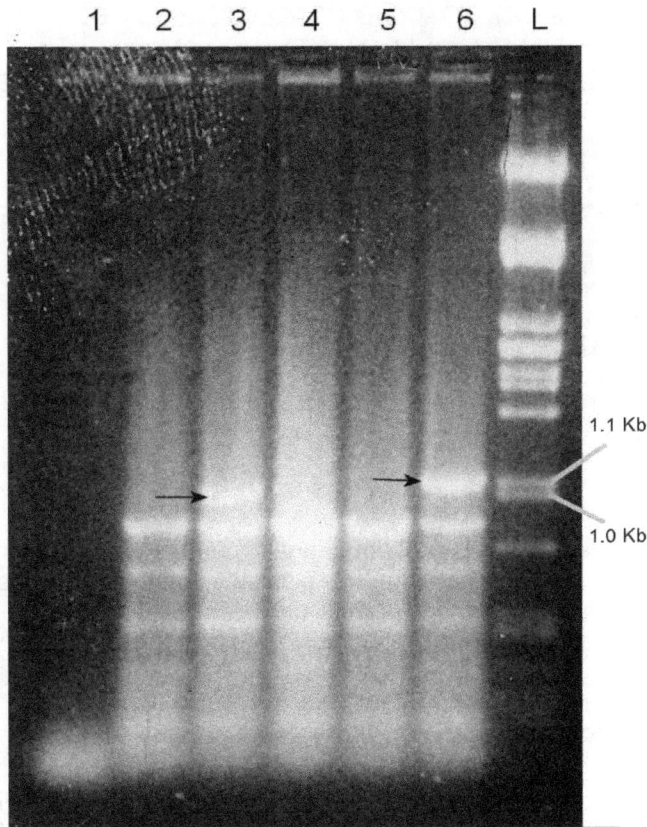

Figure 3. Agarose gel electrophoresis of IPCR products. Lane 1, IPCR no DNA control; lane 2, wild type DNA (negative control); lane 3, Ez 2508; lane 4, Ez1819; lane 5, Ez477; lane 6, Ez188; lane 7, 1 μg of λ/PstI DNA ladder (14.3 to 0.072 kb size range).

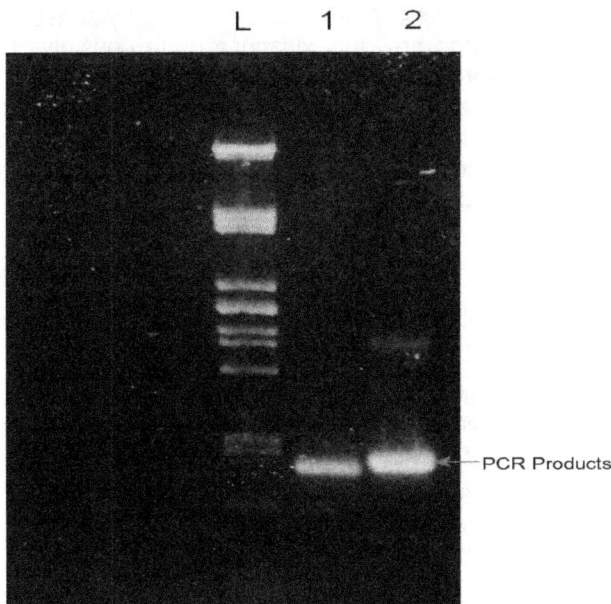

Figure 4. Agarose gel electrophoresis of PCR products. Lane 1, 1 μg of λ/PstI DNA ladder (14.3 to 0.072 kb size range); lane 2, Ez2508; lane 3, Ez188.

1990). Further, IPCR is the method of choice for amplification of unknown DNA sequences flanking a core region of known sequence (Triglia et al., 1988). The most important step in preparing the DNA for IPCR is the ligation reaction. DNA fragments are ligated in such a way that they form monomeric circles (Figure 2). The formation of circular templates is formed under dilute conditions whereby the total DNA fragment concentration is less than the local concentration of the one terminus in the neighbourhood of another on the same molecule (Dugaiczyk et al., 1975).

Iron starvation-inducible genes are under the regulation of the 17 kDa ferric uptake regulation (Fur) repressor. In *Salmonella*, Fur exerts control over a series of genes involved in the synthesis, excretion and recovery of the siderophore enterochelin (Hall and Foster, 1996).

An active Fur repressor (containing Fe^{2+}) also prevents RNA polymerase from binding to promoters of iron starvation-induced genes (Litwin and Calderwood, 1993). Thus, in order to allow transcription of iron regulated genes, 2,2'-dipyridyl inactivates the Fur-Fe^{2+} complex (Hassan and Sun, 1992; Hall and Foster, 1996). The chemical 2,2'-dipyridyl was also shown to induce anaerobic expression of a gene encoding an isozymic form of superoxide dismutase (Hassan and Sun, 1992; Dubrac and Touati, 2000). This suggests that regulation of iron starvation-induced genes by fur could be essential for biological defense against O_2 radicals and toxicity in phagocytes. In contrast, some bacterial genes such as the one encoding iron superoxide dismutase are positively regulated by Fur (Dubrac and Touati, 2000). Fur has pleitropic effects on *Salmonella* gene expression and cellular physiology (Miethke and Marahiel, 2007), but an alternative sigma factor, σ^{54}, was also shown to be responsible for recruiting core RNA polymerase during growth under low iron conditions (Cullen et al., 1994).

Our results showed that a putative xylanase gene (McClelland et al., 2001) in *S. enterica* subsp. *enterica* serovar Enteritidis mutants was induced during iron starvation conditions. Xylanases (endo-1,4-β-xylan xylanohydrolase) are *O*-glycoside hydrolases that catalyze the random hydrolysis of internal β-1,4-D-xylosidic linkages of xylan, a major component of hemicellulose found in wood, thus making these enzymes very biotechnologically significant in the paper industry (Collins et al., 2002). Xylanases are found in several fungal and bacterial species including *Salmonella*, *Escherichia coli* and *Pseudomonas* (Collins et al., 2002).

Xylanase genes were reported to be under a pH regulatory system (Eisendle et al., 2004). Iron starvation-induced genes such as those involved in siderophore biosynthesis and uptake are also governed by both iron and pH control (Eisendle et al., 2004). It was suggested that because uptake of siderophores required contransport with protons (Winkelmann, 2001); there was cross-talk between siderophore metabolism and pH sensing (Eisendle et al., 2004). Our results also suggest that xylanase genes

```
Query    7        GTCCCGGGATAGTCTCCCTTTCTGATGGACAATATGCATTAACGAGCGGGCAAACAGGTA   66
                  ||||||||||||||||||||||||||||||||||||||||||||||||||||||||||||
Sbjct    3292067  GTCCCGGGATAGTCTCCCTTTCTGATGGACAATATGCATTAACGAGCGGGCAAACAGGTA   3292126

Query    67       ACTTCTGCCCGGGTAATCCGTTACCCGACTTGATATTAACATTGTCATGAGGTGATATCA   126
                  ||||||||||||||||||||||||||||||||||| ||||||||||||||||||||||||
Sbjct    3292127  ACTTCTGCCCGGGTAATCCGTTACCCGACT-GATATTAACATTGTCATGAGGTGATATCA   3292185

Query    127      TTTG---GGCCAG--ATAATCATCCGGAACGGCGC   157
                  ||||   |||||||  |||||||||||||||||||
Sbjct    3292186  TTTGCCCGGCCAGCCATAATCATCCGGAACGGCGC   3292221
```

Figure 5. BLAST alignment results of query nucleotide sequences (Ez188 in red) and database sequences. Ez188 sequences reveal 96% similarity to a putative xylanase gene.

may be part of the molecular cross-talk network between iron starvation-induced genes and pH regulated genes. Previous observations that the role of fur in cellular physiology extends beyond that of regulatory iron utilization (Hall and Foster, 1996) also lend credence to suggestions of a cross-talk between siderophore metabolism and pH sensing. Fur was also found to regulate the acid tolerance response (ATR) in *Salmonella* (Hall and Foster, 1996).

Xylanase genes were also linked to the uptake of iron-siderophores in the phytopathogenic bacterium *Xylella fastidiosa* (Blanvillain et al., 2007). Bacterial uptake of iron-siderophore complexes depends on ABC-type transporters and in the case of gram negative bacteria, outer membrane receptors are the first gate of iron-siderophore recognition (Miethke and Marahiel, 2007). Energy for transport of iron-siderophore complexes is supplied by the TonB complex which acts as a transducer of a proton motive force (Miethke and Marahiel, 2007). TonB-dependent receptors (TBDRs) located in the outer membrane of gram negative bacteria are mainly known to transport iron-siderophore complexes into the periplasm (Blanvillain et al., 2007).

Interestingly, a carbohydrate utilization locus containing a TBDR ("CUT locus") is also involved in expression of xylanases (Blanvillain et al., 2007). The expression of genes encoding TBDRs involved in iron transport is regulated under iron depletion conditions and repressed under iron repletion conditions via the fur repressor (Blanvillain et al., 2007). In *X. fastidiosa*, two genes encoding enzymes putatively involved in xylan degradation were induced during heat stress (Koide et al., 2006). Many of the heat shock proteins were also induced under various environmental stress conditions such as nutrient starvation and changes in osmolarity or pH of the medium (Koide et al., 2006). Heat shock causes low oxygen pressure that leads to ferrous ion transport. Taken together, the induction of xylanases during iron starvation conditions is therefore well supported by previous studies. Xylan is present in agar from seaweeds (Collins et al., 2002). Therefore, it is plausible that in times of iron starvation, *Salmonella* may turn to xylan foraging as a source of carbon.

Conclusion

This study identified mini-Tn5lacZ1 *S. enterica* subsp. *enterica* serovar Enteritidis mutant fusions induced under iron deprivation conditions. IPCR enabled the amplification of genomic DNA upstream of the mini-Tn5 insertion in some mutants. Sequencing of nested PCR products revealed that iron starvation conditions induced a putative xylanase gene in *S. enterica* subsp. *enterica* serovar Enteritidis miniTn5lacZ1 mutants. Future studies will focus on the characterization of the xylanase gene and protein and how this xylanase gene possibly cross-talks with other genes that regulate responses to stresses such as pH and heat shock. Proteomic studies will also help to elucidate how whole biological systems with interconnected pathways respond to myriad environmental stimuli.

REFERENCES

Altschul SF, Gish W, Miller W, Meyers EW, Lipman DJ (1990). Basic local alignment search tool. J. Mol. Biol., 215: 403-410.

Bagg A, Neilands JB (1987). Ferric uptake regulation protein acts as a repressor, employing iron (II) as a cofactor to bind the operator of an iron transport operon in Escherichia coli. Biochemistry, 26: 5471-5477.

Blanvillain S, Meyer D, Boulanger A, Lautier M, Guynet C, Denance N, Vasse J, Lauber E, Arlat M (2007). Plant carbohydrate scavenging through TonB-dependent receptors: a feature shared by phytopathogenic and aquatic bacteria. PLos One 2: e224. Doi:10.1371/journal.pone.0000224.

Chinsembu KC (1996a). Molecular mechanisms of *Salmonella* pathogenicity: Review of recent advances. UNZA J. Sci. Technol., 1: 57-67.

Chinsembu KC (1996b). Iron and *Salmonella*: towards a new breed of vaccines. Zambian J. Med. Health Sci., 1: 2-4.

Collins T, Meuwis MA, Stals I, Claeyssens M, Feller G, Gerday C

(2002). A novel family 8 xylanase, functional and physicochemical characterization. J. Biol. Chem., 277: 35133-35139.

Cullen PJ, Foster-Hartnett D, Gabbert KK, Kranz RG (1994). Structure and expression of the alternative sigma factor, RpoN, in *Rhodobacter capsulatus*; physiological relevance of an autoactivated *nifU2-rpoN* super-operon. Mol. Microbiol., 11: 51-65.

De Bruijm FJ, Lupski JR (1984). The use of transposon *Tn5* mutagenesis in the rapid generation of correlated physical and genetic maps of DNA segments cloned into multicopy plasmids-review. Gene, 27: 131-149.

de Lorenzo V, Herrero M, Jakubzik U, Timmis NK (1990). Mini-*Tn5* transposon derivatives for insertion mutagenesis, promoter probing, and chromosomal insertion of cloned DNA in gram negative Eubacteria. J. Bacteriol., 172: 6568-6572.

Dertz EA, Raymond KN (2003). Siderophore and transferrins, pp. 141-168. In L Que, Jr, Tolman WE (eds) Comprehensive coordination chemistry II, vol.8. Elsevier Ltd, Philadelphia, PA.

Dertz EA, Raymond KN (2004). Biochemical and physical properties of siderophores, pp. 3-17. In JH Crosa, AR Mey, SM Payne (eds) Iron transport in bacteria. ASM Press, Washington, DC.

Dubrac S, Touati D (2000). Fur positive regulation of iron superoxide dismutase in Escherichia coli: functional analysis of the *sodB* promoter. J. Bacteriol., 182: 3802-3808.

Dugaiczyk A, Boyer HW, Goodman H (1995). Ligation of EcoRI endonuclease-generated DNA fragments into linear and circular structures. J. Mol. Biol., 96: 171-184.

Eisendle M, Oberegger H, Buttinger R, Illmer P, Haas H (2004). Biosynthesis and uptake of siderophores is controlled by the PacC-mediated ambient-pH regulatory system in *Aspergillus nidulans*. Eukaryotic Cell, 3(2): 561-563.

Falkow S, Mekalanos J (1990). The enteric Bacilli and Vibrios. In: Microbiology, BD Davis, R Dulbecco, HN Eisen and HS Ginsberg (eds), Lippincott Co., Philadelphia, pp. 561-587.

Hall HK, Foster JW (1996). The role of fur in the acid tolerance response of Salmonella typhimurium is biologically and genetically separable from its role in iron acquisition. J, Bacteriol., 178: 5683-5691.

Hassan H, Sun HC (1992). Regulatory roles of Fnr, Fur and Arc in expression of manganese-containing superoxide dismutase in Escherichia coli. PNAS USA 89: 3217-3221.

Herrero M, de Lorenzo V, Timmis NK (1990). Transposon vectors containing non-antibiotic resistance selection markers for cloning and stable chromosomal insertion of foreign genes in gram negative bacteria. J. Bacteriol., 172: 6557- 6567.

Jurado RL (1997). Iron, infections, and anaemia of inflammation. Clin. Infect. Dis., 25: 888-895.

Koide T, Vencio RZN, Gomes SL (2006). Global gene expression analysis of the heat shock response in the phytopathogen *Xylella fastidiosa*. J, Bacteriol., 188: 5821-5830.

Litwin CM, Calderwood SB (1993). Role of iron in regulation of virulence genes. Clin. Microbiol. Rev., 6: 137-149.

McClelland M, Sanderson KE, Speith J, Clifton SW, Latreille P, Courtney L, Porwollik S, Ali J (2001). Complete genome sequence of *Salmonella enterica* serovar Typhimurium LT2. Nature, 413: 852-856.

Miethke M, Marahiel MA (2007). Siderophore-based iron acquisition and pathogen control. Microbiol. Mole. Biol. Rev., 71(3): 413-451.

Neilands JB (1982). Microbial envelope proteins related to iron. Ann. Rev. Microbiol., 36: 285-309.

Payne SM (1988). Iron and virulence in the family Enterobacteriaceae. Crit. Rev. Microbiol., 16: 81-111.

Postle K (1990). Aerobic regulation of the *Escherichia coli* tonB gene by changes in iron availability and the fur locus. J. Bacteriol., 172: 2287-2293.

Raymond KN, Dertz EA, Kim SS (2003). Enterobactin: an archetype for microbial iron transport. PNAS USA 100: 3584-3588.

Rouault TA (2006). The role of iron regulatory proteins in mammalian iron homeostasis and disease. Nat. Chem. Biol., 2: 406-414.

Sasakawa C, Yoshikawa M (1987). A series of Tn5 variant drug-resistant markers and suicide vectors for transposon mutagenesis. Gene, 56: 283-288.

Simon R, Quandt J, Klipp W (1989). New derivatives of transposon Tn5 suitable for immobilisation of replicons, generation of operon fusions and induction of genes in gram negative bacteria. Gene, 80: 161-169.

Triglia T, Peterson MG, Kemp DT (1988). A procedure for in vitro amplification of DNA segments that lie outside the boundaries of known sequences. Nucl. Acids Res., 16: 8186.

Winkelmann G (2001). Siderophore transport in fungi, In G Winkelmann (ed) Microbial transport systems. Wiley-VCH, Weinheim, Germany. pp. 463-479.

25

Molecular assessment of *Ficus* species for identification and conservation

G. R. Rout* and S. Aparajita

Plant Biotechnology Division, Regional Plant Resource Centre, Bhubaneswar-751015, Orissa, India.

Ficus L. is widely distributed in all the climatic stages and is of great diversity. Molecular marker is used for identification of genetic resources. Inter-simple Sequence Repeat (ISSR) markers was used to assess the identification of 23 important *Ficus* species / varieties and determination of the genetic relationships among these species. Out of twenty one ISSR primers tested, five primers produced 116 detectable fragments, out of which 106 were polymorphic across the species/varieties. Each of the five primers produced fingerprint profile unique to each of the species/variety studied and thus could be solely used for their identification. Thirteen unique bands specific to nine species were detected. These may be converted into species-specific probes for identification purposes. Genetic relationships among these species/varieties were evaluated by generating a similarity matrix based on the Dice coefficient and the Unweighted Pair Group Method with Arithmetic Average (UPGMA) dendogram. The results showed a clear cut separation of the 23 *Ficus* varieties/species and were in broad agreement with the morphology. Both molecular and morphological markers will be useful for preservation of the *Ficus* germplasm.

Key words: Inter simple sequence repeat (ISSR), phylogeny, *Ficus* species, fingerprinting.

INTRODUCTION

Ficus Linn. (Moraceae) constitute one of the largest genera of flowering plants with about 800 species of free-standing trees, hemi-epiphytes and shrubs primarily occurring in subtropical and tropical regions world-wide. The genus is remarkable for the large variation in the habits of its species (Herre et al., 2008). Many species are cultivated for shade and ornament in gardens. Some species are serves as good source of latex/rubber. The fig is a very nourishing food and is used as an industrial product. It is very energizing, rich in vitamin, mineral elements, water and fats. In India, the most important among them are; *Ficus bengalensis*, *Ficus carica* and *Ficus elastica*. It is propagated by seeds/cuttings. Thus, the *Ficus* germplasm is characterized by a great diversity since a high number of varieties/ accessions has been identified (Mars, 2003). Due to industrialization and deforestation, many of the species/varieties are currently threatening. As a consequence, a lack of landraces has occurred during recent years and this constituted a constrain in the improvement of the *Ficus* varieties. Apart from its morphological, physiological and agronomic traits, the genetic analysis through molecular marker is a pre-requisite to having a deep insight of the genome organization in the wild species. Moreover, the precise number of cultivars/species is still unknown since problem of mislabeling are often detected (Gao et al., 2006). Therefore, it is imperative to establish strategies for the preservation of *Ficus* germplasm.

Many DNA based markers are also available to identify the varieties / species. These markers can be effectively used to answer the phylogenetic relationship between *Ficus* varieties/species (Khadari et al., 2005: Salhi-Hannachi et al., 2006; Chatti et al., 2007). ISSR overcomes many of the limitations faced by different marker system and has a higher reproducibility (Guasmi et al., 2006). *Ficus* species are represented by a large number of varieties/accessions which are facing genetic erosion. To save these genetic resources, the present investigation is to study the dentification and

*Corresponding author. E-mail: grrout@rediffmail.com.

hylogenetic analysis of 23 species/varieties of *Ficus* through ISSR markers.

MATERIALS AND METHODS

Plant materials

Important *Ficus* cultivars (Species/varieties) were collected from Botanical garden of Regional Plant Resource Center, Bhubaneswar and Chandaka Reserve Forest, Bhubaneswar, India. Identification of the species based on their morphological characteristics was confirmed in our laboratory and doubtful samples were excluded from the analysis. In total, twenty-three varieties / species were collected for this study. All the species/varieties were given the accession number that is: F1, *Ficus religiosa*; F2, *Ficus microcarpa* "Microcarpa"; F3, *Ficus elastica* "Rubra"; F4, *Ficus repens*; F5, *Ficus krisnae*; F6, *Ficus elmira*; F7, *Ficus petiolaris*; F8, *Ficus nervosa*; F9, *Ficus microcarpa* "Varigatae"; F10, *Ficus benjamina* "Varigatae"; F11, *Ficus elastica* "Robusta"; F12, *Ficus elastica* "varigatae"; F13, *Ficus mollis*; F14, *Ficus callosa*; F15, *Ficus rumphii* "Varigatae"; F16, *Ficus benjimina* "Nuda"; F17, *Ficus glomerata*; F18, *Ficus virens* "Glabella"; F19, *Ficus arnottina*; F20, *Ficus bhengalensis*; F21, *Ficus amplissima*; F22, *Ficus virens* "Virens"; F23, *Ficus macrophylla*.

DNA PREPARATION

DNA was isolated using CTAB method following the protocol of Doyle and Doyle (1990), with minor modification. 1.0 - 1.5 g of young non-scenence leaves were ground in liquid nitrogen. Then they were incubated in CTAB buffer (3% w/v CTAB, 100 mM Tris-HCl, 20 mM EDTA, 1.4M NaCl, 2% v/v β-mercaptoethanol, 2% w/v polyvinyl pyrrolidine, pH 8.0) for 2 h at 65°C. The homogenate was then extracted with an equal volume mixture of chloroform: isoamylalcohol (24:1) and centrifuged at 9000 rpm for 10 min. The upper aqueous layer was recovered and precipitated with prechilled isopropanol. The pallet was suspended with Tris-EDTA buffer (pH 8.0). The crude DNA was treated with RNase and incubated for 30 min at 37°C and again extracted with 1 volume phenol and subsequently with one volume of chloroform: isoamylalcohol (24:1). The supernatant were collected and precipitated with 3 M sodium acetate and prechilled ethanol. The DNA pellate was washed with 70% ethanol, dried and resuspended in TE buffer. The high molecular weight DNA was checked for quality and quantity electrophoretically using 0.8% agarose gel against a known amount of λ, DNA taken as standard. Twenty-one synthesized ISSR primers both 5'-anchored core motifs [$(AG)_8$ T, $(AG)_8$ C, $(AC)_8$ T, $(AC)_8$ G, $(GA)_8$ A, $(GA)_8$ T, $(CT)_8$ G, $(CT)_8$ A, $(TG)_8$ C, $(GT)_8$ C, $(GT)_8$ A], 3'-anchored core motifs [$CAG(CA)_7$, $TGG(AC)_7$, $ACA(TG)_6$, $ACG(GT)_7$] and Non-Anchored core motifs [$(ATG)_6$, $(ACTG)_4$, $(GACA)_4$, $(ACAG)_4$, $(GACAGATA)_2$] were used for the experiment. To ensure reproducibility, the primers generating no, weak, or complex patterns were discarded. A few well amplified fragments that were not reproducible across 2 replicate of DNA extraction were also discarded from analysis.

For ISSR study, the initial optimization of PCR was conducted including concentration of template DNA, primer, $MgCl_2$, number of PCR cycle and annealing temperature. The PCR reaction had a total volume of 25 μl containing 20 ng templates DNA, 100 mM each dNTPs, 20 ng of oligonucleotides synthesized primer (M/S Bangalore Genei, Bangalore, India), 2.5 mM $MgCl_2$, 1x Taq buffer [10 mM Tris-HCl) pH 9.0, 50 mM KCl, 0.01% gelatin] and 0.5 U Taq DNA polymerase (M/S Bangalore Genei, Bangalore, India).

DNA amplification was performed in a PTC -100 thermal cycler (M J Research, USA) programmed for a preliminary 5 min denaturation step at 94°C, followed by 40 cycles of denaturation at 94°C for 20 s, annealing temperature depending on the primer (50 - 56°C) for 30 s and extension at 72°C for 45 s and finally at 72°C for 5 min. Amplification products were separated alongside a low range molecular weight marker (M/S Bangalore Genei, India) on a 2% (w/v) agarose gel electrophoresis in 1x TAE (Tris Acetate- EDTA) buffer stained with ethidium bromide and visualized under UV light. Gel photographs were scanned through Gel Doc System (Gel Doc. 2000, BioRad, California, USA) and the amplification product sizes were evaluated using the software Quantity one (BioRad, California, USA).

Clearly defined ISSR bands that behave as dominant markers were scored for the presence (1) or absence (0) for all the species/cultivars and entered into a data matrix. The genetic relationships among the species/cultivar were determined by calculating the Dice coefficient, estimated as

$$S = 2N_{AB} / N_A + N_B$$

.Where N_{AB} is the number of amplified product common to both A and B. N_A and N_A corresponds to number of amplified product in A and B respectively. Diversity pattern were represented in the form of a dendrogram that was generated by subjecting the genetic similarity matrix to Unweighted Pair-group Method Arithmetic average (UPGMA) cluster analysis with software NTSYS-pc, Version 2.0 (Rohlf, 1995).

RESULTS AND DISCUSSION

The present investigation indicates that Twenty one ISSR primers were used for identification of genetic resources and to assess phylogenetic relationship among the *Ficus* species/ cultivars occurring in Orissa, India. Out of the 21 ISSR primers, nine primers generated clear multiplex banding profiles, among which five primers $(AG)_8$ C, $(GA)_8$ A, $(TG)_8$A, $(GT)_8$ C and $ACA(TG)$ produced the best ISSR profiles. The results showed that the most of the primers based on GA/AG and GT/TG dinucleotides core repeats generated good banding profiles. The amplification of ISSR markers was consistent across two replicate DNA extractions from three samples, over with 98% of scorable fragments reproducible. A higher concentration of $MgCl_2$ (2.5 mM) gave best results. This may be due to non-specific amplification because of reduced enzyme fidelity (Hopkins and Hilton, 2001). The concentration of $MgCl_2$ affects the specificity and yield of reaction by increasing the stringency of primer annealing or has a direct effect on Taq Polymerase (Saiki, 1989). The five primers selected, produced highly polymorphic band profiles. Furthermore, each of these primers produced fingerprints profiles unique to each of the variety. Therefore, each primer can be used separately to identify these accessions in the future. ISSR amplification for all samples resulted in multiple band fingerprint profile for the five selected ISSR primers (Figure 1). The average number of scorable fragments per primer was 23, with a range from 17 to 34, while the average number of polymorphic fragments per primer was 21 with a range from 15 to 31. Out of the 116 scorable fragments, 106

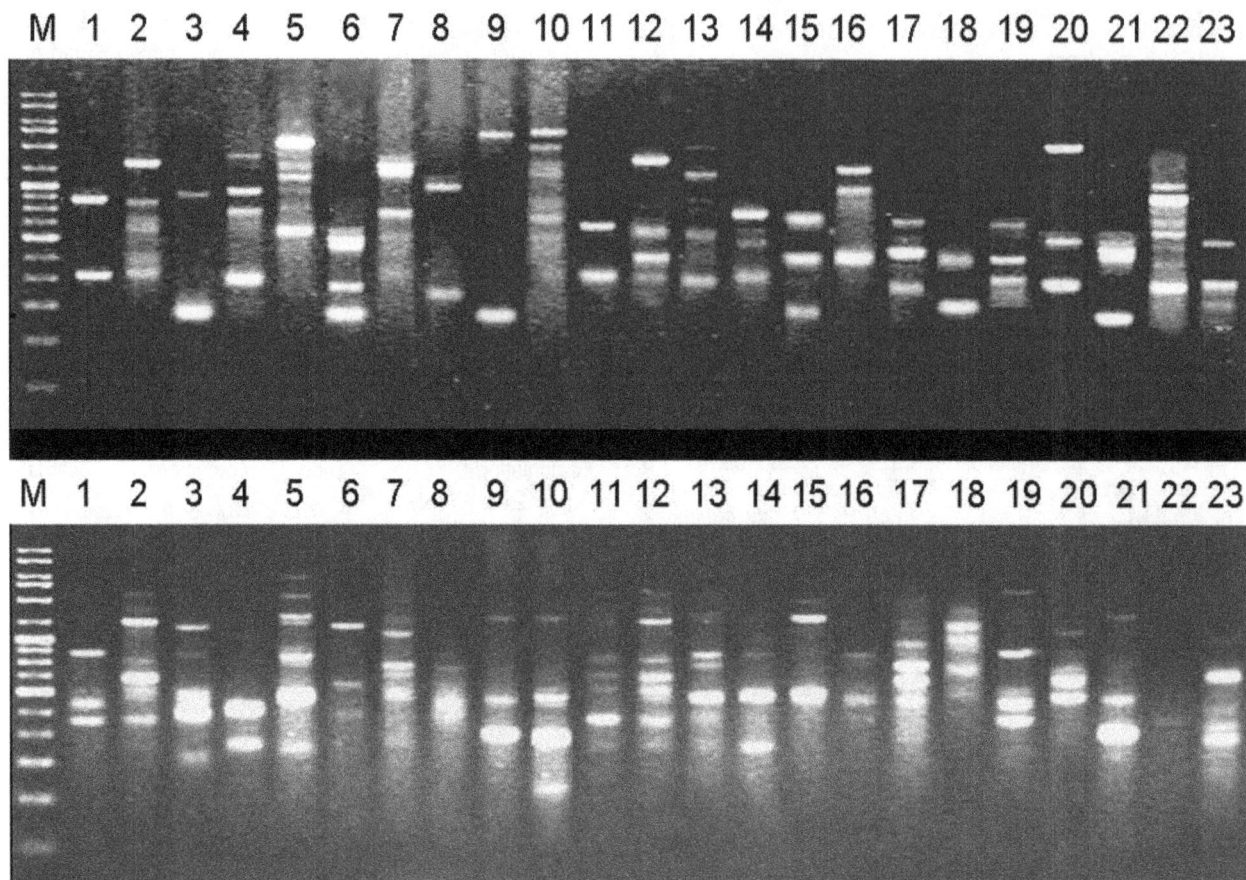

Figure 1. ISSR banding pattern of 23 species/ variety of *Ficus* using synthesis primer (GA)$_8$ A [A] and (TG)$_8$A [B]. M –low range DNA marker. **F1:** *Ficus religiosa,* **F2:** *Ficus microcarpa* "Microcarpa" , **F3:** *Ficus elastica* "Rubra", **F4:** *Ficus repens,* **F5:** *F. krisnae,* **F6:** *F. elmira,* **F7:** *F. petiolaris,* **F8:** F. nervosa, **F9:** *F. microcarpa* "Varigatae", **F10:** *F. benjamina* "Varigatae", **F11:** *F. elastica* "Robusta", **F12:** *F. elastica* "varigatae", **F13:** *F. mollis,* **F14:** *F. callosa,* **F15:** *F. rumphii* "Varigatae", **F16:** F. benjimina "Nuda", **F17:** *F. glomerata,* **F18:** *F. virens "Glabella",* **F19:** *F. arnottina,* **F20:** *F. bhengalensis,* **F21:** *F. amplissima ,* **F22:** *F. virens* "Virens", **F23:** *F. macrophylla.*

were polymorphic revealing 91.3% polymorphism across the 23 varieties/species studied. Thirteen ISSR loci were recorded as germplasm specific as they occurred in only a single species/cultivars. These may be developed into cultivar specific probes for identification purposes. In the study primers based on GA/AG and GT/TG dinucleotides core repeats generated good profiles, which seem to indicate that the more frequent microsatellite in *Ficus* contains the repeated dineucleotides (AG/GA)n and (GT/TG)n. These results are in accordance with the SSR assay. Vignes et al. (2006), developed 13 microsatellite from *Ficus insipida* by screening CT and GT genomic library. All markers revealed a broad cross-species affinity when tested in 23 other *Ficus* species. Saddouda et al. (2005) studied the genetic diversity of Tunisian fig (*F. carica*) and cultivar characterization using microsatellite markers. Khadari et al. (2005) identified eight microsatellite loci in common fig (*F. ciarica*) by screening a TC and TG enriched genomic library. Present study indicates that there was a distant variation

in DNA amplification of 23 species / varieties of *Ficus* collected from Orissa, India. The matrix calculated for all possible pair wise comparisons between accessions showed that the index value varied from a minimum of 0.15 between *F. repens* and *F. benjamina* "Variegata" to a maximum of 0.78 between *F bhengalensis* "Krishnae" and *F. bhengalensis* "Bhengalensis" with an average of 0.43 (Table 1). It seems to be the most divergent since they have presented the highest genetic distance value of 0.78. All the others ones have displayed different intermediate levels of similarity. Cluster analysis showing a dendrogram using Dice coefficient subjected to UPGMA method (Figure 2) were divided into four major clusters at 0.25 similarity level : A, B, C and D. Cluster-A had 17 accessions and cluster B, C and D, two species/accessions each. Cluster-D had two varieties/accessions of *F. virens* (Virens and Glabella) grouped together but with only 0.42 similarity. Similarly, two accessions of *F. benjamina (Varigatae* and Nuda) were also grouped together in Cluster-C with 0.41

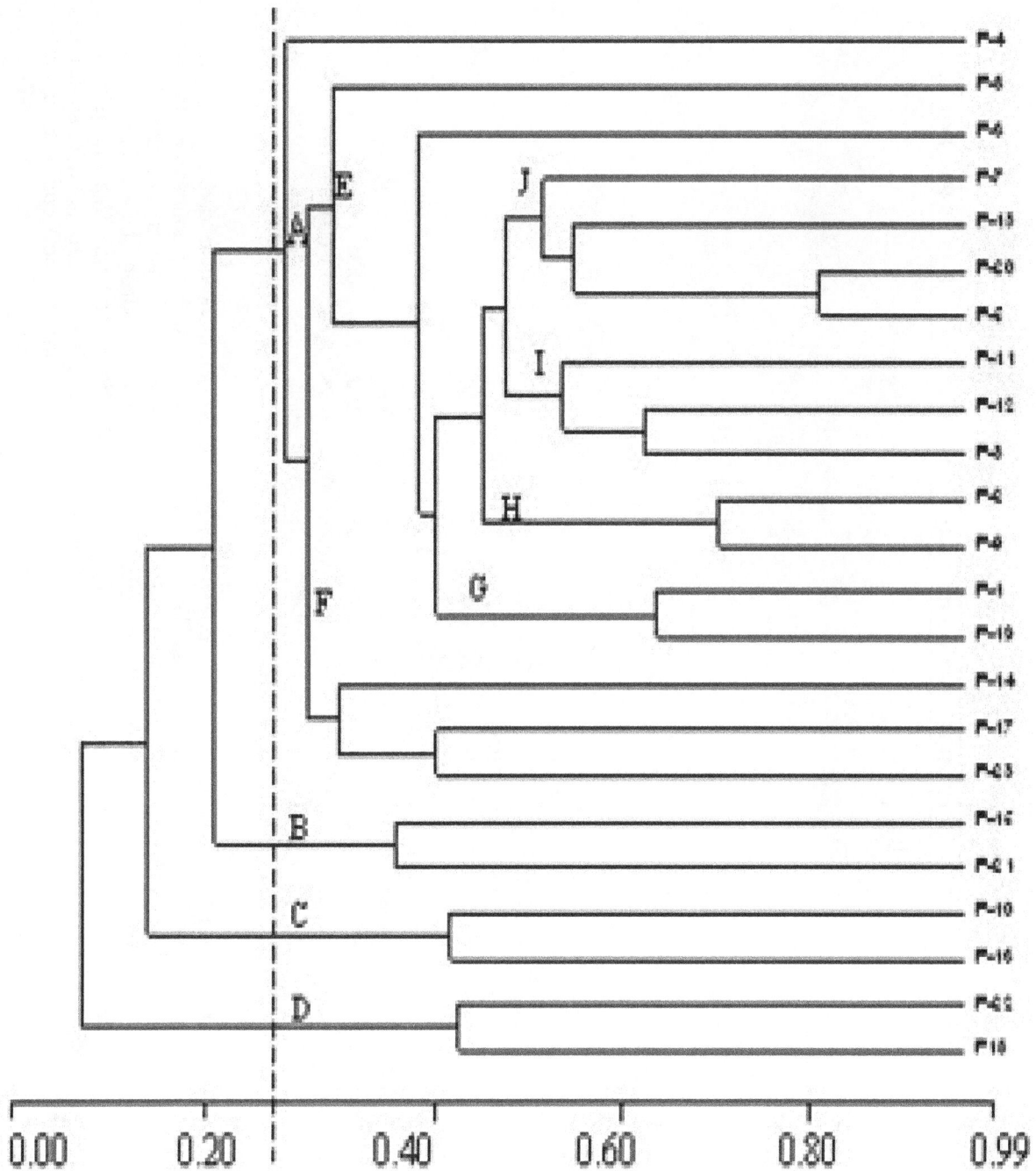

Figure 2. UPGMA dendrogram based on the similarity coefficient, showing the clustering pattern among the 23 *Ficus* species /variety F1 - F23 are the 23 *species*/variety.

similarities. *F. rumphii* "Varigata" had been grouped with *F. amplissima* at 0.36 similarity levels in Cluster B. In Cluster A, *F. repens* had isolated itself (0.27) as a unique cluster and the rest accessions had been further grouped into two sub-clusters-E and F at 0.30 levels. Sub-cluster-F had three species; *F. mollis*, *F. glomerata* and *F. callosa* whereas, sub-cluster- E had the rest 13

accessions. In sub-cluster-E, *F. nervosa* and *F. elmira* has separated themselves as single cluster at 0.32 and 0.39 similarity levels respectively. The rest 11 accessions in Cluster-E had been divided into four minor sub-clusters (G, H, I and J). *F. bhengalensis* "Bhengalensis" and *F. bhengalensis* "Krishnae" had been grouped together (0.78) in minor sub-cluster-J which was linked to *F. mollis*

Table 1. Similarity coefficient among the 23 accession of *Ficus* obtained from ISSR markers. (F1-F23 is the 23 accessions of *Ficus* as in figure 1).

	F-1	F-2	F-3	F-4	F-5	F-6	F-7	F-8	F-9	F-10	F-11	F-12	F-13	F-14	F-15	F-16	F-17	F18	F-19	F-20	F-21	F-22	F-23
F-1	1.00																						
F-2	0.61	1.00																					
F-3	0.62	0.68	1.00																				
F-4	0.25	0.33	0.38	1.00																			
F-5	0.69	0.70	0.64	0.31	1.00																		
F-6	0.44	0.45	0.55	0.32	0.52	1.00																	
F-7	0.58	0.65	0.59	0.35	0.73	0.48	1.00																
F-8	0.36	0.38	0.45	0.38	0.48	0.52	0.44	1.00															
F-9	0.48	0.75	0.57	0.37	0.62	0.52	0.62	0.44	1.00														
F-10	0.33	0.32	0.35	0.15	0.39	0.40	0.45	0.33	0.42	1.00													
F-11	0.46	0.65	0.67	0.36	0.61	0.43	0.6	0.39	0.56	0.36	1.00												
F-12	0.58	0.68	0.72	0.37	0.66	0.48	0.71	0.51	0.62	0.38	0.77	1.00											
F-13	0.50	0.60	0.58	0.48	0.60	0.48	0.65	0.41	0.56	0.28	0.55	0.63	1.00										
F-14	0.37	0.57	0.48	0.35	0.55	0.44	0.54	0.26	0.52	0.4	0.63	0.51	0.60	1.00									
F-15	0.34	0.44	0.44	0.16	0.33	0.42	0.27	0.29	0.43	0.21	0.29	0.31	0.25	0.26	1.00								
F-16	0.26	0.33	0.36	0.20	0.30	0.42	0.31	0.29	0.35	0.43	0.42	0.35	0.36	0.58	0.31	1.00							
F-17	0.50	0.47	0.51	0.27	0.48	0.64	0.47	0.56	0.47	0.38	0.39	0.5	0.44	0.43	0.40	0.46	1.00						
F18	0.23	0.24	0.30	0.26	0.24	0.41	0.26	0.33	0.29	0.33	0.27	0.26	0.24	0.41	0.55	0.36	0.32	1.00					
F-19	0.57	0.49	0.55	0.55	0.59	0.43	0.54	0.36	0.44	0.25	0.45	0.51	0.56	0.46	0.15	0.19	0.42	0.20	1.00				
F-20	0.49	0.52	0.55	0.47	0.78	0.56	0.68	0.47	0.56	0.37	0.43	0.57	0.52	0.47	0.31	0.27	0.49	0.25	0.51	1.00			
F-21	0.42	0.43	0.47	0.25	0.41	0.56	0.46	0.51	0.51	0.30	0.45	0.46	0.4	0.42	0.58	0.58	0.65	0.44	0.31	0.39	1.00		
F-22	0.35	0.33	0.37	0.16	0.38	0.54	0.44	0.55	0.44	0.30	0.30	0.40	0.3	0.32	0.48	0.40	0.53	0.48	0.19	0.6	0.67	1.00	
F-23	0.34	0.39	0.52	0.46	0.52	0.49	0.58	0.40	0.42	0.31	0.44	0.45	0.52	0.49	0.38	0.27	0.39	0.48	0.47	0.6	0.48	0.44	1.00

(0.56) and *F. petiolaris* (0.51) respectively. Similarly, *F. microcarpa* "Microcarpa" and *F. microcarpa* "Variegata" are grouped together with 75% similarity, owing to their morphological similarities with minor sub-cluster - H. All the three accessions of *F. elastica*, "Rubra", "Robusta" and "Variegata" are being grouped together in minor sub-cluster- I. Our results indicate that the genetic relationship among the *Ficus* accessions, inferred by ISSR markers, were in accordance with their morphological characters. *F. bhengalensis* "Krishnae" is morphologically similar to *F. bhengalensis* "Bhengalensis", but with most or all the leaves having the side lamina in its lower half reflexes and connate to form a cup. This modification may be due to bud-sport and most plants have been distributed in Orissa, India. Salhi-Hannachi et al (2005) assessed the genetic diversity in two Tunisian fig cultivars by using RAPD and ISSR markers. They compared two molecular markers for diversity study of Ficus. In conclusion, this study indicates that the information for species/varieties identification and the natural distribution of Ficus have been confirmed with ISSR markers. This analysis is quick and reproducible and can generate sufficient polymorphism to identify the Ficus species, although most ISSR alleles are dominant rather than co-dominant. This approach could subsequently be refined to include more markers and individual variety analysis for detailed characterization of *Ficus* taxa that would be essential for future breeding and tree improvement programme.

ACKNOWLEDGEMENTS

The authors wish to acknowledge the Department of Forest and Environment, Government of Orissa for providing the laboratory facility for conducting the experiment.

REFERENCES

Chatti K, Saddoud O, Salhi-Hannachi A, Mars Messaoud, Marrakchi M, Trifi M (2007). Analysis of genetic diversity and relationships in a Tunisian fig (*Ficus carica*) germplasm collection by Random Amplified Microsatellite Polymorphisms. J. Intgr. Plant Biol. 49: 386-391.

Doyle JJ, Doyle JL (1990). Isolation of plant DNA from fresh tissue. Focus, 12(1): 13-15.

Gao J, Zhang S, Qi L, Zhang Y, Wang C, Song W, Han S (2006). Application of ISSR markers to fingerprinting of elite cultivars (varieties / clones) from different sections of the genus *Ficus* L. Silvae Genetica. 55: 1-6.

Guasmi F, Ferchichi A, Farés K, Touil L (2006). Identification and differentiation of *Ficus carica* L. cultivars using inter simple sequence repeat markers. Afr. J. Biotechnol. 5 (15): 1370-1374.

Herre EA, Jander KC, Machado CA (2008). Evolutionary Ecology of Figs and their associates: Recent Progress and outstanding puzzles. Ann. Rev. Ecol. Evol. Syst. 39:439-458.

Khadari B, Hochu I, Santoni S, Kjellberg F (2005). Identification and characterization of microsatellite loci in the common fig (*Ficus carica* L.) and representative species of the genus *Ficus*. Mol. Ecol. Notes 1:191–193

Mars, M (2003). Fig (*Ficus carica* L.) genetic resources and breeding. Acta Hort. 605:19-27.

Rohlf FJ (1995) NTSYS-pc numerical taxonomy and multivariate analysis system. Version 1.80, Exter Software, Setauket, New York.

Salhi-Hannachi A, Chatti K, Mars M, Marrakchi M, Trifi M (2005). Comparative analysis of genetic diversity in two Tunisian collections of fig cultivars based on random amplified polymorphic DNA and Inter Simple Sequence Repeats fingerprints. Genetic Resources and Crop Evolution, 52: 563-573.

Salhi-Hannachi A, Chatti K, Saddoud O, Mars M, Rhouma A, Marrakchi M, Trifi M (2006) Genetic diversity of different Tunisian fig (*Ficus carica* L.) collection revealed by RAPD fingerprints. Hereditas. 143:15-22.

Saiki RK (1989). The design and optimization of the PCR. In: Erlich HA, editor. PCR technology: Principles and applications for DNA amplification. New York: Stockton Press. pp. 7-16.

Saddouda O, Salhi-Hannachia A, Chattia K, Marsb M, Rhoumac A, Marrakchia M, Tnifia M (2005). Tunisian fig (*Ficus carica*) genetic diversity and cultivar characterization using microsatellite markers. Fruits. 60:143-153.

26

Two genotypes of *Xanthomonas oryzae* pv. *oryzae* virulence identified in West Africa

Onasanya Amos[1], M. M. Ekperigin[2], A. Afolabi[4], R. O. Onasanya[2], Abiodun A. Ojo[1] and I. Ingelbrecht[3]

[1]Department of Chemical Sciences, Afe Babalola University, Ado-Ekiti, Ekiti State, Nigeria.
[2]Federal University of Technology, Akure, PMB 704, Akure, Ondo State, Nigeria.
[3]Institute of Plant Biotechnology for Developing Countries, Ghent University, K. L. Ledeganckstraat 35, 9000 Ghent, Belgium.
[4]Biotechnology and Genetic Engineering Advanced Laboratory, Sheda Science and Technology Complex, PMB 186, Garki, Abuja, Nigeria

Bacterial leaf blight (BLB) caused by *Xanthomonas oryzae* pv. *oryzae* (*Xoo*), is a very destructive rice disease worldwide. The aim of the present study was to examine if the *Xoo* virulence pathotypes obtained using phenotypic pathotyping could be confirmed using molecular approach. After screening of 60 Operon primers with genomic DNA of two *Xoo* isolates (virulent pathotype, *Vr* and mildly virulent pathotype, *MVr*), 12 Operon primers that gave reproducible and useful genetic information were selected and used to analyze 50 *Xoo* isolates from 7 West African countries. Genetic analysis revealed two major *Xoo* virulence molecular type (Mt) which were *Mta* and *Mtb* with *Mta* having two subgroups (*Mta1* and *Mta2*). *Mta1* (*Vr1*) subgroup genotype has occurrence in six countries and *Mta2* (*Vr2*) in three countries while *Mtb* genotype characterized mildly virulence (*MVr*) *Xoo* isolates present in five countries. The study revealed possible linkage and correlation between phenotypic pathotyping and molecular typing of *Xoo* virulence. Durable resistance rice cultivars would need to overcome both *Mta* and *Mtb Xoo* virulence genotypes in order to survive after their deployment into different rice ecologies in West Africa.
Key words: Bacterial leaf blight, *Xanthomonas oryzae* pv. *oryzae* (*Xoo*), *Xoo* virulence pathotype, molecular typing, genomic DNA, Operon primer, *Xoo* virulence genotype, *Xoo* pathogen migration, West Africa.

INTRODUCTION

Rice is one of the most widely cultivated food crop worldwide, but its production is constrained by fungal, bacterial and viral diseases. Bacterial leaf blight (BLB) caused by *Xanthomonas oryzae* pv. *oryzae* (*Xoo*), is a very destructive rice disease and its incidence has been reported from different parts of Asia, northern Australia, Africa and USA (Adhikari et al., 1995; Sere et al., 2005; Jiang et al., 2006). In West Africa, BLB disease incidence ranged from 70 to 85% and yield loss ranged from 50 to 90%, indicating a wide spread of BLB disease in farmers'

fields (Sere et al., 2005). Some selected *Xoo* isolates have shown high level of pathogenicity and virulence on the cultivated rice varieties (Sere et al., 2005; Onasanya et al., 2009; Dewa et al., 2011). Crop loss assessment studies have revealed that this disease reduces grain yield to varying levels, depending on the stage of the crop, degree of cultivar susceptibility and to a great extent, the conduciveness of the environment in which it occurs (Savary et al., 2006). The severity and significance of damage caused by infection have necessitated the development of strategies to control and manage the disease, so as to reduce crop loss and to avert an epidemic. The identification and characterization of major genes for qualitative resistance and polygenic factors controlling quantitative resistance have contributed a great deal to the success in breeding resistant cultivars and their deployment (Chen et al., 2002). Recent research has provided considerable evidence that the deployment of bacterial antagonists to *Xoo* might be an effective strategy, bringing about disease suppression by biological control (Gnanamanickam, 2009).To understand the epidemiology and ecology of *Xoo* pathogens and their potential for virulence change, various phenotypic characters as well as molecular markers have been used in studies of *Xoo* pathogen population structure (Jiang et al., 2006). Identification and classification of bacteria are normally carried out by morphology, nutritional requirements, antibiotic resistance, isozyme comparisons, phage sensitivity (Akanji et al., 2011; Chaudhary et al., 2012) and more recently by DNA based methods, particularly rRNA sequences (Anzai et al., 2002; Chandrashekar et al., 2012), strain-specific fluorescent oligonucleotides (Zhao et al., 2007) and the polymerase chain reaction (PCR) (Akanji et al., 2011). Several repetitive elements found in the *Xoo* pathogen have been used as probes in restriction fragment length polymorphism (RFLP) analysis (Gonzalez et al., 2007). However, for the large number of samples needed for ecological and virulence studies, a simpler and cheaper technology is required. PCR is increasingly becoming an important tool in population biology, because of its simplicity and potential to rapidly screen a large number of samples with a minimal amount of DNA.

In West Africa, several *Xoo* genetic studies have been conducted and different *Xoo* pathotypes identified but little information is available on *Xoo* virulence genotypes population structure and distribution (Basso et al., 2011). The virulence pathotypes of several *Xoo* isolates from West African countries based on cultivars reactions has been determined (Basso et al., 2011; Dewa et al., 2011). The main goal of this study was to determine *Xoo* virulence genotypes using the characterized *Xoo* isolates virulence pathotypes identified by Onasanya et al. (2009) using random amplified polymorphic DNA polymerase chain reaction (RAPD-PCR) assays. The identification and differentiation of different *Xoo* virulence genotypes and distribution in West Africa would greatly help in rice breeding improvement programs aiming at the effective development of rice cultivars with durable resistance to BLB disease.

MATERIALS AND METHODS

Bacterial isolates

Fifty *X. oryzae* pv. *oryzae* (*Xoo*) isolates (Table 1) used in this study

were from Onasanya et al. (2009). The identity of all the fifty *Xoo* isolates had been confirmed by oxidative biochemical test as well as their virulence pathotypes (Onasanya et al., 2009).

Isolates propagation

BLB isolates were first propagated using a modified procedure developed by Akanji et al. (2011). Nutrient broth (75 ml; pH 7.5) was prepared inside a 100 ml conical flask. Each *Xoo* isolate (100 µl) from storage was transferred into 50 ml of nutrient broth and kept under constant shaking at 30°C for 24 h for bacterial growth. The bacterial cell was removed by centrifugation, washed with 0.1

mM Tris-EDTA (pH 8.0) and kept at -20°C for DNA extraction.

Genomic DNA extraction

DNA extraction was according to Onasanya et al. (2003) with some modification. 0.3 g of washed bacterial cell was suspended in 200 µl of cetyl trimethylammonium bromide (CTAB) buffer (50 mM Tris, pH 8.0; 0.7 mM NaCl; 10 mM EDTA; 2% hexadecyltrimethylammonium bromide; 0.1% 2-mercaptoethanol), followed by 100 µl of 20% sodium dodecyl sulfate and incubated at 65°C for 20 min. DNA was purified by two extractions with chloroform and precipitated with -20°C absolute ethanol. After washed with 70% ethanol, the DNA was dried and resuspended in 200 µl of sterilized distilled water. DNA concentration was measured using DU-65UV spectrophotometer (Beckman Instruments Inc., Fullerto CA, USA) at 260 nm. DNA quality was checked on a 1% agarose gel in Tris-Acetate-EDTA (TAE) buffer (45 mM Tris-acetate, 1 mM EDTA, pH 8.0) after electrophoreses.

RAPD-PCR analysis

This analysis was performed according to Akanji et al. (2011). DNA primers used were purchased from Operon Technologies (Alameda, CA, USA) and each was ten nucleotides long. Two concentrations of each DNA (25 and 95 ng per reaction) were used to test reproducibility and eliminate sporadic amplification products from the analysis. Sixty primers (OPP, OPQ, OPR, OPS, OPT, OPV, OPX and OPY series) were screened with DNA of two *Xoo* isolates (virulence, *Vr*, and mildly virulence, *MVr*, isolates) for their ability to amplify the *Xoo* genomic DNA. Primers that gave useful polymorphisms were selected and used in amplifying the DNA from all *Xoo* isolates. Amplifications was performed in 25 µl reaction mixture consisting of genomic DNA, reaction buffer (Promega), 100 µM each of dATP, dCTP, dGTP and dTTP, 0.2 µM Operon random primer, 2.5 µM MgCl₂ and 1U of Taq polymerase (Boehringer, Germany). A single primer was used in each reaction. Amplification was performed in a thermowell microtiter plate (Costa Corporation) using a MJ Research programmable thermal controller. The cycling program was (i) 1 cycle of 94°C for 3 min; (ii) 45 cycles of 94°C for 1 min for denaturation, 40°C for 1 min for annealing of primer and 72°C for 2 min for extension; and (iii) a final extension at 72°C for 7 min. Amplification products were maintained at 4°C until electrophoresis.

Electrophoresis of PCR products

The amplification products were resolved by electrophoresis in a

Table 1. Identity of *X. oryzae* pv. *oryzae* isolates used for the study.

S/N	Isolates code*	Host plant	Country
1	XN-1	D52-37	Niger
2	XN-2	D52-37	Niger
3	XN-3	IR15296829	Niger
4	XN-4	IR15296829	Niger
5	XN-5	WITA 8	Niger
6	XN-6	WITA 8	Niger
7	XB-7	Local	Benin
8	XB-8	Local	Benin
9	XB-9	Local	Benin
10	XB-10	Local	Benin
11	XB-11	Local	Benin
12	XNG-12	WITA9	Nigeria
13	XNG-13	WITA9	Nigeria
14	XNG-14	WITA 4	Nigeria
15	XNG-15	WITA 4	Nigeria
16	XNG-16	WITA 8	Nigeria
17	XBF-17	TS2	Burkina Faso
18	XBF-18	TS2	Burkina Faso
19	XBF-19	FKR14	Burkina Faso
20	XBF-20	FKR19	Burkina Faso
21	XBF-21	FKR14	Burkina Faso
22	XBF-22	Chinese	Burkina Faso
23	XM-23	Adventices	Mali
24	XM-24	Kogoni	Mali
25	XM-25	Kogoni	Mali
26	XM-26	Kogoni	Mali
27	XM-27	Kogoni	Mali
28	XM-28	Kogoni	Mali
29	XM-29	Jamajigi	Mali
30	XM-30	Nionoka	Mali
31	XG-31	Weed	Guinea
32	XG-32	Weed	Guinea
33	XG-33	Weed	Guinea
34	XG-34	Local	Guinea
35	XG-35	Local	Guinea
36	XG-36	Local	Guinea
37	XG-37	Local	Guinea
38	XG-38	Local	Guinea
39	XG-39	Local	Guinea
40	XG-40	Local	Guinea
41	XTG-41	Local	The Gambia
42	XTG-42	Local	The Gambia
43	XTG-43	Local	The Gambia
44	XTG-44	Local	The Gambia
45	XTG-45	Local	The Gambia
46	XTG-46	Local	The Gambia

Table 1. Cont.

47	XTG-47	Local	The Gambia
48	XTG-48	Local	The Gambia
49	XTG-49	Weed	The Gambia
50	XTG-50	Weed	The Gambia

* = *X. oryzae* pv. *oryzae* isolates obtained from Onasanya et al. (2009).

1.4% agarose gel using Tris-Acetate-EDTA (TAE) buffer (45 mM Tris-acetate, 1 mM EDTA, pH 8.0) at 100 V for 2 h. A 1 kb ladder (Life Technologies, Gaithersburg, MD, USA) was included as molecular size marker. Gels were visualized by staining with ethidium bromide solution (0.5 µg/ml) and banding patterns were photographed over UV light using using UVP-computerized gel photo documentation system.

Cluster analysis

Positions of scorable amplified DNA bands were transformed into a binary character matrix ("1" for the presence and "0" for the absence of a band at a particular position). Pairwise distance matrices were compiled by the Numerical Taxonomy System (NTSYS) 2.0 software (Rohlf, 2000) using the Jaccard coefficient of similarity (Ivchenko and Honov, 1998). Cluster dendrogram was created by unweighted pair-group method arithmetic (UPGMA) cluster analysis (Éena et al., 2009). Principal component analysis with GGEbiplot was carried out on 50 *Xoo* isolates using genetic data generated from twelve Operon primers (Ebdon and Gauch, 2002).

RESULTS

Genetic analysis of fifty *X. oryzae* pv *oryzae* (*Xoo*) isolates from West Africa have been carried out. After screening of 60 Operon primers with genomic DNA of two *Xoo* isolates (virulent pathotype, *Vr* and mildly virulent pathotype, *MVr*), only 12 primers gave reproducible polymorphism and useful genetic information that differentiated the fifty *Xoo* isolates. Amplification with the 12 primers generated 210 bands from which 136 (64.8%) of them were polymorphic (Table 2) with sizes ranging between 0.5 and 4.0 kb (Figure 1). Using the 136 RAPD markers (Table 2) in cluster and principal component analyses revealed two major (*Mta* and *Mtb*) molecular typing (Mt) virulence genotypes among fifty *Xoo* isolates (Figures 2 and 3). *Mta* genotype was made up of 42 virulence (*Vr*) *Xoo* isolates with two subgroup genotypes (*Mta1* and *Mta2*). *Mta1* (*Vr1*) subgroup genotype was typical of 25 *Xoo* isolates with 50% occurrence in six countries (Niger, Benin Republic, Nigeria, Burkina Faso, Mali and Guinea) (Table 3). *Mta2* (*Vr2*) subgroup genotype was typical of 17 *Xoo* isolates with 34% occurrence in three countries (Mali, Guinea and The Gambia) (Table 3). *Mtb* genotype characterized 8 mildly virulence (*MVr*) *Xoo* isolates with 16% occurrence in five

Table 2. Oligonucleotide primers that showed genetic polymorphism among the *X. oryzae* pv. *oryzae* isolates using random amplified polymorphic DNA polymerase chain reaction analysis.

Operon primer	Nucleotide sequence 5' to 3'	No. of fragments amplified	No. of polymorphic bands	Polymorphism (%)
OPP-17	TGACCCGCCT	18	16	88.9
OPP-18	GGCTTGGCCT	14	11	78.6
OPR-07	ACTGGCCTGA	20	11	55.0
OPS-08	TTCAGGGTGG	23	13	56.5
OPS-10	ACCGTTCCAG	20	13	65.0
OPS-13	GTCGTTCCTG	16	9	56.3
OPT-09	CACCCCTGAG	16	10	62.5
OPT-12	GGGTGTGTAG	13	7	53.8
OPT-15	GGATGCCACT	18	10	55.6
OPV-05	TCCGAGAGGG	19	12	63.2
OPY-06	AAGGCTCACC	16	11	68.8
OPY-08	AGGCAGAGCA	17	13	76.5
Total		210	136	64.8

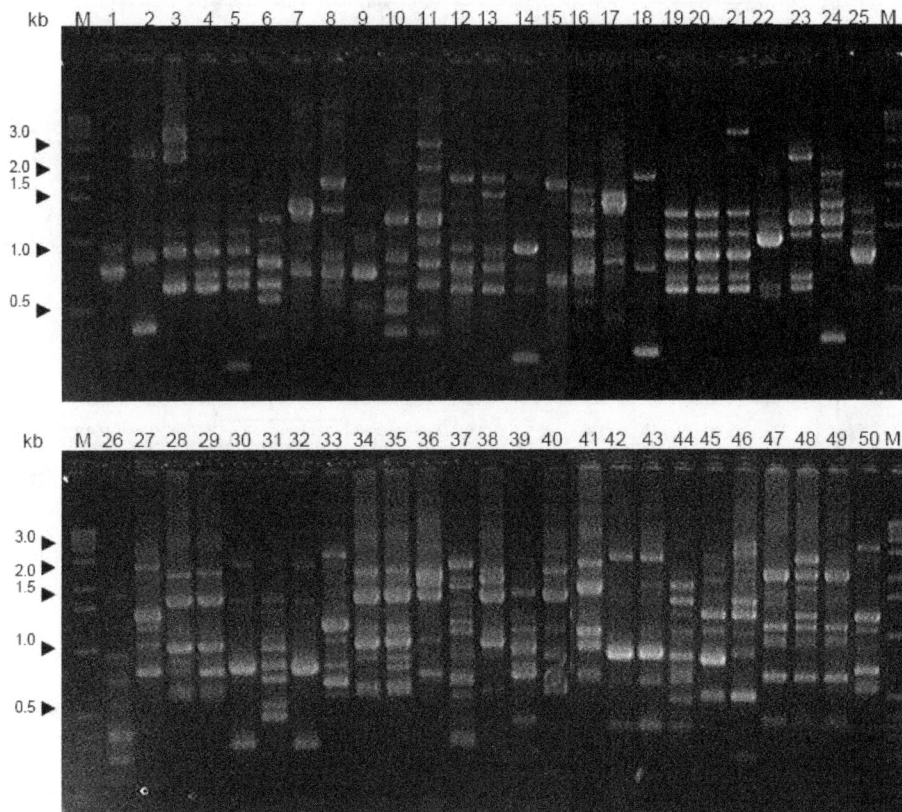

Figure 1. DNA fingerprinting patterns of 50 *X. oryzae* pv. *oryzae* (*Xoo*) isolates using OPS-08 random amplified polymorphic DNA primer. M: 1kb molecular size marker; kb: kilobase pair. *Xoo* isolate: 1 = XN-1; 2 = XN-2; 3 = XN-3; 4 = XN-4; 5 = XN-5; 6 = XN-6; 7 = XB-7; 8 = XB-8; 9 = XB-9; 10 = XB-10; 11 = XB-11; 12 = XNG-12; 13 = XNG-13; 14 = XNG-14; 15 = XNG-15; 16 = XNG-16; 17 = XBF-17; 18 = XBF-18; 19 = XBF-19; 20 = XBF-20; 21 = XBF-21; 22 = XBF-22; 23 = XM-23; 24 = XM-24; 25 = XM-25; 26 = XM-26; 27 = XM-27; 28 = XM-28; 29 = XM-29; 30 = XM-30; 31 = XG-31; 32 = XG-32; 33 = XG-33; 34 = XG-34; 35 = XG-35; 36 = XG-36; 37 = XG-37; 38 = XG-38; 39 = XG-39; 40 = XG-40; 41 = XTG-41; 42 = XTG-42; 43 = XTG-43; 44 = XTG-44; 45 = XTG-45; 46 = XTG-46; 47 = XTG-47; 48 = XTG-48; 49 = XTG-49; 50 = XTG-50.

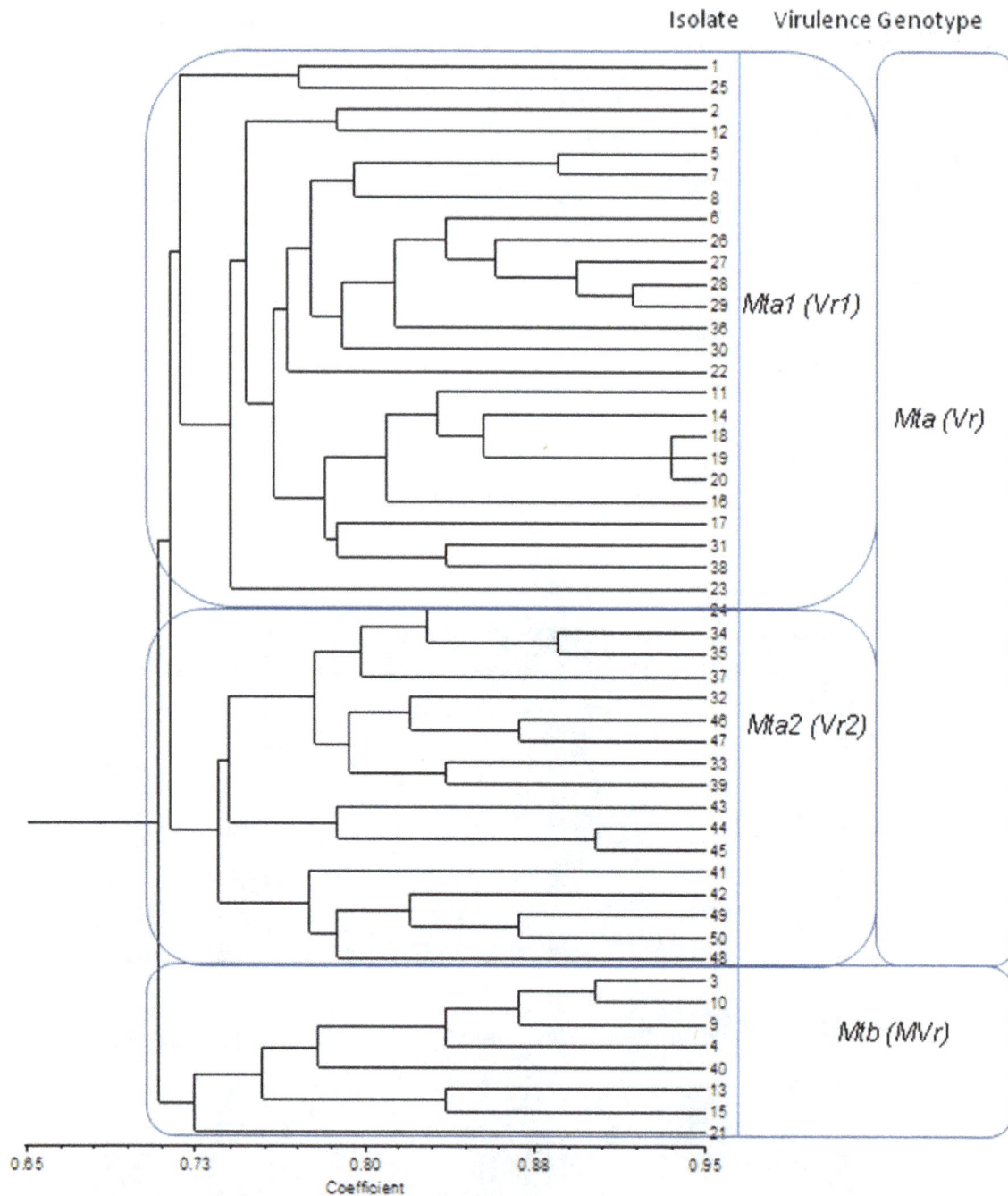

Figure 2. Molecular typing of 50 *Xanthomonas oryzae* pv. *oryzae* (*Xoo*) virulence as revealed by 136 random amplified polymorphic DNA markers. *Mta* = Molecular type a; *Mtb* = Molecular type b; *Vr* = Virulence; *MVr* = Mildly virulence. *Xoo* isolate: 1 = XN-1; 2 = XN-2; 3 = XN-3; 4 = XN-4; 5 = XN-5; 6 = XN-6; 7 = XB-7; 8 = XB-8; 9 = XB-9; 10 = XB-10; 11 = XB-11; 12 = XNG-12; 13 = XNG-13; 14 = XNG-14; 15 = XNG-15; 16 = XNG-16; 17 = XBF-17; 18 = XBF-18; 19 = XBF-19; 20 = XBF-20; 21 = XBF-21; 22 = XBF-22; 23 = XM-23; 24 = XM-24; 25 = XM-25; 26 = XM-26; 27 = XM-27; 28 = XM-28; 29 = XM-29; 30 = XM-30; 31 = XG-31; 32 = XG-32; 33 = XG-33; 34 = XG-34; 35 = XG-35; 36 = XG-36; 37 = XG-37; 38 = XG-38; 39 = XG-39; 40 = XG-40; 41 = XTG-41; 42 = XTG-42; 43 = XTG-43; 44 = XTG-44; 45 = XTG-45; 46 = XTG-46; 47 = XTG-47; 48 = XTG-48; 49 = XTG-49; 50 = XTG-50.

countries (Niger, Benin Republic, Nigeria, Burkina Faso, and Guinea) (Table 3). Thus in Niger, Benin Republic, Nigeria and Burkina Faso molecular typing revealed the presence of *Mta1* (*Vr1*) and *Mtb* (*MVr*) *Xoo* genotypes, *Mta1* (*Vr1*) and *Mta2* (*Vr2*) genotypes in Mali, *Mta1* (*Vr1*), *Mta2* (*Vr2*) and *Mtb* (*MVr*) genotypes in Guinea, and

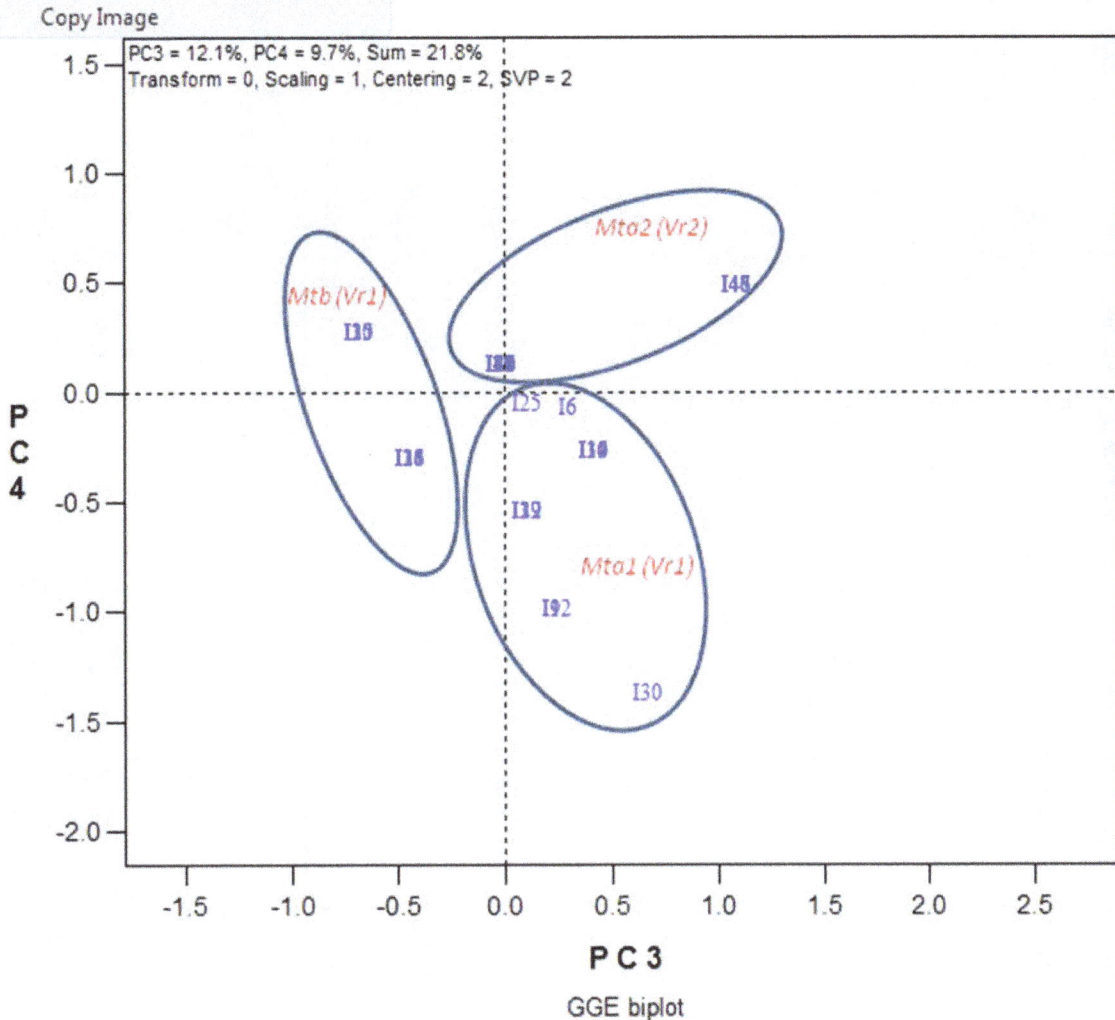

Figure 3. Principal component analysis that revealed subgroup virulence genotypes among 50 *X. oryzae* pv. *oryzae* isolates using genetic data generated from twelve random amplified polymorphic DNA (RAPD) primers.

Mta2 (*Vr2*) genotype in The Gambia (Figure 4 and Table 3).

DISCUSSION

Molecular basis for African *Xoo* virulence identification is a prerequisite into understanding the genetics of *Xoo* virulence population structure in West Africa and deployment of durable resistance cultivars (Sere et al., 2005; Adhikari et al., 1999; Adhikari et al., 1995). The present study examined if the two *Xoo* virulence pathotypes (*Pta* and *Ptb*) obtained using phenotypic pathotyping by Onasanya et al. (2009) could be confirmed using molecular approach. Molecular typing using random amplified polymorphic (RAPD) markers has revealed two major (*Mta* and *Mtb*) virulence genotypes

among the 50 *Xoo* isolates in which *Mta* was virulence (*Vr*) and *Mtb* mildly virulence (*MVr*). This report supports recent isozyme fingerprints of 30 *Xoo* isolates from 5 countries (Mali, Burkina Faso, Niger, Benin Republic and Nigeria) in West Africa that revealed two major genetic groups (Onasanya et al., 2008). These two genotypes of *Xoo* virulence identified by molecular typing were very identical to *Xoo* virulence pathotypes (*Pta* and *Ptb*) obtained using phenotypic pathotyping indicating possible linkage and correlation between phenotypic pathotyping and molecular typing of *Xoo* virulence (Adhikari et al., 1999; Lalitha et al., 2010).

The high distinction pattern of each isolates in this study suggests possible high level of genetic variation among *Xoo* isolates in different host cells (Innes et al., 2001; Mongkolsuk et al., 2000). The genetic analyses revealed that *Mta* virulence genotype might cover about 84% of BLB population across Niger, Benin Republic,

Table 3. *X. oryzae* pv. *oryzae* isolate group, virulence and distribution relative to country of origin.

Typing		Virulence	Isolate origin and distribution							Occurrence (%)
Main group	Subgroup		Niger	Benin	Nigeria	Burkina Faso	Mali	Guinea	The Gambia	
Pathotype*										
Pta	Pta1		–	–	–	4	1	4	1	20
	Pta2	Vr	3	–	–	–	–	–	–	16
	Pta3		–	2	3	–	2	3	–	22
Ptb	Ptb1		2	2	1	1	1	1	2	20
	Ptb2	MVr	1	1	1	–	1	–	4	22
Molecular type										
Mta	Mta1	Vr1	4	3	3	5	7	3	–	50
	Mta2	Vr2	–	–	–	–	1	6	10	34
Mtb	–	MVr	2	2	2	1	–	1	–	16

* = Onasanya et al., 2009; *Pta* = pathotype a; *Ptb* = pathotype b; *Mta* = molecular type a; *Mtb* = molecular type b; *Vr* = virulence; *MVr* = mildly virulence.

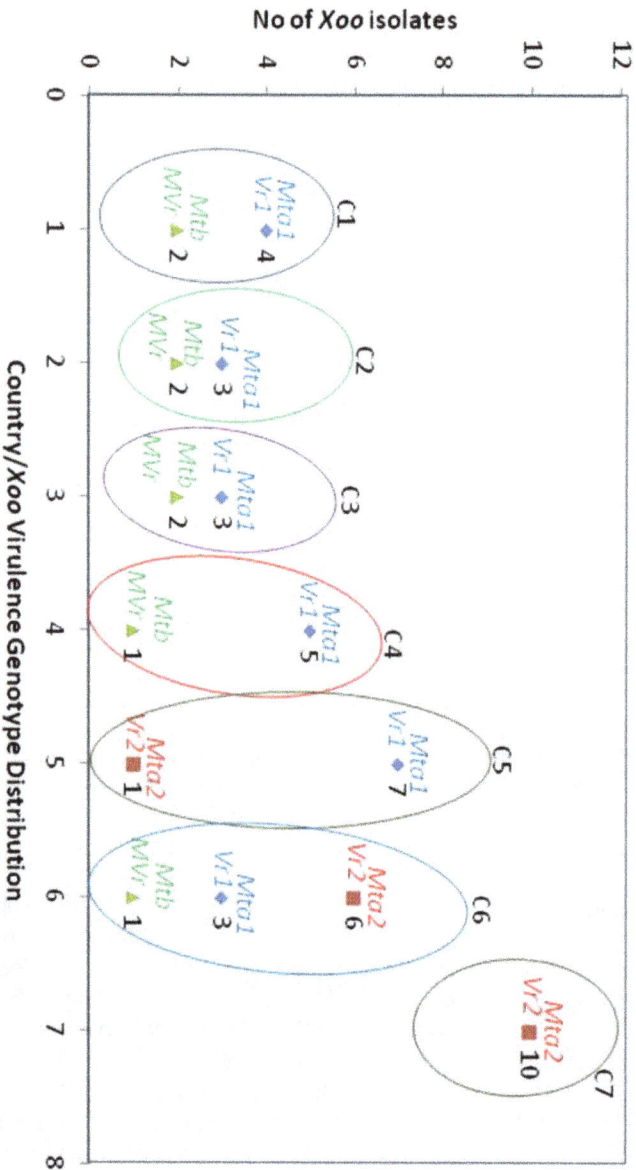

No of *Xoo* isolates

Country/*Xoo* Virulence Genotype Distribution

C1: *Mta1 Vr1* 4; *Mtb MVi* 2
C2: *Mta1 Vr1* 3; *Mtb MVi* 2
C3: *Mta1 Vr1* 3; *Mtb MVi* 2
C4: *Mta1 Vr1* 5; *Mtb MVi* 1
C5: *Mta1 Vr1* 7; *Mta2 Vr2* 1
C6: *Mta1 Vr1* 3; *Mta2 Vr2* 6; *Mtb MVi* 1
C7: *Mta2 Vr2* 10

Figure 4. *X. oryzae* pv. *oryzae* isolates virulence genotype population structure status in West Africa. *Mta* = Molecular type a; *Mtb* = molecular type b; *Vr* = virulence; *MVr* = mildly virulence. Country: C1 = Niger; C2 = Benin Republic; C3 = Nigeria; C4 = Burkina Faso; C5 = Mali; C6 = Guinea; C7 = The Gambia.

Nigeria, Burkina Faso, Mali, the Gambia and Guinea and possibly be responsible for most sporadic cultivars infestation and epidemics in these countries. Also, the existence of *Mta1* and *Mta2* subgroups were likely due to interactions among isolates and strains that originally constituted *Mta* genotype (Innes et al., 2001; Chisholm et al., 2006). *Mtb* genotype existed in over 16% of BLB population across Niger, Benin Republic, Nigeria, Burkina Faso and Guinea, and might be responsible for most sporadic cultivars infestation and epidemics in these countries. *Mta1* (*Vr1*) and *Mtb* (*MVr*) genotypes were found to exist in Niger, Benin Republic, Nigeria, and Burkina Faso, *Mta1* and *Mta2* in Mali, *Mta1*, *Mta2* and *Mtb* in Guinea, and *Mta2* in The Gambia suggesting possible *Xoo* pathogen migration between these countries and long-term *Xoo* pathogen survival (Adhikari et al., 1995; Dewa et al., 2011).

Distinct phenotypes usually consist of isolates that are genetically less related and such identification of isolates using cultural and morphological techniques often lack consistency and precision (Onasanya et al., 2009). Molecular typing of *Xoo* virulence has proven particularly useful in situations where it is necessary to differentiate virulence among two or more bacterial pathogens (Lalitha et al., 2010). In the current study, it was discovered that identification of virulence in *Xoo* depends on different host origins and occurrence of mutants. For instance, *Mta* virulence genotype might cover about 84% of BLB population across Niger, Benin Republic, Nigeria, Burkina Faso, Mali, The Gambia and Guinea and *Mtb* genotype existed in over 16% of BLB population across Niger, Benin Republic, Nigeria, Burkina Faso and Guinea but isolates virulence distributions vary within subgroups. Based on phylogenetic study, it was discovered that after prolonged season-to-season interactions among isolates of *Mta* or *Mtb* genotype in different cultivated rice and weed hosts, different subgroup virulence genotypes (*Mta1* and *Mta2*) may emerge as a result of mutation (Mongkolsuk et al., 2000). The emerged subgroup virulence genotypes might result in occurrence of highly virulent isolates and strains with very broad interaction and pathogenicity across wide range of cultivated rice varieties across West African countries.

Conclusion

The present molecular study of *Xoo* virulence identified two major *Xoo* virulence genotypes (*Mta* and *Mtb*) and two subgroups (*Mta1* and *Mta2*). Existence of different *Xoo* virulence genotypes suggests high level of *Xoo* pathogen interaction with host cells and mutation. The study revealed possible linkage between *Xoo* virulence pathotype and *Xoo* virulence genotype. Different *Xoo* virulence genotypes were known to exist within country and there was evidence of *Xoo* pathogen migration between countries. Durable resistance rice cultivars

would need to overcome both *Mta* and *Mtb Xoo* virulence genotypes in order to survive after their deployment into different rice ecologies in West Africa.

ACKNOWLEDGEMENT

We are very grateful to the Central Biotechnology Laboratory, International Institute of Tropical Agriculture (IITA) Nigeria for funding this research in its laboratory.

REFERENCES

Adhikari TB, Vera Cruz CM, Zhang Q, Nelson RJ, Skinner DZ, Mew TW, Leach JE (1995). Genetic diversity of *Xanthomonas oryzae* pv. *oryzae* in Asia. Appl. Environ. Microbiol. 61: 966-971.

Adhikari TB, Basnyat RC, Mew TW (1999). Virulence of *Xanthomonas oryzae* pv. *oryzae* on rice lines containing single resistance genes and gene combinations. Plant Dis. 83(1): 46-50.

Akanji BO, Ajele JO, Onasanya A, Oyelakin O (2011). Genetic Fingerprinting of *Pseudomonas aeruginosa* Involved in Nosocomial Infection as Revealed by RAPD-PCR Markers. Biotechnol. 10(1): 70-77.

Anzai Y, Kim H, Park JY, Wakabayashi H, Oyaizu H (2002). Phylogenetic affiliation of the Pseudomonads based on 16S rRNA sequence. Int. J. Syst. Evol. Microbiol. 50(4): 1563-1589.

Basso A, Onasanya A, Issaka S, Sido AY, Haougui A, Adam T, Sere Y, Saadou M (2011). Bacterial leaf blight of rice in Niger: Pathological diversity of isolates collected on irrigated lands. J. Applied Biosci. 38: 2551-2563.

Chandrashekar S, Umesha S, Chandan S (2012). Molecular detection of plant pathogenic bacteria using polymerase chain reaction single-strand conformation polymorphism. Acta. Biochim. Biophys. Sin. 44(3): 217-223.

Chaudhary SU, Javed I, Muzzammil H (2012). Effectiveness of different fungicides and antibiotics against bacterial leaf blight in rice. J. Agric. Res. (2012). 50(1): 109-117.

Chen H, Wang S, Zhang Q (2002). New gene for bacterial blight resistance in rice located on chromosome 12 identified from Minghui 63, an elite restorer line. Phytopathology 92: 750–754.

Chisholm ST, Coaker G, Day B, Staskawicz BJ (2006). Host-microbe interactions: Shaping the evolution of the plant immune response. Cell 124: 803–814.

Dewa KMK, Banito A, Onasanya A, Kpemoua KE, Sere Y (2011). Rice Bacterial Blight in Togo: Importance of the Disease and Virulence of the Pathogen. Current Research in Bacteriology 4: 94-100.

Ebdon JS, Gauch HG (2002). AMMI analysis of national turfgrass performance trials: I. Interpretation of genotype by environment interaction. Crop Sci. 42:489–496.

Éena J, Eszter A, Péter I, Arnold H, János P (2009). BOOL-AN: A method for comparative sequence analysis and phylogenetic reconstruction. Mol. Phylogenet. Evol. 52(3): 887-897.

Gnanamanickam SS (2009). An Overview of Progress in Biological Control. Biomedical and Life Sciences 8: 43-51.

Gonzalez C, Szurek B, Manceau C, Mathieu T, Sere Y, Verdier V (2007). Molecular and pathotypic characterization of new *Xanthomonas oryzae* strains from West Africa. Mol. Plant Microbe Interact. 20(5): 534-546.

Innes D, Beacham IR, Beven CA, Douglas M, Laird MW, Joly JC, Burns DM (2001). The cryptic *ushA* gene (*ushAc*) in natural isolates of *Salmonella enterica* (serotype Typhimurium) has been inactivated by a single missense mutation. Microbiology 147: 1887–1896.

Ivchenko GI, Honov SA (1998). On The Jaccard Similarity Test. J. Mathematical Sci. 88(6): 789-794.

Jiang GH, Xia ZH, Zhou YL, Wan J, Li DY, Chen RS, Zhai WX, Zhu LH (2006). Testifying the rice bacterial blight resistance gene xa5 by genetic complementation and further analyzing xa5 (Xa5) in comparison with its homolog TFIIAγ1. Mol. Gen. Genomics 275: 354–

366.

Lalitha SM., Varma CMK, Premalatha P, Devi GL, Zehr U, Freeman W (2010). Understanding the Bacterial Blight Pathogen-Combining Pathotyping and Molecular Marker Studies. International J. Plant Pathol. 1 (2): 58 – 68.

Mongkolsuk S, Whangsuk W, Fuangthong M, Loprasert S (2000). Mutations in oxyR resulting in peroxide resistance in *Xanthomonas campestris*. J. Bacteriol. 182: 3846- 3849.

Onasanya A, Mignouna HD, Thottappilly G (2003). Genetic fingerprinting and phylogenetic diversity of isolates of *Staphylococcus aureus* from Nigeria. Afri. J. Biotechnol. 2(8): 246 – 250.

Onasanya A, Ekperigin MM, Nwilene FE, Sere Y, Onasanya RO (2009). Two pathotypes of *Xanthomonas oryzae* pv. *oryzae* virulence identified in West Africa. Current Res. Bacteriol. 2(2): 22-35.

Onasanya A, Ekperigin MM, Sere Y, Nwilene FE, Ajele JO (2008). Enzyme polymorphism and genetic diversity in *Xanthomonas oryzae* pv. *oryzae* isolates causing rice bacterial leaf blight disease in West Africa. Int. J. Agric. Res. 3(3): 227-236.

Rohlf FJ (2000). NTSys pc, Version 2.02j, Exeter Software, Setauket, New York.

Savary S, Teng PS, Willocquet L, Jr. Nutter FW (2006). Quantification and modeling of 25 crop losses: a review of purposes. Ann. Rev. Phytopathol. 44:89-112.

Sere Y, Onasanya A, Verdier V, Akator K, Ouédraogo LS, Segda Z, Mbare MM, Sido AY, Basso A (2005). Rice Bacterial Leaf Blight in West Africa: Preliminary Studies on Disease in Farmers` Fields and Screening Released Varieties for Resistance to the Bacteria. Asian J. Plant Sci. 4: 577-579.

Zhao WJ, Zhu SF, Liao XL, Chen HY, Tan TW (2007). Detection of *Xanthomonas oryzae* pv. *oryzae* in seeds using a specific TaqMan probe. Molecular Biotechnology 35(2): 119-127.

27

Lack of DLAD mutations in age-related nuclear cataract

Nitza Goldenberg-Cohen[1,5,6]*, **Bat Chen R. Avraham-Lubin**[1], **Dorina Calles**[1],
Olga Dratviman-Storobinsky[1], **Rita Ehrlich**[4], **Boris Paritiansky**[4], **Yoram Cohen**[3] and
Dov Weinberger[4,6]

[1]The Krieger Eye Research Laboratory, Felsenstein Medical Research Center, Petah Tiqwa, Israel.
[2]The Mina and Everard Goodman Faculty of Life Sciences, Bar Ilan University, Ramat Gan, Tel Aviv, Israel.
[3]Sheba Cancer Research Center, Sheba Medical Center, Tel Hashomer, Israel.
[4]Department of Ophthalmology, Rabin Medical Center, Petach Tiqwa, Israel.
[5]Pediatric Unit, Rabin Medical Center, Petach Tiqwa, Israel.
[6]Sackler Faculty of Medicine, Tel Aviv University, Tel Aviv, Israel.

DNase II like acid DNase (DNase IIβ, DLAD) is expressed in human and murine cells in the lens. Studies in mice have reported that abnormal degeneration of cellular organelles by DLAD reduced lens transparency and that the *DLAD* gene may be involved in cataract formation. The aim of the present study was to search for possible genetic alterations in the *DLAD* gene in human senile cataract. Anterior lens capsule material was collected during surgery from 55 patients with senile cataract, with or without a subcapsular component. Total DNA was extracted, amplified by polymerase chain reaction and sequenced for exon 3 (n = 51) exon 4 (n = 40) and all 6 exons of the *DLAD* gene (n = 27). No mutation was found. There were genomic polymorphisms in all exons except 3 and 4. Nonsynonymous genomic polymorphisms were detected in exon 1 (rs738573) and exon 2 (rs3754274) and synonymous polymorphisms were detected in exon 5 (rs7511984) and exon 6 (rs3768250). In contrast to findings in mice, based on the limited samples analyzed, this study suggests that human age-related nuclear cataract is not associated with *DLAD* mutations.

Key words: DNAase II like acid DNAase- DNAse IIβ- *DLAD*- senile cataract-nuclear cataract.

INTRODUCTION

The lens is an avascular tissue that focuses light onto the retina. It consists of closely packed fibers that are produced continuously by the differentiation of epithelial cells located near its equator. During their proliferation, differentiation and migration toward the nucleus, the epithelial cells elongate and their light-scattering organelles, including the endoplasmic reticulum, mitochondria, golgi apparatus and nucleus, disappear, together with housekeeping enzymes (RNA polymerase and DNA polymerase, etc.) (Bassnett, 2002).

Concomitantly, the production of lens-specific proteins and crystallins increases. The lens nucleus is composed of fully differentiating fiber-like cells (Nakahara et al., 2007). Because there is no cell turnover, the lens grows throughout life (McAvoy et al., 1999; Piatigorsky, 1981).

Accordingly, the cortex of the adult lens has two populations of fiber cells: a cortical layer of differentiating cells that contain organelles and the nonnucleated fiber cells that do not contain organelles. Several organelle-degradation mechanisms have been proposed, most prominently apoptosis (Dahm, 1999; Ishizaki et al., 1998; Zandy and Bassnett, 2007), cytosolic degradation (van Leyen et al., 1998) and autophagy (Vrensen et al., 1991). Autophagy has been extensively studied in yeast and several molecules involved in this process have been identified (Nagata, 2005). However, the pathways involved in the differentiation of fiber cells are still elusive.

The loss of organelles during fiber cell differentiation apparently occurs to ensure the transparency of the lens (Nagai et al., 2006). Cataract is an alteration in the optical homogeneity of the lens or a decrease in its transparency. It leads to decreased vision and even blindness.

---

*Corresponding author. Email: ncohen1@gmail.com.

Abbreviations: DLAD, DNase II like acid DNase; **LOCS,** lens opacities classification scale; **PCR,** polymerase chain reaction.

Table 1. Primers and polymerase chain reaction (PCR) reaction conditions.

Exon	Primers	Product size bp	Annealing temperature(°C)
1 F	5'-CGCCTTGAAACTCAGACTCC	239	56
1 R	5'-ATGCAGCAGTCCTTCCATTT		
2 F	5'-CTGCCACAGGAAAACAGTCA	307	56
2 R	5'-ACTCTGGTGAAGGCTTTGGA		
3 F	5'-GGGCGTTTGTAAGTGGAAGA	174	56
3 R	5'-TTCCAGTAGGAAAGAGAAACCAA		
4 F	5'-TGTTGCCACTGTGAACCCTA	300	60
4 R	5'-TTTCCAAATCATTTCTTCAACA		
5 F	5'-TGCCTTTGTTTTTGTGTTGTTT	343	56
5 R	5'-CAGCATGTTGTTTGATGATGG		
6 F	5'-CATCCACATATCAGGGGTGA	718	56
6 R	5'-TTGGAGTTGACTAATGTGGAAAAA		

Although cataract may result from normal aging, multiple factors can contribute to this process, including the precipitation and degradation of proteins, exposure to ultraviolet light, exposure to oxidative and other toxic agents, genetic factors, diabetes, smoking, poor nutrition and alterations in endocrinic or enzymatic equilibrium (Cekic et al., 1999; Dilsiz et al., 1999; Garland et al., 1988; Garland, 1999; Tumminia et al., 2001). DNase II-like acid DNase (DLAD; also known as DNase IIβ) is expressed in human and murine cells, specifically in the lens, and in the liver, salivary glands, and lungs (Nakahara et al., 2007; Nagai et al., 2006). Its role in degrading DNA during lens cell differentiation was reported in a study of mice deficient in the *DLAD* gene (Nishimoto et al., 2003). The failure to degrade DNA led to the development of cataract and the accumulation of undegraded DNA in the fiber cells. Nishimoto et al. (2003) identified mutations in the activated enzymatic site of exons 3 and 4 of the *DLAD* gene in cataractous mice.

Recently, Nakahara et al. (2007) found that the DLAD protein is localized to lysosomes of the cortical fiber cells, close to the organelle-free zone. Its absence from epithelial cells indicated that DLAD expression is induced during differentiation of the epithelial cells into fiber cells. Immunohistochemical analysis suggested that this process involves the fusion of the nuclei with the lysosomes and DLAD digestion of the chromosomal DNA.

We sought to determine if findings in murine models could be extended to humans. The aim of the present study was to explore the role of genetic variations in the *DLAD* gene in human age-related cataract formation.

METHODS

Patients

The study group included 55 patients undergoing routine cataract surgery at a major tertiary medical center. All patients signed an informed consent form. All were examined preoperatively by slit-lamp and categorized by cataract type according to the Lens Opacities Classification Scale (LOCS) Background data were derived from the medical files. Anterior lens capsule material excised during surgery (one sample per patient) was analyzed for mutations in exons 1 to 6 of the *DLAD* gene. The study was approved by the institutional and national review boards.

DNA isolation

The capsules containing single-layer lens epithelial cells were suspended in 5 ml of conservation medium until isolation of genomic DNA. DNA was extracted using standard sodium dodecylsulfate (SDS)/proteinase K digestion followed by phenol-chloroform extraction and ethanol precipitation.

Detection of DLAD mutations

The 6 exons of the *DLAD* gene were analyzed using polymerase chain reaction (PCR) amplification and direct sequencing. PCR primer sequences were designed with the Primer 3 Program (http://frodo.wi.mit.edu/rgi-bin/primer3/) and are listed in Table 1. PCR amplification was performed in a 50 µL reaction volume containing 100 ng of sample DNA as a template. Details of the conditions for each reaction are summarized in Table 1. The PCR parameters were as follows: denaturation at 95°C for 5 min, 35 cycles of 1 min at 95°C, annealing at 56 - 60°C for 1 min (Table 1) and extension of 1 min at 72°C with Taq polymerase. The PCR product was amplified on 2% agarose gel and visualized with ethidium bromide staining. Direct sequencing of the PCR products was performed with Big Dye Terminator Cycle Sequencing reagents using the ABI PRISM 3700 DNA Analyzer (Applied Biosystems, Foster City, CA).

RESULTS

The patient group consisted of 33 women and 22 men aged 47 to 93 years (mean 75.8 years ± 8.5). All were Caucasian Jews. All cataracts had a nuclear component,

Table 2. Comparison of genotypic variations and frequencies in Caucasian Jews in our study and other ethnic populations.

Exon	SNP ID	Alleles	Amino acid change	Amino acid co-ordinate	Our results	European	Chinese	Afro-American	Japanese
				SNP Data			SNP Frequency (%)		
1	rs3738573	C/G	Q/H	3 (3)	CC- 11 CG- 45 GG- 45	CC- 12.5 CG- 50 GG- 37.5	CC- 8.3 CG- 45.8 GG- 45.8	CC- 0 CG- 30 GG- 70	CC- 18.2 CG- 43.2 GG- 38.6
2	rs3754274	A/G	R/K	51(2)	AA- 13.5 AG- 32 GG- 54.5	AA- 8.3 AG- 41.7 GG- 50	AA- 11.1 AG- 42.2 GG- 46.7	AA- 0 AG- 22.7 GG- 77.3	AA- 13.6 AG- 45.5 GG- 40.9
5	rs7511984	C/T	Y	248(3)	CC- 74 TC- 26 TT- 0	CC- 71.9 TC- 24.6 TT- 3.5	CC- 97.8 TC- 2.2 TT- 0	No data No data No data	CC- 95.5 TC- 4.5 TT- 0
6	rs3768250	C/T	H	305(3)	CC- 15.7 TC- 47.3 TT- 37	CC- 12.5 TC- 58.3 TT- 29.2	CC- 15.6 CT- 51.1 TT- 33.3	CC- 9.1 CT- 40.9 TT- 50	CC- 18.2 CT- 47.7 TT- 34.1

and 7 also had a cortical component.

No mutations were detected on analysis of the *DLAD* gene. A number of genomic polymorphisms were found in exons 1, 2, 5, and 6; all of them were previously reported in the Genome Database (http://www.ensembl.org/index.html) (Table 2). The PCR amplification products for each exon are presented in Figure 1. DNA alterations and base-pair positions related to the sequences were deposited in the Genebank (Homosapiens DLAD chromosome 1, NM 058248).

The polymorphisms and their frequencies are described in Table 2. Nonsynonymous single-nucleotide polymorphisms (SNPS) were identified in exons 1 and 2 (Table 2, Figure 2). Exon 1 was characterized by a glutamine to histidine change; SNP frequencies were as follows: CC 12, CG 44 and GG 44%. Exon 2 had an arginine-to-lysine change; SNP frequencies were GG 54.5, GA 37.5 and AA 8%. Synonymous polymorphisms were detected in exons 5 and 6 (Table 2, Figure 2): in exon 5 for aspartate, with frequencies of 74% for CC, 26% for CT and 0 for TT, and in exon 6 for histidine, with frequencies of 37% for TT, 47% for TC, and 16% for CC. No polymorphisms were found in exons 3 and 4 in any of the samples tested.

DISCUSSION

We investigated the possible relationship of epithelial cell organelle degradation by lens DLAD and senile nuclear cataract in humans. To our knowledge, this is the first study to explore *DLAD* mutations in this setting.

Our study was prompted by findings in a *DLAD*-deficient mouse model wherein no nuclear DNA degradation occurred during lens cell differentiation (Nishimoto et al., 2003). Specifically, in the *DLAD*-null fiber cells, naked DNA remained in the cytoplasm, but the mitochondria, the endoplasmic reticulum and the nuclear membrane were completely lost. The mice acquired weak cataracts and their response to light was severely reduced (Nishimoto et al., 2003). The data suggested that the nuclei disappear from fiber cells to ensure the transparency of the lens and that cataract develops as a direct consequence of the accumulated DNA. If other organelles are left undigested in the lens, they may also cause cataract. This observation led to the assumption that human cataract, too, might be caused by a genetic defect that impairs degradation of the organelles during lens cell differentiation (Nagata, 2005). The findings were supported by the study of Nakahara et al. (2007), who reported that *DLAD* was a major acid DNase in nuclear cataract in the mouse.

We analyzed DNA extracted from the capsule of nuclear cataracts, with or without cortical components, of 53 patients. No truncating mutations or inactivating mutations were found. By contrast, studies in transgene mice with cataract identified mutation in exons 3 and 4 of the *DLAD* gene (Nishimoto et al., 2003). We speculate that despite the sequence similarity of human and mouse *DLAD* and the apparently similar mechanism that regulates lens transparency in the two species, different genes or different genomic variations in some genes are involved in the process. Interestingly, our results are in line with the conclusion of Nagai et al. (2006) that undigested DNA is not located in the nuclear portion of the lens.

Hawse et al. (2005) analyzed the gene expression profiles of epithelial and fiber cells in the human lens and found global differences between them, although their

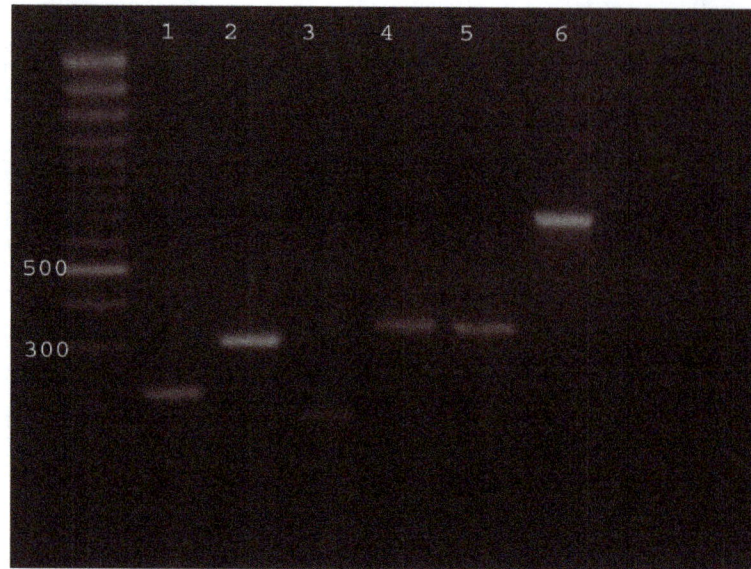

Figure 1. Gel electrophoresis of the PCR products of the 6 *DLAD* exons.

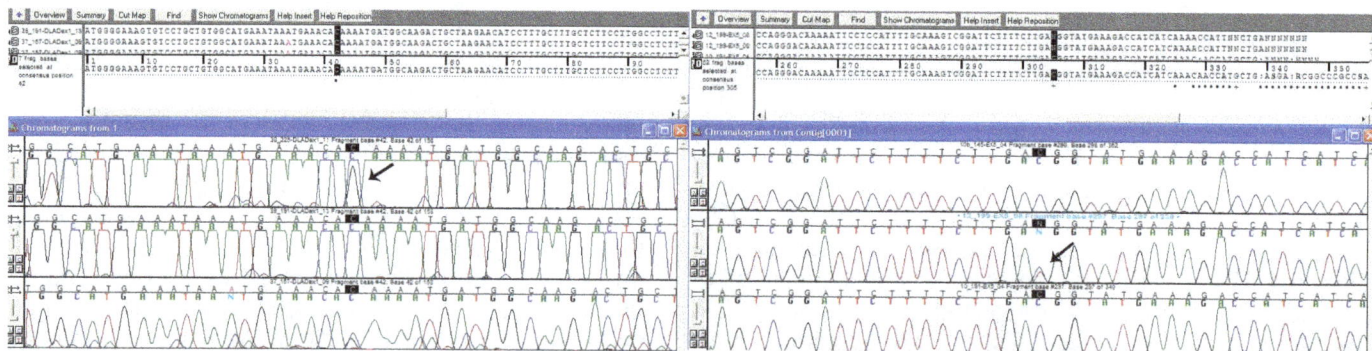

Figure 2. DNA sequence for exon 1 and 5 in a few samples. Note heterozygosity (arrow).

analysis did not clearly point to genes other than *DLAD* that might play a role in organelle degradation. Given that DLAD is active in acid conditions and is probably localized to the lysosomes, the process of DNA degradation may involve a canonical autophagic mechanism (Nakahara et al., 2007). However, Matsui et al. (2006) failed to detect any abnormalities in lens cell differentiation in mice lacking Atg5, a protein essential for autophagosome formation.

The genotype variations and the distribution of the SNP in exons 1, 2, 5 and 6 were similar to those reported previously in studies of the genomic polymorphism of *DLAD* in Chinese and Japanese populations, but different from those in African and African-American populations (Genome Browser). Although we found no association between cataract formation and genotype, the different distributions in the Africans/African-Americans are in line with the different type of cataract (cortical rather than nuclear) that characterizes these populations, which occurs at an earlier age.

The present study was restricted to senile cataract and we cannot extrapolate the findings to other types of cataract, such as cortical or congenital cataract or acquired cataract in younger groups. However, even when the nuclear cataract had a cortical component, no mutation could be detected. Many studies have presented evidence of a genetic component in the risk of congenital cataract, but none has explored the role of *DLAD*.

Our study was also limited by the small size of the sample. Furthermore, although we screened all exons, we did not examine mRNA expression or levels of the protein itself, or enzymatic dysfunction in the lens. Nevertheless, the results may suggest that *DLAD* mutations are not associated with nuclear cataract in elderly people. The question of whether animals are a good model to define genes involved in acquired human disease is still unresolved. The widespread occurrence of cataract in the elderly may point to multiple causative factors. Moreover, given that some eyes are spared, it is possible that a

DLAD mutation is not enough to account for the sequence of events leading to the development of cataract. Recent studies report an increase in the annual incidence of cataract surgery in younger patients (Francis and Moore, 2004), which may coincide with changes in other risk factors.

Further studies in larger populations are needed to evaluate the role of DLAD in different types of the cataract and the possible involvement of protein dysfunction.

ACKNOWLEDGMENT

This study was partially supported by the Zanvyl and Isabelle Krieger Fund, Baltimore, Maryland, USA, and the Claire and Amedee Maratier Institute for the Study of Visual Disorders and Blindness, Tel Aviv University, Tel Aviv, Israel.

This work was presented in part at the annual meeting of the Israel Association for Research in Vision and Ophthalmology, Neve Ilan, Israel, March 2005

REFERENCES

Bassnett S (2002). Lens organelle degradation. Exp. Eye Res. 74: 1-6.

Cekic O, Bardak Y, Totan Y, Akyol O, Zilelioglu G (1999). Superoxide dismutase, catalase, glutathione peroxidase and xanthine oxidase in diabetic rat lenses. Ophthalmic. Res. 31: 346-350.

Dahm R (1999). Lens fibre cell differentiation - A link with apoptosis? Ophthalmic. Res. 31: 163-183.

Dilsiz N, Olcucu A, Cay M, Naziroglu M, Cobanoglu D (1999). Protective effects of selenium, vitamin C and vitamin E against oxidative stress of cigarette smoke in rats. Cell Biochem. Funct. 17: 1-7.

Francis PJ, Moore AT (2004). Genetics of childhood cataract. Curr. Opin. Ophthalmol. 15: 10-15.

Garland D (1999). Role of site-specific, metal-catalyzed oxidation in lens aging and cataract: a hypothesis. Exp. Eye Res. 50: 677-682.

Garland D, Russell P, Zigler JS Jr (1988). The oxidative modification of lens proteins. Basic Life Sci. 49: 347-352.

Hawse JR, DeAmicis-Tress C, Cowell TL, Kantorow M (2005). Identification of global gene expression differences between human lens epithelial and cortical fiber cells reveals specific genes and their associated pathways important for specialized lens cell functions. Mol. Vis. 11: 274-283.

Ishizaki Y, Jacobson MD, Raff MC (1998). A role for caspases in lens fiber differentiation. J. Cell Biol. 140: 153-158.

Matsui M, Yamamoto A, Kuma A, Ohsumi Y, Mizushima N (2006). Organelle degradation during the lens and erythroid differentiation is independent of autophagy. Biochem. Biophys. Res. Commun. 339: 485-489.

McAvoy JW, Chamberlain CG, de Iongh RU, Hales AM, Lovicu FJ (1999). Lens development. Eye (Lond) 13(Pt 3b): 425-437.

Nagai N, Takeuchi N, Kamei A, Ito Y (2006). Involvement of DNase II-like acid DNase in the cataract formation of the UPL rat and the Shumiya cataract rat. Biol. Pharm. Bull. 29: 2367-2371.

Nagata S (2005). DNA degradation in development and programmed cell death. Annu. Rev. Immunol. 23: 853-875.

Nakahara M, Nagasaka A, Koike M, Uchida K, Kawane K, Uchiyama Y, Nagata S (2007). Degradation of nuclear DNA by DNase II-like acid DNase in cortical fiber cells of mouse eye lens. FEBS J. 274: 3055-3064.

Nishimoto S, Kawane K, Watanabe-Fukunaga R, Fukuyama H, Ohsawa Y, Uchiyama Y, Hashida N, Ohguro N, Tano Y, Morimoto T, Fukuda Y, Nagata S (2003). Nuclear cataract caused by a lack of DNA degradation in the mouse eye lens. Nature 424: 1071-1074.

Piatigorsky J (1981). Lens differentiation in vertebrates. A review of cellular and molecular features. Differentiation 19: 134-153.

Tumminia SJ, Clark JI, Richiert DM, Mitton KP, Duglas-Tabor Y, Kowalak JA, Garland DL, Russell P (2001). Three distinct stages of lens opacification in transgenic mice expressing the HIV-1 protease. Exp. Eye Res. 72: 115-121.

van Leyen K, Duvoisin RM, Engelhardt H, Wiedmann M (1998). A function for lipoxygenase in programmed organelle degradation. Nature 395: 392-395.

Vrensen GF, Graw J, De Wolf A (1991). Nuclear breakdown during terminal differentiation of primary lens fibres in mice: a transmission electron microscopic study. Exp. Eye Res. 52: 647-659.

Zandy AJ, Bassnett S (2007). Proteolytic mechanisms underlying mitochondrial degradation in the ocular lens. Invest. Ophthalmol. Vis. Sci. 48: 293-302.

28

Deletion of derivative ABL, BCR or ABL-BCR fusion gene is associated with shorter disease free survival in CML patients

Beena P. Patel*, Pina J. Trivedi, Manisha M. Brahmbhatt, Sarju B. Gajjar, Ramesh R. Iyer, Esha N. Dalal, Shilin N. Shukla and Pankaj M. Shah

The Gujarat Cancer and Research Institute, Asarwa, Ahmedabad -380 016, India.

Chronic myeloid leukemia (CML) is characterized by formation of the BCR/ABL fusion gene, usually as a consequence of the Philadelphia (Ph) translocation between chromosomes 9 and 22. However, deletions of the derivative 9 chromosome [der (9)] in 10 - 15% of CML patients with a standard Ph translocation as well as > 30% of CML patients with a variant Ph translocation may have verse prognosis. The study shed light on prognostic effect of submicroscopic deletions of the derivative chromosome 9 in CML in untreated patients and their follow up samples to correlate with survival. The study included blood and/or bone marrow (BM) samples of 65 untreated CML patients (PT) and 76 follow-up samples, classified as cytogenetic responders (CyR, n = 42), non cytogenetic responders (NCyR, n = 25) and partial cytogenetic responder (PCyR, n = 9). Karyotype analysis was performed on metaphases obtained through short term cultures of BM and blood. Detection of BCR–ABL fusion gene was performed using dual colour dual fusion (D-FISH) translocation probes. Data were analyzed using SPSS statistical software. CyR showed significantly elevated Hemoglobin (p = 0.0001) and decreased in total WBC (p = 0.0001) and Platelet counts (p = 0.0001) as compared to pretreatment levels. 61.5, 30.8 and 7.7% of the PT showed CyR, NCyR and PCyR respectively. Kaplan-Meier survival curve showed the patient with CyR, NCyR and PCyR as well as patients with different stage of the disease did not find difference in survival time. ABL-BCR deletion on derivative 9 was seen in 9.2% of PT, while ABL-BCR, ABL or BCR deletion on derivative 9 was found around 7.7% of PT. Patients with deletion of ABL-BCR on derivative 9 and deletion of ABL or BCR and/or ABL-BCR on derivative 9 have significantly reduced survival (log rank = 14.54; p = 0.001) than non deleted patients. Deletion in *ABL*, *BCR* or *ABL-BCR* on derivative 9 could predict over the survival of all CML patients.

Key words: CML, *BCR-ABL* fusion Gene, FISH, derivative 9 deletion.

INTRODUCTION

The Philadelphia (Ph) chromosome, resulting from the

---

*Corresponding author. E-mail: patel.beena@yahoo.com.

Abbreviations: AP, Accelerated phase; **BAC,** bacterial artificial chromosome; **BC,** blast crisis; **BCR-ABL fusion gene,** break point Cluster Region-Abelson fusion gene; **CML,** chronic myeloid leukemia; **CP,** chronic phase; **CyR,** cytogenetic responders; **D-FISH,** dual colour dual fusion fluorescence *in situ* hybridization; **FISH,** Fluorescence *in situ* hybridization; **NCyR,** non cytogenetic responders; **PCyR,** partial cytogenetic responder; **PT,** untreated patients.

balanced translocation, t(9;22)(q34;q11.2), which fuses the 5' sequences of the BCR gene on chromosome 22 with 3' sequences of the ABL gene on chromosome 9, is the diagnostic hallmark of chronic myeloid leukaemia (CML).

This translocation generates two fusion genes, BCR–ABL on derivative chromosome 22q, known as the Ph chromosome and a reciprocal ABL–BCR fusion gene on derivative chromosome 9q. The technique of fluorescence *in situ* hybridisation (FISH), using probes that bind to specific sequences at the Ph translocation breakpoints, has been applied extensively at initial diagnosis of CML to investigate cases with failed cytogenetics, to detect cryptic BCR–ABL gene fusion and to decipher complex

Table 1. Patients and samples details.

Untreated CML patients (PT)	N = 65
Gender	
Male (%)	28 (43)
Female (%)	37 (57)
Age (Years)	
Mean	38
Median	37
Range	11 - 70
Diagnosis (%)	
Chronic Phase (CML-CP)	61 (93.8)
Accelerated Phase (CML-AP)	3 (4.7)
Blast Crisis (CML-BC)	1 (1.5)
Final response (%)	
Responded	40 (61.5)
Partially Responded	20 (30.8)
Not Responded	5 (7.7)
Follow-up in months	
Mean	24.86
Median	23.01
Range	10.82 - 45.57
Follow-up samples (%)	76
Cytogenetic Responder (CytR)	42 (55.3)
Non cytogenetic responder (NcyR)	25 (32.9)
Partial cytogenetic responder (PCyR)	9 (11.8)

Ph rearrangements (Wan et al., 20003).

The quantification of BCR-ABL positive cells is achievable through interphase FISH and hence facilitates the monitoring of disease response to treatment.

A dual colour dual fusion (D-FISH) BCR–ABL probe system has been developed (Dewald et al., 1998; Buño et al., 1998), which are large probes designed to span the translocation breakpoints and are labelled with different fluorochromes. In addition to one signal each from the normal BCR and ABL genes, two fusion signals are created in a Ph positive cell: a BCR–ABL fusion signal on derivative chromosome 22q and an ABL–BCR fusion signal on derivative chromosome 9q. Variant signal patterns are encountered in three way Ph translocations and the addition of an extra Ph chromosome when investigated by the D-FISH system (Dewald et al., 2000). It has been reported that *ABL* deletion on derivative 9 was associated with poor prognosis while *BCR* deletion did not affect survival in CML (Sinclair et al., 2000; Gonza´lez et al., 2001). However, role of *ABL-BCR* fusion gene is not well documented (Huntly et al., 2002).

Therefore, the present study aimed to evaluate prognostic effect of sub-microscopic deletions of the derivative chromosome 9 in untreated CML patients and their follow up samples. We investigated the frequencies of ABL, BCR or ABL-BCR deletion on the derivative chromosome 9 using D-FISH and analyzed the association of deletion on derivative 9 with disease free survival in CML patients.

MATERIALS AND METHODS

Subjects

The study included 65 samples from untreated CML patients (PT) from the medical oncology department of The Gujarat Cancer and Research Institute. Among them 28 were male and 37 female.

Follow-up samples

The patients were evaluated at various intervals after initiation of therapy. Blood and/or bone marrow (BM) samples were collected at the time of diagnosis and during follow-up visit of these patients. Seventy six follow-up samples from these patients, classified as cytogenetic responders (CcyR, n = 42), non cytogenetic responders (NCyR, n = 25) and partial cytogenetic responder (PCyR, n = 9) were also collected. Imatinib (Gleevec) was given to all the patients as per the protocol. Clinical details of the patients are given in Table 1. Clinical, radiological and other necessary examinations like molecular cytogenetics were performed during the follow-up study. Patients were monthly followed-up for first 3 months after initiation of the treatment and then after every 3 months based on heamatological and cytogenetic response. As the patients were not regular for follow-up, follow-up number and samples were not identical in all the patients. Clinical status of the patients during/after therphy was evaluated by the clinicians treating them by international criteria of heamatological and cytological parameters.

Study ethics

The study design and patients consent to participate in the study was ethically approved by hospital based ethical committee of The Gujarat Cancer and Research Institute.

Methods

Conventional cytogenetic analysis

Cytogenetic analysis was performed on Giemsa banded (G banded) metaphases obtained through short term cultures of bone marrow and/or peripheral blood cells using standardized protocols. Karyotypes were reported in minimum of 30 metaphases as per ISCN 2005 (Shaffer and Tommerup, 2005). Metaphase cells were captured and analyzed using automated karyotyping system consisting of Axioplan universal epifluorescence microscope (Carl Zeiss) and IKAROS software (Metasystems, Germany).

Fluorescence in situ hybridization (FISH)

Detection of BCR–ABL fusion gene was performed using BCR/ABL dual colour dual fusion (D-FISH) translocation probe (Vysis, Downers Grove, Illinois, USA), according to the manufacturer's instructions. This probe mixture contained directly labelled Spectrum Orange™ probe that spanned the ABL locus at 9q34 ("O"

Table 2. Comparison of haematological parameters between untreated samples and their paired three groups of follow-up samples.

Paired t test		Haemoglobin (gm/dl) Mean (S.D.)	Leukocyte count (counts/cumm.) Mean (S.D.)	Platlets count (counts/cumm.) Mean (S.D.)
PT vs. CR	PT	10.2444 (1.78)	131022.22 (127638.64)	495363.63 (261726.58)
	CR	11.7333 (1.18)	6672.22 (1924.82)	244333.33 (75636.33)
significance	p =	0.0001	0.0001	0.0001
PT vs. PR	PT	10.45 (1.848)	186914.28 (134101.01)	488000.0 (240806.28)
	PR	10.8 (2.65)	6228.57 (1943.98)	190142.86 (81462.52)
significance	p =	0.696	0.012	0.033
PT vs. NR	PT	9.94 (1.89)	134576.12 (121281.19)	425500.0 (303006.01)
	NR	10.35 (2.03)	15454.17 (31851.5)	383954.54 (451010.98)
significance	p=	0.466	0.0001	0.732

denotes Orange labelled ABL gene) and directly labelled Spectrum Green™ probe that spanned the BCR locus at 22q11.2 ("G" denotes Green labelled BCR gene). 400 interphase/metaphase nuclei were analyzed for the presence of fusion signals. The OGFF pattern is the typical pattern for CML and indicates no gross sub-microscopic deletions. ("F" denotes yellow Fusion signal of orange and green probe indicative of BCR-ABL fusion on derivative 22 and ABL-BCR fusion on derivative 9). Atypical patterns of D-FISH include OGGF, OOGF and OGF which are indicative of deletion of ABL, BCR and ABL-BCR respectively on derivative 9. Image acquisition was performed either on automated Olympus epifluorescence microscope and Cytovision software (version 3.7, Applied Imaging System) or Carl Zeiss with ISIS software (Metasystems, Germany).

Treatment response criteria

Treatment response in follow-up samples was evaluated based on bone marrow pathology/morphology report and cytogenetic analysis for Philadelphia chromosome. The CcyR was defined as disappearance of signs and symptoms of disease, including palpable splenomegaly, normalization of peripheral blood counts and differentials (white blood cell [WBC] count < 10 × 10^9/L; no peripheral blasts or promyelocytes; < 5% myelocytes -metamyelocytes; platelets < 450 × 10^9/L) as well as absence of Ph chromosome in conventional cytogenetics and/or absence of BCR-ABL fusion gene by FISH analysis, with BM morphology report as remission. The complete cytogenetic response was confirmed by qualitative BCR-ABL PCR outside the institute (Data not shown). Minor cytogenetic response having Ph and/or BCR-ABL fusion 35 to 90% with no morphology change/relapse in BM report was considered as NCyR. PCyR was defined as presence of Ph positive (1 to 34%) cells by conventional cytogenetic and/or presence of mixed clone of BCR-ABL fusion gene positive by FISH aided with BM pathology report being controlled CML activity.

Statistical analysis

Data were statistically analyzed using the SPSS statistical software (version 15.0; SPSS, Inc., Chicago, IL, USA). Students paired't' test

was performed to compare haematological parameters like haemoglobin, total white blood cells (WBC) and platelet counts between untreated and their follow-up samples. Kaplan-Meier survival curves were plotted using log rank test to compare disease free survival between patients with deletion of ABL or BCR and/or ABL-BCR on derivative 9 with patients with non deleted derivative 9. Statistical significance was considered when 'p' values were less then 0.05.

RESULTS

Variations in haematological parameters between untreated samples and follow-up samples Paired t test was performed between untreated samples and their paired follow-up samples for haemoglobin, leukocyte and platelet counts (Table 2). CcyR showed significantly elevated Hemoglobin (p = 0.0001) and decreased in leukocyte (p=0.0001) and Platelet counts (p = 0.0001) as compared to their paired PT levels. PcyR samples showed significantly lower leukocyte and platelet counts as compared to their paired PT levels (p = 0.012 and p = 0.033 respectively). NcyR samples showed significantly elevated leukocyte counts (p = 0.0001) as compared to their paired PT levels.

Conventional cytogenetic and FISH

Table 3 shows frequency of different clone in karyotype and FISH at PT. Seventy two percentage of PT showed t (9; 22) at the time of diagnosis, while 24.6% of PT showed non informative karyotype results due to no metaphase or poor metaphase preparations. One patient showed normal karyotype with no t(9; 22), which was confirmed for BCR-ABL negative by FISH. Only one patient showed variant translocation having t (9; 22)?add9q.

Table 3. Karyotype and FSIH results.

Karyotype	(PT = 65) (%)
t (9:22)	47 (72.31)
Normal (N)	01 (1.54)
Non Informative (NI)	16 (24.61)
Variant translocations other than t(9:22)	01 (1.54)
FISH for BCR-ABL DCDF	**(PT=65)**
Negative (OOGG) (no fusion of BCR-ABL)	01 (1.54)
Positive (OGFF) (no deletion of derivative 9)	36 (55.39)
Positive (OGF) (Deletion of derivative 9)	06 (9.23)
Positive mix clone (OGFF + OGF)	17 (26.15)
Variant mix clone (OGF + OOGF + OGGF) (Deletion on derivative 9 for ABL/BCR/ABL-BCR)	05 (7.69)

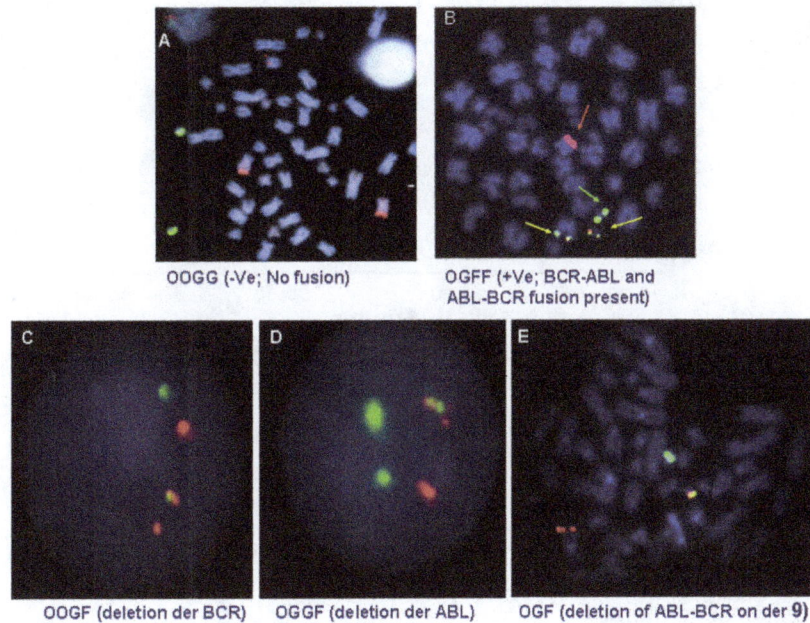

OOGG (-Ve; No fusion)   OGFF (+Ve; BCR-ABL and ABL-BCR fusion present)

OOGF (deletion der BCR)   OGGF (deletion der ABL)   OGF (deletion of ABL-BCR on der 9)

Figure 1. Representative FISH signal pattern using DCDF BCR-ABL FISH probe.

Figures 1A - E shows representative FISH signal pattern observed in the study. Sample with no fusion of BCR-ABL showed OOGG signal pattern (Figure 1A), samples with BCR-ABL fusion with no deletion in derivative 9 showed OGFF signal pattern (Figure 1b). Deletion in BCR, ABL and ABL-BCR on derivative 9 was seen as OOGF, OGGF and OGF respectively (Figure 1C, D and E respectively). Fifty-five percent of untreated patients showed presence of BCR-ABL fusion gene with no deletion of ABL-BCR on derivative 9 chromosome (Table 3). Deletion of ABL-BCR on derivative 9 was seen in 9.23% of PT while 26.15% of PT showed mixed clone for non-deleted and deleted ABL-BCR on derivative 9 showing signal pattern of OGF+OGFF. Mix clone for deletion of ABL or BCR or ABL-BCR (OGF + OOGF + OGGF) was seen in 7.69% of PT.

Frequency of FISH variant signals in t(9;22)

As mentioned in Table 3, 72.31% of untreated samples showed t(9;22) by conventional cytogenetic. When D-FISH was performed in these patients, they showed wide range of FISH signal pattern indicating deletion of ABL, BCR or ABL-BCR on derivative 9 chromosome which was not possible to diagnose by conventional cyto-genetic. Table 4 shows frequency of patients showing response to therapy and their D-FISH signal pattern. Sixty percentages of patients having deletion of ABL-BCR on derivative 9 showed no response to therapy

Table 4. FISH signal pattern analysis in patients with t(9;22) by conventional cytogenetics (karyotyping).

FISH signal pattern	Final response		
	Response (%)	No response (%)	Partial response
Der 9 non deleted (OGFF) (n=28)	20 (71.43)	06 (21.43)	02 (7.14)
Der 9 deleted (OGF) (n=05)	02 (40)	03 (60)	-
Mix (OGF/OGFF) (n=11)	07 (63.64)	04 (36.36)	-
Other variants (OGGF/OOGF/OGF/OGFF) (n=03)	02 (66.66)	-	01 (33.33)

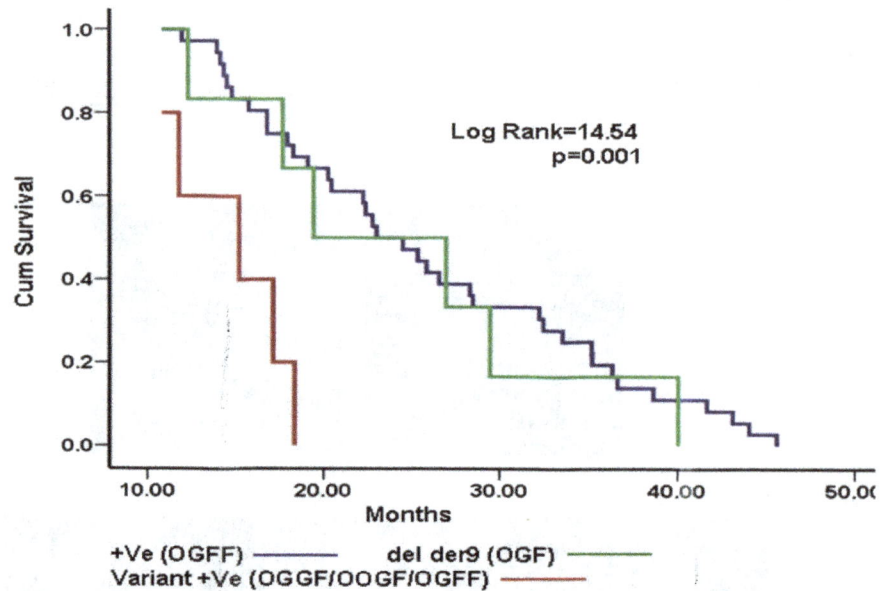

Figure 2. Kaplan-Meier survival analysis between derivative 9 deletion and non deletion.

while 40% showed response. While 71.43% of patients with non deleted ABL-BCR (OGFF) showed response and 21.43% showed no response to therapy.

Kaplan-Meier survival analysis

Kaplan-Meier disease free survival curve showed patients with deletion of ABL-BCR on derivative 9 and deletion of ABL or BCR and/or ABL-BCR on derivative 9 have significantly reduced survival (log rank = 14.54; p = 0.001) than non deleted patients (Figure 2).

DISCUSSISON

The present study aimed to evaluate prognostic effect of sub-microscopic deletions of the derivative chromosome 9 in CML in untreated patients and their follow up samples to correlate with disease free survival. The submicroscopic deletion can not be detected by conventional cytogenetic method using karyotype analysis.

D-FISH *BCR/ABL* probe used in the present study is useful to reveal the locations of 3' *ABL* and 5'*BCR* as well as 5'*ABL* and 3'*BCR* on metaphase chromosomes making this technique capable of detecting minimal residual disease in CML (Wan et al., 20003; Qiu et al., 2009). However, it is widely accepted that the clinical, prognostic and haematological features of CML patients with complex variant translocations are not different from those with the classical t(9; 22) translocation because it is accepted that the key pathological event is the formation of the BCR/ABL fusion gene (Johansson et al., 2002). In the present study only one patient showed complex variant translocation at the time of diagnosis showing t (9; 22)? add9, however at the end of 21 months, the patient showed complete response to therapy.

For each patient, pre treatment and follow-up haematological evaluation was also carried out to confirm haematological response. We found that the standard haematological response criteria in responders, non responder and partial responders (Table 2). However, there was no difference in leukocyte and platelet counts were seen between derivative 9 deleted and non deleted patients (data not shown). Lee et al. (2006) have shown that patients with ABL deletion showed significantly higher leukocyte count then deletion in ABL-BCR together. However, other clinical and haematological parameters were similar in deleted and non deleted group (Lee et al., 2006). FISH plays a complementary role in providing information in which cytogenetic studies are inadequate because of poor metaphase yield (Tefferi et al., 2005; Qiu et al., 2009). In this study, 16 (24.61%) patients at the time of diagnosis showed non informative metaphase. In these patients FISH showed not only positive results but also be useful to pin point deletion on derivative 9. Out of these 16 patients, 8 showed mix clone for deletion in ABL-BCR and non deleted derivative 9. Similarly, patients with t(9;22) at the time of diagnosis also showed different deletion status in derivative 9 by D-FISH (Table 4). Majority of patients with non deleted derivative 9 showed response as compared to deletion in ABL, BCR or ABL-BCR on derivative 9.

However, the association of poor prognosis and poor response to therapeutic modalities for patients harbouring the deletion of 9q34 is controversial. Several studies have suggested that 9q34 deletions confer an adverse prognosis and a poor response to therapeutic modalities (Li et al., 2008; Wu et al., 2007; Kreil et al., 2007; Gorusu et al., 2007; Cohen et al., 2001; Huntly et al., 2001; Storlazzi et al., 2002; Huntly et al., 2003) whereas others did not observe this association (Dong et al., 2008; Yoong et al., 2005; Quintas-Cardama et al., 2005).

We have observed significantly reduced disease free survival in patients with deletion in ABL, BCR or ABL-BCR on derivative 9 as compared to non deleted derivative 9. The survival time was shorter in those patients with deletions, regardless of the deletion patterns. These findings suggest that the gene segments around the breakpoint are deleted in heterogeneous patterns when the translocation occurs. In addition, the genes located near the breakpoint may play a role in the delayed tumor progression and a loss of this sequence may reduce the survival time.

Conclusion

In this study deletion on derivative 9 was heterogeneous involving either *ABL* or *BCR*, or *ABL-BCR*. Similar finding were observed by Lee et al. (2006) and Wu et al. (2006). This raised the question which region of the derivative 9 is important for disease prognosis. Many tumor-related genes are located near the translocation breakpoints (Huntly et al., 2001). It has been reported that p21rac acts on cell growth and proliferation associated with *RAS*, which moves along with 3'*BCR* region on derivative 9 during translocation. GTPase-activating protein binds with p21rac and inhibits its activity. Therefore, a loss of this region can induce abnormal cell growth and proliferation (Diekmann et al., 1991). If these genes are deleted during *BCR-ABL* gene rearrangement and the residue

allele is injured by "two hit" events, the tumor suppressor functions are destroyed and the disease can progress. Similarly Argininosuccinate Synthetase (*ASS*) gene located on 9q34 region adjacent to *ABL* and Immunoglobulin light chain (*IGL1*) gene located near *BCR* gene on chromosome 22 might be candidate genes to analyze along with deletion in *ABL-BCR* on derivative 9 (Storlazzi et al., 2002). Therefore, an array of adjacent genes at the breakpoint and fusion regions needs to be analyzed using Bacterial Artificial Chromosome (BAC) FISH clones. We are currently working on the BACs for different chromosome 9 and 22 breakpoint and fusion regions in cases with deletion in derivative 9.

ACKNOWLEDGEMENT

The authors acknowledge financial support provided by Gujarat Council of Science and Technology (GUJCOST), Government of Gujarat State, India.

REFERENCES

Buño I, Wyatt WA, Zinsmeister AR Band JD, Silver RT, Dewald GW (1998). A special fluorescent in situ hybridization technique to study peripheral blood and assess the effectiveness of interferon therapy in chronic myeloid leukemia. Blood, 92: 2315-2321.

Cohen N, Rozenfeld-Granot G, Hardan I, Brok-Simoni F, Amariglio N, Rechavi G, Trakhtenbrot L (2001). Subgroup of patients with Philadelphia-positive chronic myelogenous leukemia characterized by a deletion of q proximal to ABL gene: expression profiling, resistance to interferon therapy, and poor prognosis. Cancer Genet. Cytogenet., 128: 114-9.

Dewald G, Stallard R, Alsaadi A Arnold S, Blough R, Ceperich TM, Rafael Elejalde B, Fink J, Higgins JV, Higgins RR, Hoeltge GA, Hsu WT, Johnson EB, Kronberger D, McCorquodale DJ, Meisner LF, Micale MA, Oseth L, Payne JS, Schwartz S, Sheldon S, Sophian A, Storto P, Van Tuinen P, Wenger GD, Wiktor A, Willis LA, Yung JF, Zenger-Hain J (2000). A multicenter investigation with D-FISH BCR/ABL1 probes. Cancer Genet. Cytogenet., 116: 97-104.

Dewald GW, Wyatt WA, Juneau AL, Carlson RO, Zinsmeister AR, Jalal SM, Spurbeck JL, Silver RT (1998). Highly sensitive fluorescence in situ hybridization method to detect double BCR/ABL fusion and monitor response to therapy in chronic myeloid leukemia. Blood, 91: 3357-3365.

Diekmann D, Brill S, Garrett MD, Totty N, Hsuan J, Monfries C,et al Hall C, Lim L, Hall A (1991). BCR encodes a GTPase-activating protein for p21rac. Nature, 351: 400-402.

Dong Hwan, Dennis K, Gizelle P, Lakshmi S, Suzanne KR, Hong C, HA, Jeffrey HL (2008). No significance of derivative chromosome 9 deletion on the clearance kinetics of BCR/ABL fusion transcripts, cytogenetic or molecular response, loss of response, or treatment failure to imatinib mesylate therapy for chronic myeloid leukemia. Cancer, 113: 772-81.

Gonza´lez FA, Anguita E, Mora A, Asenjo S, Lo´pez I, Polo M Villegas A (2001). Deletion of BCR region 3' in chronic myelogenous leukemia. Cancer Genet. Cytogenet., 130: 68-74.

Gorusu M, Benn P, Li Z, Fang M (2007). On the genesis and prognosis of variant translocations in chronic myeloid leukemia. Cancer Genet. Cytogenet., 173: 97-106.

Huntly BJ, Bench AJ, Delabesse E, Reid AG, Li J, Scott MA, Campbell L Byrne J, Pinto E, Brizard A, Niedermeiser D, Nacheva EP, Guilhot F, Deininger M, Green AR (2002). Derivative chromosome 9 deletions in chronic myeloid leukemia: poor prognosis is not associated with loss of ABL-BCR expression, elevated BCR-ABL levels, or karyotypic instability. Blood, 99: 4547-4553.

Huntly BJ, Guilhot F, Reid AG, Vassiliou G, Hennig E, Franke C, Byrne J, Brizard A, Niederwieser D, Freeman-Edward J, Cuthbert G, Bown N, Clark RE, Nacheva EP, Green AR, Deininger MW (2003). Imatinib improves but may not fully reverse the poor prognosis of patients with CML with derivative chromosome 9 deletions. Blood, 102: 2205-12.

Huntly BJ, Reid AG, Bench AJ, Campbell LJ, Telford N, Shepherd P, Szer J, Prince HM, Turner P, Grace C, Nacheva EP, Green AR (2001). Deletions of the derivative chromosome 9 occur at the same time of the Philadelphia translocation and provide a powerful and independent prognostic indicator in chronic myeloid leukemia. Blood, 98: 1732-1738.

Johansson B, Fioretos T, Mitelman F (2002). Cytogenetic and molecular genetic evolution of Philadelphia-chromosome-positive chronic myeloid leukaemia. In: Chronic myeloproliferative disorders. Cytogenetic and molecular genetic abnormalities. Bain BJ, editor. Basel: S. Kargerpp. 44-61.

Kreil S, Pfirrmann M, Haferlach C, Waghorn K, Chase A, Hehlmann R, Reiter A, Hochhaus A, Cross NC (2007). Heterogeneous prognostic impact of derivative chromosome 9 deletions in chronic myelogenous leukemia. Blood, 110: 1283-1290.

Lee YK, Kim YR, Min HC, Oh BR, Kim TY, Kim YS Cho HI, Kim HC, Lee YS, Lee DS (2006). Deletion of any part of the BCR or ABL gene on the derivative chromosome 9 is a poor prognostic marker in chronic myelogenous leukemia. Cancer Genet. Cytogenet., 166: 65-73.

Li JY, Xu W, Wu W, Zhu Y, Qiu HR, Zhang R, Zhang SJ, Qian SX (2008). The negative prognostic impact of derivative 9 deletions in patients who received hydroxyurea treatment for chronic myelogenous leukemia in the chronic phase. Onkologie, 31: 585-589.

Qiu HR, Miao KR, Wang R, Qiao C, Zhang JF, Zhang SJ, Qian SX, Xu W, Li JY (2009). The application of fluorescence in situ hybridization in detecting chronic myeloid leukaemia. Zhonghua Yi Xue Yi Chuan Xue Za Zhi. 26: 207-210.

Quintas-Cardama A, Kantarjian H, Talpaz M, O'Brien S, Garcia-Manero G, Verstovsek S, Rios MB, Hayes K, Glassman A, Bekele BN, Zhou X, Cortes J (2005). Imatinib mesylate therapy may overcome the poor prognostic significance of deletions of derivative chromosome 9 in patients with chronic myelogenous leukemia. Blood, 105: 2281-6.

Shaffer LG, Tommerup N (2005). (ed) ISCN: An international system for human cytogenetic nomenclature. Basel: S Karger, 2005.

Sinclair PB, Nacheva EP, Leversha M, Telford N, Chang J, Reid A Bench A, Champion K, Huntly B, Green AR (2000). Large deletions at the t(9;22) breakpoint are common and may identify a poor-prognosis subgroup of patients with chronic myeloid leukemia. Blood 2000; 95: 738-744.

Storlazzi CT, Specchia G, Anelli L, Albano F, Pastore D, Zagaria A, Rocchi M, Liso V (2002). Breakpoint characterization of der (9) deletions in chronic myeloid leukemia patients. Genes Chromosomes Cancer, 35: 271-6.

Tefferi A, Dewald GW, Litzow ML, Cortes J, Mauro MJ, Talpaz M, Kantarjian HM (2005). Chronic myeloid leukemia: Current application of cytogenetics and molecular testing for diagnosis and treatment. Mayo Clinic Proceedings. Rochester, 80(3): 390-403.

Wan TSK, Ma SK, Au WY, Chan LC (2003). Derivative chromosome 9 deletions in chronic myeloid leukaemia: interpretation of atypical D-FISH pattern J. Clin. Pathol., 56: 471-474.

Wu W, Li JY, Shen YF, Cao XS, Qiu HR, Xu W (2007). The prognostic significance of derivative chromosome 9 deletions in chronic myeloid leukaemia. Zhonghua Nei Ke Za Zhi., 46: 386-388.

Wu W, Xue YQ, Wu YF, Pan JL, Shen J (2006). Study of deletion of derivative chromosome 9 in patients with Ph+ chronic myeloid leukaemia. Zhonghua Xue Ye Xue Za Zhi. 27: 183-186.

Yoong Y, VanDeWalker TJ, Carlson RO, Dewald GW, Tefferi A (2005). Clinical correlates of submicroscopic deletions involving the ABL-BCR translocation region in chronic myeloid leukemia. Eur. J. Haematol., 74: 124-7.

Permissions

All chapters in this book were first published in IJGMB, by Academic Journals; hereby published with permission under the Creative Commons Attribution License or equivalent. Every chapter published in this book has been scrutinized by our experts. Their significance has been extensively debated. The topics covered herein carry significant findings which will fuel the growth of the discipline. They may even be implemented as practical applications or may be referred to as a beginning point for another development.

The contributors of this book come from diverse backgrounds, making this book a truly international effort. This book will bring forth new frontiers with its revolutionizing research information and detailed analysis of the nascent developments around the world.

We would like to thank all the contributing authors for lending their expertise to make the book truly unique. They have played a crucial role in the development of this book. Without their invaluable contributions this book wouldn't have been possible. They have made vital efforts to compile up to date information on the varied aspects of this subject to make this book a valuable addition to the collection of many professionals and students.

This book was conceptualized with the vision of imparting up-to-date information and advanced data in this field. To ensure the same, a matchless editorial board was set up. Every individual on the board went through rigorous rounds of assessment to prove their worth. After which they invested a large part of their time researching and compiling the most relevant data for our readers.

The editorial board has been involved in producing this book since its inception. They have spent rigorous hours researching and exploring the diverse topics which have resulted in the successful publishing of this book. They have passed on their knowledge of decades through this book. To expedite this challenging task, the publisher supported the team at every step. A small team of assistant editors was also appointed to further simplify the editing procedure and attain best results for the readers.

Apart from the editorial board, the designing team has also invested a significant amount of their time in understanding the subject and creating the most relevant covers. They scrutinized every image to scout for the most suitable representation of the subject and create an appropriate cover for the book.

The publishing team has been an ardent support to the editorial, designing and production team. Their endless efforts to recruit the best for this project, has resulted in the accomplishment of this book. They are a veteran in the field of academics and their pool of knowledge is as vast as their experience in printing. Their expertise and guidance has proved useful at every step. Their uncompromising quality standards have made this book an exceptional effort. Their encouragement from time to time has been an inspiration for everyone.

The publisher and the editorial board hope that this book will prove to be a valuable piece of knowledge for researchers, students, practitioners and scholars across the globe.

List of Contributors

Ajit Kumar Saxena
Human Molecular Cytogenetic Laboratory, Centre of Experimental Medicine and Surgery, Institute of Medical Sciences, Banaras Hindu University, Varanasi-221005, India

S. Pandey
Human Molecular Cytogenetic Laboratory, Centre of Experimental Medicine and Surgery, Institute of Medical Sciences, Banaras Hindu University, Varanasi-221005, India

L. K. Pandey
Department of Obstetrics and Gynecology, Institute of Medical Sciences, Banaras Hindu University, Varanasi-221005, India

M. C. Ogwu
Plant Conservation Unit, Department of Plant Biology and Biotechnology, Faculty of Life Sciences, University of Benin, Benin City, Edo State, Nigeria

M. E. Osawaru
Plant Conservation Unit, Department of Plant Biology and Biotechnology, Faculty of Life Sciences, University of Benin, Benin City, Edo State, Nigeria

C. M. Ahana
Plant Conservation Unit, Department of Plant Biology and Biotechnology, Faculty of Life Sciences, University of Benin, Benin City, Edo State, Nigeria

Haddad El Rabey
Genetic Engineering and Biotechnology Institute, Minufiya University, P. O. Box 79 Sadat City, Egypt

Francesca Barale
Faculty of Agriculture, Milan University, Milan, Italy

Radha Saraswathy
Division of Biomolecules and Genetics, School of Biosciences and Technology, VIT University, Vellore 632014, Tamilnadu, India

V. G. Abilash
Division of Biomolecules and Genetics, School of Biosciences and Technology, VIT University, Vellore 632014, Tamilnadu, India

G. Manivannan
Division of Biomolecules and Genetics, School of Biosciences and Technology, VIT University, Vellore 632014, Tamilnadu, India

Alex George
Division of Biomolecules and Genetics, School of Biosciences and Technology, VIT University, Vellore 632014, Tamilnadu, India

K. Thirumal Babu
Heart line Medical and Research Centre, 72, Thennamaram Street, Vellore-1, India

J. M. Aparicio-Rodríguez
Department of Genétics, Hospital para el Nino Poblano, Mexico
Department of Estomatology, Benemérita Universidad Autónoma de Puebla, Mexico

M. L. Hurtado-Hernández
Department of Cytogenetics, Hospital para el Nino Poblano, Mexico
Department of Perinatology, Hospital de la Mujer S.S.A. Puebla, Mexico

M. Barrientos-Perez
Department of Endocrinology, Hospital para el Nino Poblano, Mexico

S. I. Assia-Robles
Department of Pediatrics, Hospital para el Nino Poblano, Mexico

N. C. Gil-Orduña
Department of Estomatology, Hospital para el Nino Poblano, Mexico
Department of Estomatology, Benemérita Universidad Autónoma de Puebla, Mexico

R. Zamudio-Meneses
Department of Cardiology, Hospital para el Nino Poblano, Mexico

J. S. Rodríguez-Peralta
Department of Neurosurgery, Hospital para el Nino Poblano, Mexico

S. M. Brieke-Walter
Department of Estomatology, Hospital para el Nino Poblano, Mexico

F. Almanza-Flores
Department of Pediatrics, Hospital para el Nino Poblano, Mexico

C. Silva-Xilotl
Department of Perinatology, Hospital de la Mujer S.S.A. Puebla, Mexico

Pooja Chadha
Department of Zoology, Guru Nanak Dev University, Amritsar, Punjab, Pin Code: 143005, India

Anupam Mehta
Department of Zoology, Guru Nanak Dev University, Amritsar, Punjab, Pin Code: 143005, India

Peter Twumasi
Department of Biochemistry and Biotechnology, Kwame Nkrumah University of Science and Technology, Kumasi, Ghana

Eric Warren Acquah
Department of Biochemistry and Biotechnology, Kwame Nkrumah University of Science and Technology, Kumasi, Ghana
Crops Research Institute (CRI) of Council for Scientific and Industrial Research (CRI-CSIR)-Fumesua, Ghana, P.O.
Box 3785, Kumasi, Ghana

Marian D. Quain
Crops Research Institute (CRI) of Council for Scientific and Industrial Research (CRI-CSIR)-Fumesua, Ghana, P.O.
Box 3785, Kumasi, Ghana

Elizabeth Y. Parkes
Crops Research Institute (CRI) of Council for Scientific and Industrial Research (CRI-CSIR)-Fumesua, Ghana, P.O.
Box 3785, Kumasi, Ghana

Aimola Idowu Asegame
Department of Biochemistry, Ahmadu Bello University, Zaria, Nigeria

Inuwa Hajia Mairo
Department of Biochemistry, Ahmadu Bello University, Zaria, Nigeria

Nok Andrew Jonathan
Centre for Biotechnology Research and Training, Ahmadu Bello University, Zaria, Nigeria

Mamman I. Aisha
Hematology Department, Ahmadu Bello University Teaching Hospital, Zaria, Nigeria

Khaled Mohamed Anwar Aboshanab
Department of Microbiology and Immunology, Faculty of Pharmacy, Ain Shams University, Organization of African Unity St., POB: 11566, Abbassia, Cairo, Egypt

Mostafa Mahmoud Elshafey
Department of Biochemistry, Faculty of Pharmacy, Al-Azhar University (Boys), Nasr City, Cairo, Egypt

Elodie Richard
Laboratoire PsyNuGen, INRA, UMR 1286, CNRS 5226, Université de Bordeaux, 146 rue Léo Saignat, F-33077 Bordeaux, France

José-Manuel Fernandez-Real
Unitat de Diabetologia, Endocrinologia i Nutricio, University Hospital of Girona, "Dr. Josep Trueta" and CIBER
Fisiopatología de la Obesidad y Nutrición CB06/03/010, 17007 Girona, Spain

Abel Lopez-Bermejo
Unitat de Diabetologia, Endocrinologia i Nutricio, University Hospital of Girona, "Dr. Josep Trueta" and CIBER
Fisiopatología de la Obesidad y Nutrición CB06/03/010, 17007 Girona, Spain

Wifredo Ricart
Unitat de Diabetologia, Endocrinologia i Nutricio, University Hospital of Girona, "Dr. Josep Trueta" and CIBER
Fisiopatología de la Obesidad y Nutrición CB06/03/010, 17007 Girona, Spain

Henri Déchaud
Fédération d'Endocrinologie, INSERM U863, IFR62, Groupe Hospitalier Est F-69437 Lyon, France

Michel Pugeat
Fédération d'Endocrinologie, INSERM U863, IFR62, Groupe Hospitalier Est F-69437 Lyon, France

Marie-Pierre Moisan
Laboratoire PsyNuGen, INRA, UMR 1286, CNRS 5226, Université de Bordeaux, 146 rue Léo Saignat, F-33077 Bordeaux, France

Isaac Kofi Bimpong
Africa Rice Centre, BP 96. St Louis, Senegal
International Rice Research Institute (IRRI), DAPO Box 7777, Metro Manila, Philippines

Joong Hyoun Chin
International Rice Research Institute (IRRI), DAPO Box 7777, Metro Manila, Philippines

Joie Ramos
International Rice Research Institute (IRRI), DAPO Box 7777, Metro Manila, Philippines

Hee-Jong Koh
Department of Plant Science, College of Agriculture and Life Sciences, Seoul National University, Seoul, 151-921, Korea

Nada Babiker Hamza
Department of Molecular Biology, Commission for Biotechnology and Genetic Engineering, National Centre for Research, P. O. Box 2404, Khartoum, Sudan
Nile Basin Research Programme, UNIFOB-Global, University of Bergen, P. O. Box 7800, N-5020 Bergen, Norway

Tafere Mulualem
Pawe Agricultural Research Center, Pawe, Ethiopia

Tadesse Dessalegn
Pawe Agricultural Research Center, Pawe, Ethiopia

Yigzaw Dessalegn
Amhara Regional Agricultural Research Institute, Bahir Dar, Ethiopia

SK. Kiplagat
Department of Biochemistry and Molecular Biology, Egerton University, Kenya

M. Agaba
International Livestock Research Institute (ILRI), Nairobi, Kenya

IS. Kosgey
Department of Animal Sciences, Egerton University, Kenya

M. Okeyo
International Livestock Research Institute (ILRI), Nairobi, Kenya

D. Indetie
Kenya Agricultural Research Institute (KARI), National Beef Research Centre-Lanet, Kenya

O. Hanotte
School of Biology, University of Nottingham, UK

MK. Limo
Department of Biochemistry and Molecular Biology, Egerton University, Kenya

Ambreen Ijaz
Department of Bioinformatics and Biotechnology, GC University, Faisalabad, Pakistan

Sadia Ali
Department of Bioinformatics and Biotechnology, GC University, Faisalabad, Pakistan

Usman Ijaz
Ayub Agriculture Research Institute, Faisalabad, Pakistan

Smiullah
Ayub Agriculture Research Institute, Faisalabad, Pakistan

Tayyaba Shaheen
Department of Bioinformatics and Biotechnology, GC University, Faisalabad, Pakistan

Praveen Mamidala
Department of Biotechnology, Kakatiya University, Warangal, A.P- 506009, India
Department of Biotechnology, Vaagdevi College of Engineering, Warangal, 506009, India

Rama Swamy Nanna
Department of Biotechnology, Kakatiya University, Warangal, A.P- 506009, India

Zhang Kejin
Key Laboratory of Resource Biology and Biotechnology in Western China (Northwest University) Ministry of Education, College of Life Science, Institute of Population and Health, Northwest University, Xi'an 710069, China

He Bo
Key Laboratory of Resource Biology and Biotechnology in Western China (Northwest University) Ministry of Education, College of Life Science, Institute of Population and Health, Northwest University, Xi'an 710069, China

Gong Pingyuan
Key Laboratory of Resource Biology and Biotechnology in Western China (Northwest University) Ministry of Education, College of Life Science, Institute of Population and Health, Northwest University, Xi'an 710069, China

Gao Xiaocai
Key Laboratory of Resource Biology and Biotechnology in Western China (Northwest University) Ministry of Education, College of Life Science, Institute of Population and Health, Northwest University, Xi'an 710069, China
Institute of Applied Psychology, College of Public Management, Northwest University, Xi'an 710069, China

Zheng Zijian
Key Laboratory of Resource Biology and Biotechnology in Western China (Northwest University) Ministry of Education, College of Life Science, Institute of Population and Health, Northwest University, Xi'an 710069, China
Institute of Applied Psychology, College of Public Management, Northwest University, Xi'an 710069, China

Huang Shaoping
The 2nd Affiliated Hospital, Xi'an Jiaotong University, Xi'an 710004, China

Zhang Fuchang
Key Laboratory of Resource Biology and Biotechnology in Western China (Northwest University) Ministry of Education, College of Life Science, Institute of Population and Health, Northwest University, Xi'an 710069, China Institute of Applied Psychology, College of Public Management, Northwest University, Xi'an 710069, China

V. G. Abilash
Division of Biomolecules and Genetics, School of Bio Sciences and Technology, VIT University, Vellore 632014, Tamilnadu, India

Radha Saraswathy
Division of Biomolecules and Genetics, School of Bio Sciences and Technology, VIT University, Vellore 632014, Tamilnadu, India

K. M. Marimuthu
Emeritus Professor, Dept of Genetics, University of Madras, Chennai 600113, Tamilnadu, India

Emil José Hernández-Ruz
Laboratório de Zoologia, Faculdade de Ciências Biológicas, Universidade Federal do Pará/UFPA, Campus Universitário de Altamira, Rua Coronel José Porfírio, 2515 CEP 68.372-040 - Altamira - PA, Brazil

Evonnildo Costa Gonçalves
Laboratório de Tecnologia Biomolecular, Universidade Federal do Pará, Instituto de Ciências Biológicas, Rua Augusto Corrêa, 01, Guamá, Belém, Pará, Brasil, 66075-900

Artur Silva
Laboratório de Polimorfismo de DNA, Universidade Federal do Pará, Instituto de Ciências Biológicas, Rua Augusto Corrêa, 01, Guamá, Belém, Pará, Brasil, 66075-900

Maria Paula Cruz Schneider
Laboratório de Polimorfismo de DNA, Universidade Federal do Pará, Instituto de Ciências Biológicas, Rua Augusto Corrêa, 01, Guamá, Belém, Pará, Brasil, 66075-900

M. Alavi
Department of Plant Breeding and Biotechnology, Gorgan University of Agricultural Sciences and Natural Resources, Gorgan, Iran

A. Ahmadikhah
Department of Plant Breeding and Biotechnology, Gorgan University of Agricultural Sciences and Natural Resources, Gorgan, Iran

B. Kamkar
Department of Agronomy, Gorgan University of Agricultural Sciences and Natural Resources, Gorgan, Iran

M. Kalateh
Agricultural Selection Station, Gorgan, Iran

R. S. Bhat
Department of Biotechnology, University of Agricultural Sciences, Dharwad 580 005, India

S. S. Biradar
Department of Biotechnology, University of Agricultural Sciences, Dharwad 580 005, India

R. V. Patil
Division of Horticulture, of Agricultural Sciences, Dharwad 580 005, India

V. U. Patil
Department of Biotechnology, University of Agricultural Sciences, Dharwad 580 005, India

M. S. Kuruvinashetti
Department of Biotechnology, University of Agricultural Sciences, Dharwad 580 005, India

Manal H. Eid
Botany Department, Faculty of Agriculture, Suez Canal University, 41522, Ismailia, Egypt

Mutiu Oyekunle Sifau
Department of Cell Biology and Genetics, University of Lagos, Lagos, Lagos State, Nigeria
Molecular Biology Laboratory, Biotechnology Unit, National Centre for Genetic Resources and Biotechnology (NACGRAB), PMB 5382, Ibadan, Oyo State, Nigeria

Liasu Adebayo Ogunkanmi
Department of Cell Biology and Genetics, University of Lagos, Lagos, Lagos State, Nigeria

Khalid Olajide Adekoya
Department of Cell Biology and Genetics, University of Lagos, Lagos, Lagos State, Nigeria

Bola Olufunmilayo Oboh
Department of Cell Biology and Genetics, University of Lagos, Lagos, Lagos State, Nigeria

Oluwatoyin Temitayo Ogundipe
Department of Botany, University of Lagos, Lagos, Lagos State, Nigeria

Kazhila C. Chinsembu
Department of Biological Sciences, Faculty of Science, University of Namibia, P/Bag 13301, Windhoek, Namibia

André Faul
Department of Biological Sciences, Faculty of Science, University of Namibia, P/Bag 13301, Windhoek, Namibia

G. R. Rout
Plant Biotechnology Division, Regional Plant Resource Centre, Bhubaneswar-751015, Orissa, India

S. Aparajita
Plant Biotechnology Division, Regional Plant Resource Centre, Bhubaneswar-751015, Orissa, India

Onasanya Amos
Department of Chemical Sciences, Afe Babalola University, Ado-Ekiti, Ekiti State, Nigeria

M. M. Ekperigin
2Federal University of Technology, Akure, PMB 704, Akure, Ondo State, Nigeria

A. Afolabi
Biotechnology and Genetic Engineering Advanced Laboratory, Sheda Science and Technology Complex, PMB 186, Garki, Abuja, Nigeria

R. O. Onasanya
Federal University of Technology, Akure, PMB 704, Akure, Ondo State, Nigeria

Abiodun A. Ojo
Department of Chemical Sciences, Afe Babalola University, Ado-Ekiti, Ekiti State, Nigeria

I. Ingelbrecht
Institute of Plant Biotechnology for Developing Countries, Ghent University, K. L. Ledeganckstraat 35, 9000 Ghent, Belgium

Nitza Goldenberg-Cohen
The Krieger Eye Research Laboratory, Felsenstein Medical Research Center, Petah Tiqwa, Israel
Pediatric Unit, Rabin Medical Center, Petach Tiqwa, Israel
Sackler Faculty of Medicine, Tel Aviv University, Tel Aviv, Israel

Bat Chen R. Avraham-Lubin
The Krieger Eye Research Laboratory, Felsenstein Medical Research Center, Petah Tiqwa, Israel

Dorina Calles
The Krieger Eye Research Laboratory, Felsenstein Medical Research Center, Petah Tiqwa, Israel

Olga Dratviman-Storobinsky
The Krieger Eye Research Laboratory, Felsenstein Medical Research Center, Petah Tiqwa, Israel

Rita Ehrlich
Department of Ophthalmology, Rabin Medical Center, Petach Tiqwa, Israel

Boris Paritiansky
Department of Ophthalmology, Rabin Medical Center, Petach Tiqwa, Israel

Yoram Cohen
Sheba Cancer Research Center, Sheba Medical Center, Tel Hashomer, Israel

Dov Weinberger
Department of Ophthalmology, Rabin Medical Center, Petach Tiqwa, Israel
Sackler Faculty of Medicine, Tel Aviv University, Tel Aviv, Israel

Beena P. Patel
The Gujarat Cancer and Research Institute, Asarwa, Ahmedabad -380 016, India

Pina J. Trivedi
The Gujarat Cancer and Research Institute, Asarwa, Ahmedabad -380 016, India

Manisha M. Brahmbhatt
The Gujarat Cancer and Research Institute, Asarwa, Ahmedabad -380 016, India

Sarju B. Gajjar
The Gujarat Cancer and Research Institute, Asarwa, Ahmedabad -380 016, India

Ramesh R. Iyer
The Gujarat Cancer and Research Institute, Asarwa, Ahmedabad -380 016, India

Esha N. Dalal
The Gujarat Cancer and Research Institute, Asarwa, Ahmedabad -380 016, India

Shilin N. Shukla
The Gujarat Cancer and Research Institute, Asarwa, Ahmedabad -380 016, India

Pankaj M. Shah
The Gujarat Cancer and Research Institute, Asarwa, Ahmedabad -380 016, India

www.ingramcontent.com/pod-product-compliance
Lightning Source LLC
Chambersburg PA
CBHW080655200326
41458CB00013B/4865

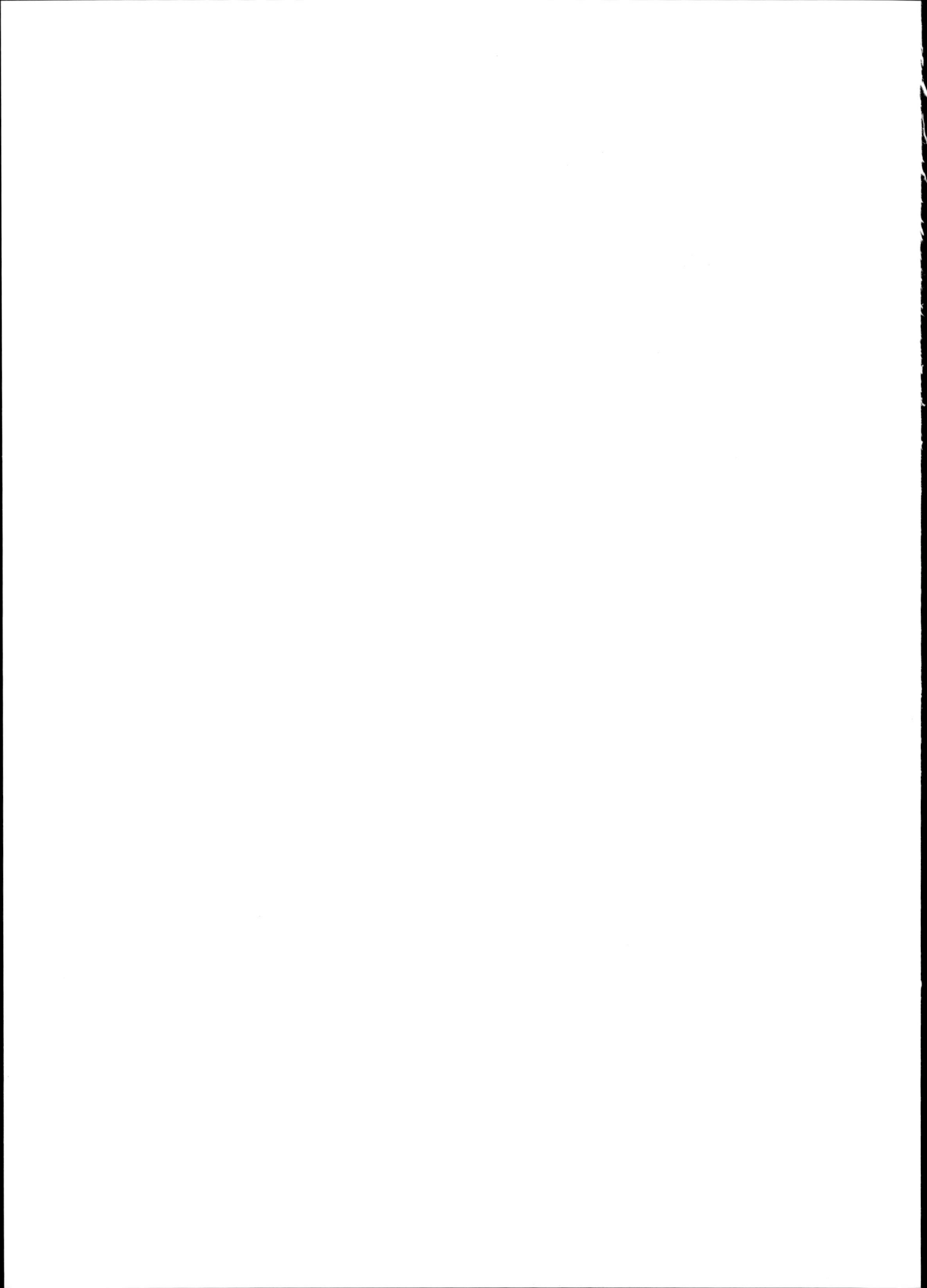